SpringerWienNewYork

CISM COURSES AND LECTURES

Series Editors:

The Rectors
Friedrich Pfeiffer - Munich
Franz G. Rammerstorfer - Wien
Jean Salençon - Palaiseau

The Secretary General
Bernhard Schrefler - Padua

Executive Editor
Paolo Serafini - Udine

The series presents lecture notes, monographs, edited works and proceedings in the field of Mechanics, Engineering, Computer Science and Applied Mathematics.
Purpose of the series is to make known in the international scientific and technical community results obtained in some of the activities organized by CISM, the International Centre for Mechanical Sciences.

INTERNATIONAL CENTRE FOR MECHANICAL SCIENCES

COURSES AND LECTURES - No. 540

WAVE PROPAGATION IN LINEAR AND NONLINEAR PERIODIC MEDIA

ANALYSIS AND APPLICATIONS

EDITED BY

FRANCESCO ROMEO
UNIVERSITY "LA SAPIENZA", ROME, ITALY

MASSIMO RUZZENE
GEORGIA INSTITUTE OF TECHNOLOGY, ATLANTA, USA

SpringerWienNewYork

This volume contains 175 illustrations

All contributions have been typeset by the authors.

ISBN 978-3-7091-1308-0 SpringerWienNewYork

PREFACE

Periodic structural configurations are ubiquitous: many heterogeneous structures and materials, both man-made and naturally occurring, feature geometry, micro-structural and/or materials properties that vary periodically in space. The classes of periodic materials and structures span a wide range of length scales, and a broad range of applications. Periodic trusses, periodically stiffened plates, shells and beam-like assemblies can be found for example in many civil, aerospace, mechanical and ship constructions. Their introduction is mostly motivated by structural strength and weight requirements, however recent studies have shown how the periodicity can be exploited to attenuate, isolate and localize vibrations. Such studies explore the unique ability of periodic assemblies to impede the propagation of elastic waves over specified frequency bands, within which strong attenuation of vibration and radiated noise can be achieved. The attenuation levels that can be obtained through tailored structural periodicity far exceed the performance of most energy dissipation and damping mechanisms. For this reason, passive, active and hybrid periodic structural configurations are being proposed for the reduction of vibration transmission and structure-born as well as airborne noise. In addition, the understanding of the dynamics of periodic structures is essential for the analysis of bladed disc assemblies which are found in turbo machinery and in turbines for energy generation where failure mechanisms due to localization phenomena may occur. At much smaller scales, extensive research is being devoted to the analysis and design of phononic metamaterials for a variety of applications. Phononic metamaterials are essentially periodic structural configurations obtained through composite designs featuring periodic modulations of mass and stiffness properties, or elastic lattice structures. Gigahertz communication devices, such as mobile phones, use phononic-based systems for their low-power filtering characteristics. Many sensing devices based on resonators, acoustic logic ports, and surface acoustic wave-based filters rely on the unique band gap characteristics of periodic phononic materials. These properties are associated with the destructive and constructive interference of acoustic waves originating at the periodic interfaces, which produce frequency band of strong attenuation of acoustic waves (band gaps). Depending on the inclusions, geometry, and elastic properties, one can design for specific band gaps. In photonic crystals, periodic modulations of the dielectric properties of a medium allow guiding, focusing and steering of electromagnetic waves. Properties modulations and engineered anisotropy in heterogeneous media can also produce negative refractive indexes, both in photonics as well as in phononic metamaterials, which lead to super-lensing or super-focusing characteristics. Dispersion of waves in media with boundaries or built-in micro-structure is a fundamental phenomenon, which occurs in problems of acoustics, models of water waves, simple systems involving onedimensional harmonic oscillators, as well as complex elastic systems. Moreover, enhanced transmission based on formation of defect modes within an interface plays a crucial role in a wide range of practical applications involving filtering and polarisation of waves of different physical origins. Other potential implications of the acoustic wave guiding technology include active sensing of structural integrity, smart sensing of environment, dissipation of high frequency modes of vibration to enhance vehicle performance or stealth, as well as applications to the medical field for sensing or diagnostic applications. Complex dynamical phenomena may be encountered in nonlinear periodic media such as the existence of energy- depended nonlinear propagation regions. Depending on substructure coupling, distinctly nonlinear dynamical phenomena can arise, such as wave localization and solitary waves. Potential applications of such phenomena are currently investigated in the MEMS context, where arrays of coupled micro/nanoresonators have been recently proposed. The CISM course "Wave propagation in linear and nonlinear periodic media: analysis and applications" was an opportunity to combine the material issued from such a variety of applications aiming to present both the theoretical background and an overview of the state-of-the art in wave propagation in linear and nonlinear periodic media in a consistent lecture format. The course is intended for doctoral and postdoctoral researchers in civil and mechanical engineering, applied mathematics and physics, academic and industrial researchers, which are interested in conducting research in the topic. The opening chapter, by A.B. Movchan, M. Brun and N.V. Movchan, gives an overview of some mathematical models of wave propagation in structured media, with the emphasis on dispersion and enhanced transmission through structured interfaces.

Chapter 2, by L. Airoldi, M. Senesi and M. Ruzzene, presents two examples of internally resonating metamaterials whose behavior is based on multi-field coupling. The first part of the chapter is devoted to the analysis of a one-dimensional waveguide with a periodic array of shunted piezoelectric patches. The second part presents the study of piezoelectric superlattices as an additional example of an internally resonant metamaterial. Topology optimization relevant to problems in vibration and wave propagation is then addressed in Chapter 3 by J. S. Jensen; in this chapter the steady-state optimization procedure for dynamical systems is extended to a nonlinear wave propagation problem and then the topology optimization procedure is applied to transient dynamic simulations for which the optimized material distributions may vary in time. In Chapter 4, by F. Romeo, the dynamic behaviour of continuous and discrete models of both linear and nonlinear periodic mechanical systems are dealt with by means of maps. At first, linear problems consisting of general multi-coupled periodic systems are presented and they are handled with linear maps, namely the transfer matrices of single units. Afterwards, a perturbation method is applied to the transfer matrix of a chain of continuous nonlinear beams while nonlinear maps are considered to address chains of nonlinear oscillators. In the closing chapter, by A. Vakakis, analytical methodologies for analyzing waves in weakly or strongly nonlinear periodic media are discussed. The generation of spatial chaos in ordered granular media, which are a special class of spatially periodic, highly discontinuous and strongly nonlinear media is eventually highlighted.

CONTENTS

Waves and defect modes in structured media

A.B. Movchan[a], M. Brun[b], N.V. Movchan[a]

[a] Department of Mathematical Sciences,
University of Liverpool, Liverpool L69 7ZL, U.K.

[b] Department of Structural Engineering,
University of Cagliari, Cagliari I-09123, Italy

Abstract. The paper gives an overview of some mathematical models of wave propagation in structured media, with the emphasis on dispersion and enhanced transmission through structured interfaces. We begin with conventional definitions and classical models of dispersive waves, discuss lattice approximations and then show the results of the recent work on modelling of waves interacting with a structured stack acting as a polarizer of elastic waves. Special attention is given to the resonance modes which may enhance transmission across the structured stack. The analytical work is accompanied by numerical simulations.

0.1 Introduction

Dispersion of waves in periodic inhomogeneous media is important in many problems of physics, mechanics and engineering. A special mention should be given to theoretical and experimental studies of photonic band gap materials (see, for example, Yablonovitch (1987, 1993)). Scalar models for Bloch waves in periodic lattices are discussed in Brillouin (1953), Maradudin *et al.* (1963) and Kunin (1975). This theory naturally extends to vector problems of elasticity. In particular, dispersion of elastic waves in lattice systems of different types has been analysed in Martinsson and Movchan (2003), Jensen (2003), Cai *et al.* (2005).

Dynamic lattice Green's functions in the propagating frequency range were analysed in Martin (2006), whereas the stop band Green's functions and exponentially localised "defect modes" were studied in Movchan & Slepyan (2007).

In wave propagation problems for structured media, the classical homogenisation approaches have limitations when the wavelength is comparable with the characteristic size of elements of the structure, such as a typical size of the elementary cell. In particular, this appears to be important in the analysis of wave interaction with finite width structured interfaces. Several classes of *non-local structured interfaces* were analysed in Bigoni and Movchan (2002) in static and dynamic configurations, which also included semi-discrete systems incorporating both continuous parts and a lattice. Brun et al. (2010a) have addressed anomalies in transmission of slow waves through dynamic structured interface of finite width. Brun et al. (2010b) addressed the case of a shear-type structured interface excited by a plane pressure wave at an oblique incidence, and this paper also shows that the shear-type interface could serve as a polariser.

In the present paper we aim to discuss the dispersion, polarisation and filtering of waves from a common perspective, with the emphasis on the dynamics of structured interfaces and connections between discrete and high-contrast continuous systems. The structure of the paper is as follows. We begin with the discussion of the fundamental phenomena, such as wave dispersion, and in Section 0.2 we show a very intuitive textbook example of dispersive waves, derived in the framework of the linear water wave theory. Section 0.3 includes the classical notion of Bloch-Floquet waves and an explicit analytical study of wave dispersion in one-dimensional lattice systems. Furthermore, this section also introduces the high-contrast continuous systems, with the emphasis on the lattice approximations of the dispersion equations in such systems. An example of the transmission problem for a one-dimensional finite width structured interface is discussed in Section 0.5, where it is shown that an irregularity within the structure of the interface may lead to an enhanced transmission for waves of certain frequency. A full vector problem of elasticity is described in Section 0.6 for plane elastic waves interacting with a shear-type structured interface; the finite width interface contains several parallel elastic bars, whose motion is preferential in the direction along the interface. This creates an effect of polarization, and we also describe an enhanced transmission for waves of certain frequencies.

0.2   Wave dispersion

This introductory section discusses some classical examples of dispersive waves, i.e. the waves whose speeds are different for different frequencies. Here we refer to the linear water wave theory, which is very well known in linearised models of fluid flow (see, for example, Billingham & King (2001), Ockendon *et al.* (2003)). The second example, included in Section 0.3, is a one-dimensional lattice system, known from classical texts (see, for example, Kittel (1996) and Brillouin (1953)). The dispersion relations are written in an explicit form, and formation of a stop band is discussed for heterogeneous systems. An asymptotic model is given for a high-contrast stratified structure in Section 0.3. Following Movchan *et al.* (2002a) we show the lattice approximation of such a system, which is capable of reproducing the dynamic response of the heterogeneous continuum system in the low frequency range.

Formulation for linear water waves

One of the straightforward examples is based on the linear theory of waves propagating along the surface of an incompressible inviscid fluid of the uniform density. For convenience, we call such a fluid "water". The continuity equation and the equation of motion for the velocity $\mathbf{u}$ and pressure p have the form

$$\nabla \cdot \mathbf{u} = 0 \quad , \tag{1}$$

$$\frac{\partial \mathbf{u}}{\partial t} + (\mathbf{u} \cdot \nabla)\mathbf{u} + \frac{1}{\rho}\nabla p = \mathbf{F} \quad , \tag{2}$$

where $\mathbf{F}$ is the body force density, ρ is the mass density and t is time. In particular, if the body force represents gravity we have $\mathbf{F} = -g\mathbf{e}^{(3)}$, and

$$\mathbf{F} = \nabla\Xi \text{ with } \Xi = -gx_3, \tag{3}$$

where g is a positive constant (normalised gravitational acceleration).

Assuming that the fluid flow is irrotational and using the notation ϕ for the velocity potential we have $\mathbf{u} = \nabla\phi$. Hence according to (1) the function ϕ is harmonic

$$\nabla^2\phi = 0. \tag{4}$$

The non-linear term in (2) becomes

$$\{(\mathbf{u} \cdot \nabla)\mathbf{u}\}_i = \sum_j u_j u_{i,j} = \sum_j \phi_{,j}\phi_{,ij} = \frac{1}{2}|\nabla\phi|^2_{,i} \tag{5}$$

Thus, (2) may be written as

$$\nabla\left(\frac{\partial\phi}{\partial t} + \frac{1}{2}|\nabla\phi|^2 + \frac{p}{\rho} + gx_3\right) = 0,$$

which leads to Bernoulli's equation

$$\frac{\partial\phi}{\partial t} + \frac{1}{2}|\nabla\phi|^2 + \frac{p}{\rho} + gx_3 = f(t), \tag{6}$$

where $f(t)$ is a function of t only.

Next, we discuss the boundary conditions. The schematic diagram and relevant notations are shown in Fig. 1. It is assumed that the bottom surface S_1 of the channel is fixed and hence the normal component of the velocity equals zero, that is

$$\mathbf{u} \cdot \mathbf{n} = 0 \quad \text{on} \quad S_1. \tag{7}$$

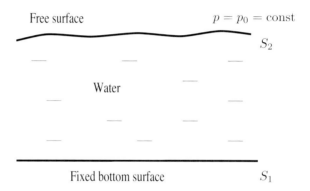

Figure 1. Surface water waves in a flow of finite depth.

It is also assumed that the upper surface S_2 is characterised by the equation

$$x_3 = \zeta(x_1, x_2, t),$$

and there is no variation in pressure on S_2:

$$p = p_0 = \text{const} \quad \text{on} \quad S_2.$$

Hence, according to (6) on the free surface we have

$$\frac{\partial \phi}{\partial t} + \frac{1}{2}|\nabla \phi|^2 + \frac{p_0}{\rho} + g\zeta(x_1, x_2, t) = f(t), \quad \text{as } x_3 = \zeta(x_1, x_2, t). \tag{8}$$

This is accompanied by the identity

$$u_3 = \frac{dx_3}{dt} = \frac{d\zeta}{dt} = \frac{\partial \zeta}{\partial t} + u_1 \frac{\partial \zeta}{\partial x_1} + u_2 \frac{\partial \zeta}{\partial x_2} \quad \text{on} \quad x_3 = \zeta(x_1, x_2, t). \quad (9)$$

In the linear approximation, we simplify the problem by addressing the case when $|\mathbf{u}|$ and the surface fluctuations are small.

In particular, if the unperturbed surface is $x_3 = h$, then for the free surface we can write

$$\phi(\mathbf{x}, t) = \phi(x_1, x_2, \zeta(x_1, x_2, t), t) =$$
$$= \phi(x_1, x_2, h, t) + (\zeta - h)\phi_{,3}(x_1, x_2, h, t) + \ldots$$

To the leading order approximation, the relations (8), (9) are set on the unperturbed surface $x_3 = h$. The second-order term $\frac{1}{2}|\nabla \cdot \phi|^2$ is neglected, and equation (8) becomes

$$\frac{\partial \phi}{\partial t} + \left(\frac{p_0}{\rho} - f(t) + gh \right) + g(\zeta - h) = 0.$$

Since the term $\left(\frac{p_0}{\rho} - f(t) + gh \right)$ is a function of t only, it does not give any contribution to $\mathbf{u}$, and may be "absorbed" into $\frac{\partial \phi}{\partial t}$. Thus, the velocity potential ϕ can be chosen in such a way that

$$\frac{\partial \phi}{\partial t} + g(\zeta - h) = 0 \quad \text{on} \quad x_3 = h. \quad (10)$$

By neglecting the second-order terms $\phi_{,1}, \zeta_{,1}$ and $\phi_{,2}, \zeta_{,2}$ in (9), involving the partial derivatives of ζ and ϕ, we deduce that to the leading order

$$\frac{\partial \phi}{\partial x_3} = \frac{\partial \zeta}{\partial t} \quad \text{on} \quad x_3 = h, \quad (11)$$

and therefore equations (10), (11) yield

$$\frac{\partial^2 \phi}{\partial t^2} + g \frac{\partial \phi}{\partial x_3} = 0 \quad \text{on} \quad x_3 = h. \quad (12)$$

Thus, in the "ocean" with the flat bottom surface $x_3 = 0$ and the unperturbed upper surface $x_3 = h$, the linearized formulation for the velocity potential ϕ is comprised of Laplace's equation (4) within the fluid layer $0 < x_3 < h$, the boundary condition $\mathbf{n} \cdot \nabla \phi = 0$ at the fixed bottom surface $x_3 = 0$, and the "free surface" boundary condition (12). Note that this formulation involves the second-order time derivative of the velocity potential on the upper free surface. In turn, after evaluation of ϕ, the "wave profile" on the free upper surface is defined by (10).

Deep water waves versus waves in shallow water

Consider a surface water wave, with the straight front perpendicular to the x_1−axis. We seek solutions independent of x_2 with the velocity potential

$$\phi = \phi(x_1, x_3, t). \tag{13}$$

It is convenient to work with the complex potential Ψ of the form

$$\Psi = W(x_3)e^{-i(\omega t - kx_1)},$$

and assume that $\phi = \Re e\Psi$. Considering a harmonic solution (i.e. $\nabla^2\Psi = 0$) we obtain

$$W'' - k^2 W = 0,$$

which yields

$$W = C_1 \cosh(kx_3) + C_2 \sinh(kx_3).$$

The boundary condition at the bottom surface ($x_3 = 0$) leads to $W'(0) = 0$, and hence $C_2 = 0$ and

$$\Psi = C_1 \cosh(kx_3)e^{-i(\omega t - kx_1)}. \tag{14}$$

In turn, the free surface condition

$$\left. \left(\frac{\partial^2 \Psi}{\partial t^2} + g\frac{\partial \Psi}{\partial x_3} \right) \right|_{x_3=h} = 0 \tag{15}$$

becomes

$$\{-\omega^2 C_1 \cosh(kh) + gkC_1 \sinh(kh)\}e^{-i(\omega t - kx_1)} = 0,$$

which simplifies to the form

$$gk\tanh(kh) = \omega^2. \tag{16}$$

Equation (16) relates the wave number k and the radian frequency ω, and thus the wave speed $c = \frac{\omega}{k}$ is in general frequency dependent. This shows that the water surface waves are *dispersive*. The relation (16) is called the *dispersion equation.*

We would like to note two important particular cases, which are

 (a) Deep water, when the non-dimensional quantity $kh \gg 1$ is large, and hence $\tanh(kh) \simeq 1$. Thus to leading order the dispersion equation (16) yields

$$\omega^2 = gk.$$

In this case, the wave speed increases when the frequency decreases, i.e. $c = g\omega^{-1}$, which is fully consistent with the physical observation that low frequency waves in the ocean propagate faster than the high frequency ripples.

(b) Shallow water, when $kh \ll 1$, and hence $\tanh(kh) \sim kh$. In this case, to leading order the dispersion equation (16) takes the form

$$\omega^2 = gk^2 h,$$

which implies $c = \sqrt{gh}$. In the framework of this approximation, the wave speed in shallow water is frequency independent and hence the shallow water waves can be treated as non-dispersive.

It is noted that the dispersion of surface water waves in the above illustrative example is linked to the boundary condition (12), which involves the second-order time derivative of the velocity potential.

The phenomenon of wave dispersion can also be observed in simple periodic structures. Conventionally, dispersion diagrams are used to describe the change in frequency with the wave number, which will be illustrated in the next section.

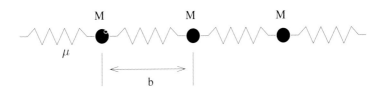

Figure 2. The one-dimensional spring-mass periodic structure.

0.3 Bloch-Floquet waves in periodic structures

A Bloch-Floquet wave in a periodic system is an object discussed in the classical texts, such as Kittel (1996) and Brillouin (1953). We give an outline of elementary examples incorporating a time-harmonic motion of a one-dimensional periodic lattice, consisting of rigid particles connected by

massless springs. The sketch of such a system is shown in Fig. 2. First, we
consider the case when all particles have the same mass M, which will be
followed by the comparative analysis of the inhomogeneous lattice involving
particles of different mass. The separation between neighbouring masses is
set to be unity, and the stiffness of springs is assumed to be μ.

Let u_n be the displacement of the n-th node within the chain, with
the unit interatomic spacing ($b = 1$). Then the equations of motion take
the form

$$M\ddot{u}_n = \mu(u_{n+1} + u_{n-1} - 2u_n) \quad \text{with } n \text{ being integer.} \tag{17}$$

If the motion is time harmonic then $u_n = U_n \exp(-i\omega t)$, and therefore

$$-M\omega^2 U_n = \mu(U_{n+1} + U_{n-1} - 2U_n). \tag{18}$$

In the case of travelling waves, we have

$$U_n = Ue^{inK}, \quad \text{with } U = \text{const}, \tag{19}$$

and the equation of motion becomes

$$-\omega^2 M U_n = 2\mu U_n(\cos(K) - 1).$$

The quantity K is said to be the Bloch parameter, and a solution u_n of
(17), which satisfies the condition

$$u_{s+n} = u_s e^{inK},$$

is referred to as the *Bloch-Floquet wave*. A non-trivial solution U_n of (18),
(19) exists, provided ω and K satisfy the following dispersion equation

$$\omega^2 - \frac{2\mu}{M}(1 - \cos(K)) = 0,$$

whose non-negative roots are

$$\omega = 2\sqrt{\frac{\mu}{M}}|\sin(K/2)|. \tag{20}$$

The corresponding dispersion curve, representing a periodic function of period 2π, is shown in Fig. 3. When the separation between the neighbouring
particles equals b, the period for the dispersion curve becomes $2\pi/b$. The
interval $(-\pi/b, \pi/b)$ is known as the *irreducible Brillouin zone*. Conventionally the dispersion diagrams are displayed for the values of K from the
irreducible Brillouin zone. In the case when the "interatomic spacing" in
the lattice equals b, the dispersion equation (20) is modified as follows:

$$\omega = 2\sqrt{\frac{\mu}{M}}|\sin\left(\frac{Kb}{2}\right)|. \tag{21}$$

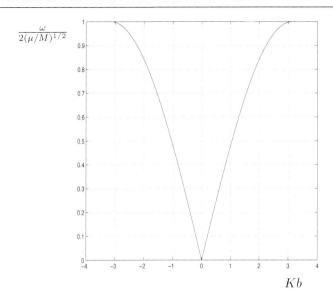

$$\frac{\omega}{2(\mu/M)^{1/2}}$$

$$Kb$$

Figure 3. Dispersion diagram for a uniform spring-mass structure

Standing waves and "long-wave" asymptotic estimates

It is convenient to refer to the general case of an arbitrary interatomic separation b within the lattice. In particular, taking $b \to 0$ one obtains the continuum limit. We note that

$$U_{n+1}/U_n = e^{iKb},$$

and within the irreducible Brillouin zone we have $K_{max} = \pi/b$. In the continuum limit, as $b \to 0$, we deduce $K_{max} \to \infty$.

The transmission velocity of a wave packet is called the *group velocity*, and it is given as

$$v_g = \frac{d\omega}{dK} = \sqrt{\frac{\mu}{M}} b \cos(Kb/2).$$

On the boundaries $K = \pm\pi/b$ of the Brillouin zone, we have $v_g = 0$; the solution U_n represents a *standing wave* with zero net transmission velocity, and $U_n = (-1)^n U$.

In many physical applications, it is useful to have asymptotic approximations corresponding to the *long wave limit*. In particular, this will give the slope of the curve $\omega = \omega(K)$ as $K \to 0+$, which is also referred to as the effective group velocity v_{eff} often used in *homogenization* approximations of wave phenomena in structured media. When $Kb \ll 1$, we have $\cos(Kb) \simeq 1 - \frac{1}{2}(Kb)^2$, and hence $\omega^2 \simeq \frac{\mu}{M}(Kb)^2$. Hence, $v_{\text{eff}} = (\mu/M)^{1/2}b$, and thus v_{eff} is frequency-independent in this limit.

A bi-atomic chain and stop bands

As a classical example (see, for example, Kittel (1996) and Brillouin (1953)), the one-dimensional periodic lattice consisting of two types of particles, of different masses, gives a very good illustration of filtering properties of structured media.

   Compared to the previous section, the new periodic system has two types of masses, M_1 and M_2, and springs of the normalized stiffness μ (see Fig. 4).

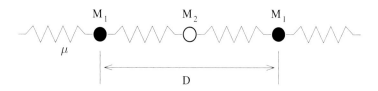

Figure 4. The one-dimensional "bi-atomic" periodic structure.

   In this case the elementary cell of the periodic lattice includes two different particles, with displacements u_n and v_n, and one needs two equations of motion as follows

$$M_1 \frac{d^2 u_n}{dt^2} = \mu(v_n + v_{n-1} - 2u_n), \ M_2 \frac{d^2 v_n}{dt^2} = \mu(u_{n+1} + u_n - 2v_n), \quad (22)$$

Let D be the size of the elementary cell, which is equal to the distance between the nearest particles of the same mass. Travelling waves are defined by

$$u_n = U e^{i(nKD - \omega t)}, \quad v_s = V e^{i(nKD - i\omega t)},$$

where K is the wave number. On substitution into the equations of motion, we have

$$\omega^2 \mu^{-1} M_1 U + V(1 + e^{-iKD}) - 2U = 0, \quad \omega^2 \mu^{-1} M_2 V + U(1 + e^{iKD}) - 2V = 0.$$

This is a homogeneous system of linear algebraic equations with respect to U and V, and it has a non-trivial solution if and only if

$$\frac{M_1 M_2}{\mu^2}\omega^4 - \frac{2(M_1 + M_2)}{\mu}\omega^2 + 2(1 - \cos(KD)) = 0. \tag{23}$$

This is the *dispersion equation* providing the relation between ω and K. For the present simple system, this equation has the explicit solution

$$\omega^2 = \mu \frac{M_1 + M_2 \pm \sqrt{(M_1 + M_2)^2 - 2M_1 M_2 (1 - \cos(KD))}}{M_1 M_2}. \tag{24}$$

This formula suggests that the corresponding dispersion diagram, constructed for equation (23) has two branches. They are referred to as the "acoustic branch" (for the sign "-" in (24)) and the "optical branch" (for the sign "+" in (24)), as shown in Fig. 5.

We note that for the non-homogeneous system, when $M_1 \neq M_2$, there is a non-zero separation between the dispersion curves, a *stop band*. The width of the stop band can be computed by evaluating the frequencies at the end points of the Brillouin zone. At the boundary of the Brillouin zone, when $K = \pm\pi/a$, the roots of the dispersion equation are defined by

$$\omega^2 = \frac{\mu}{M_1 M_2}\{M_1 + M_2 \pm |M_1 - M_2|\}.$$

Assuming that $M_1 < M_2$, one can deduce that the width of the band gap is equal to $\sqrt{2\mu/M_1} - \sqrt{2\mu/M_2}$. If we fix M_2 and let the contrast parameter $r = M_2/M_1$ increase, then the width of the stop band will increase. No propagating wave exists within the interval of frequencies $(\sqrt{2\mu/M_2}, \sqrt{2\mu/M_1})$.

By looking at the dispersion diagram in Fig. 5, we can also identify the *high frequency stop band* defined as the interval $(\omega_*, +\infty)$, where $\omega_* = 2\mu(\frac{1}{M_1} + \frac{1}{M_2})$ is the upper limit for frequencies corresponding the optical branch of the dispersion diagram, obtained as $K \to 0$.

For the bi-atomic lattice system, the Bloch-Floquet waves exist only within the frequency intervals $[0, \sqrt{2\mu/M_2}]$ and $[\sqrt{2\mu/M_1}, \sqrt{2\mu(\frac{1}{M_1} + \frac{1}{M_2})}]$, which are also called the *pass-band* intervals; one can control the size of the pass-band intervals by changing the stiffness of springs and the masses of particles.

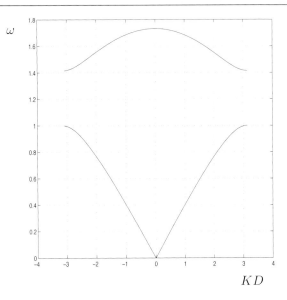

KD

Figure 5. Dispersion curves for the "bi-atomic" spring-mass structure; in this computation $M_1 = 1, M_2 = 2, \mu = 1$.

0.4   Lattice approximation for continuous structured media

As we have demonstrated in the above section, the whole pass-band set, incorporating all pass-band intervals for a discrete periodic lattice, is bounded. In contrast, a continuum system would support propagation of high-frequency waves. For high-contrast periodic structures it is sometimes possible to construct a "lattice approximation", which would describe adequately the dispersion properties of Bloch-Floquet waves in the low frequency range. An example of such a structure is discussed in this section, which is based on the paper by Movchan *et al.* (2002b).

Instead of a discrete system of masses connected by weightless springs, we consider here a one-dimensional periodic array of elastic rods of different stiffnesses μ_j, $j = 1, 2$, and non-zero mass density, as shown in Fig. 6. For convenience, it is assumed that the linear mass density ρ is the same for all elements of this structure. Thus, the elementary cell contains two types of elastic rods. We use the notations $S_1^{(n)} = (-b + nD, nD)$, $S_2^{(n)} = (nD, a + nD)$, where n is integer, and $D = a + b$ is the total size of the elementary cell.

Assuming that the waves are time-harmonic, of the radian frequency ω, we deduce that the amplitudes of the longitudinal displacements $U_j(x)$, $j = 1, 2$, satisfy the equations

$$\mu_j U_j'' + \omega^2 \rho U_j = 0, \quad x \in S_j^{(n)}, \; j = 1, 2. \tag{25}$$

The ideal contact conditions at the interface between two neighbouring rods imply

$$U_1 = U_2, \quad \mu_1 U_1' = \mu_2 U_2'. \tag{26}$$

The solution is sought in the class of Bloch-Floquet waves, for which we can write

$$U_j(x + D) = e^{iKD} U_j(x), \; j = 1, 2, \; |K| < \pi/D, \tag{27}$$

where K is the Bloch parameter.

The general solution of (25) is $U_j = A_j e^{ik_j x} + B_j e^{-ik_j x}$, $x \in S_j^{(n)}$, $j = 1, 2$, where A_j, B_j, $j = 1, 2$, are constant coefficients, and $k_j(\omega) = \omega \sqrt{\rho/\mu_j}$, $j = 1, 2$, are linear functions of ω. The interface conditions (26) and the Bloch-Floquet's condition (27), written for displacements within the elementary cell, lead to a homogeneous system of linear algebraic equations with respect to A_j, B_j, $j = 1, 2$:

$$\mathbf{Q}(K, \omega) \begin{pmatrix} A_1 \\ B_1 \\ A_2 \\ B_2 \end{pmatrix} = 0, \tag{28}$$

where

$$\mathbf{Q}(K, \omega)$$

$$= \begin{pmatrix} 1 & 1 & -1 & -1 \\ \mu_1 k_1 & -\mu_1 k_1 & -\mu_2 k_2 & \mu_2 k_2 \\ -e^{i(KD-k_1 b)} & -e^{i(KD+k_1 b)} & e^{ik_2 a} & e^{-ik_2 a} \\ k_1 e^{i(KD-k_1 b)} & -k_1 e^{i(KD+k_1 b)} & -\frac{\mu_2}{\mu_1} k_2 e^{ik_2 a} & \frac{\mu_2}{\mu_1} k_2 e^{-ik_2 a} \end{pmatrix}. \tag{29}$$

This system has a non-trivial solution, provided

$$\det \mathbf{Q}(K, \omega) = 0, \tag{30}$$

which relates the radian frequency ω to the Bloch parameter K and hence gives the dispersion equation for Bloch-Floquet waves propagating along the periodic structure. It is convenient to introduce the notation $\varepsilon = \mu_1/\mu_2$, and in this case $k_2 = k_1\sqrt{\epsilon}$. Then (30) leads to

$$(\epsilon+1)\sin(k_1 b)\sin(k_1 a\sqrt{\epsilon})-2\sqrt{\epsilon}(\cos(k_1 b)\cos(k_1 a\sqrt{\epsilon})-\cos(KD)) = 0. \quad (31)$$

The dispersion diagram, which shows ω versus K, is given in Fig. 7.

Let $\epsilon \ll 1$ and $k_1 b \ll 1$, i.e. the structure has the high-contrast in stiffness for different components, and the elastic rods of length b are relatively "short". Then the corresponding trigonometric parts in the left-hand side of (31) can be expanded into power series, and after truncation we obtain a polynomial expression in powers of ω.

Retaining the terms up to ω^4, we obtain the *approximate dispersion equation*

$$\omega^4 \mathcal{P} - \omega^2 \mathcal{Q} + 2(1 - \cos(KD)) = 0, \quad (32)$$

where the quantities $\mathcal{P}$ and $\mathcal{Q}$ are positive and have the form

$$\mathcal{P} = \frac{\rho^2}{12/\mu_1^2}(\epsilon a^2 + 2\epsilon ba + b^2)(\epsilon a^2 + b^2 + 2ba), \quad \mathcal{Q} = \frac{\rho}{\mu_1}(a+b)(\epsilon a + b).$$

Remarkably, equation (32) has the same structure as (23) derived for a discrete system incorporating a one-dimensional array of masses M_1, M_2 connected by elastic springs of stiffness μ. This observation suggests that a discrete lattice system can be used to approximate a high-contrast periodic continuous systems to obtain its dynamic response within the low frequency range.

This discrete model can be characterised by the two parameters M_1/μ and M_2/μ, and the equations (32) and (23) become the same when

$$\mathcal{P} = M_1 M_2/\mu^2, \quad \mathcal{Q} = 2(M_1 + M_2)/\mu.$$

Of course, in this case the dispersion diagrams, within the range of frequencies covering the acoustic and optical branches, also agree. In the dispersion diagram of Fig. 7 we show the dependance of k_1 versus K and note the presence of stop bands, which occur for Bloch waves in the high-contrast bi-material periodic system. For this computational example we used the following numerical values of the geometrical and materials parameters: $D = 1, b = 0.5, \epsilon = 0.1$.

It is noted that an infinite number of dispersion curves corresponds to the bi-material continuum system, whereas the lattice approximation covers only first two dispersion curves adjacent to the origin.

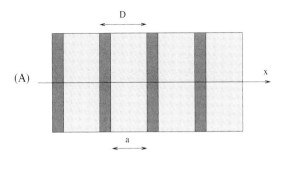

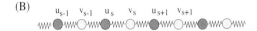

(B) u_{s-1} v_{s-1} u_s v_s u_{s+1} v_{s+1}

Figure 6. One-dimensional arrays of two layers (A) and the approximating periodic lattice system of two types of particles connected by springs (B).

0.5 Transmission through a finite thickness interface

We can consider a one-dimensional "structured interface", which incorporates a finite number of elastic rods aligned with the $x-$axis and joined together sequentially. The rods themselves may have different lengths, densities and elastic stiffness. A wave propagating in the positive direction of the $x-$axis interacts with the interface, and some of the energy gets reflected whereas the remaining energy is transmitted through the interface. The percentage of the reflected energy is frequency dependent. For the case when there is a repeating pattern, and the size of such an interface increases to infinity to form a periodic structure, it is appropriate to make a connection with analysis of Bloch-Floquet waves and their dispersion properties in the periodic system, as explained in Lekner (1994) and Brun *et al.* (2010a). However, when the thickness of the interface structure is not large, the analysis follows a different pattern. The emphasis is on trapped modes, which may exist within the structured interface and which may enhance its transmittance properties. The transmission matrix technique, as in Lekner (1994), is an efficient tool outlined below.

The displacement amplitude u for the time-harmonic motion of ra-

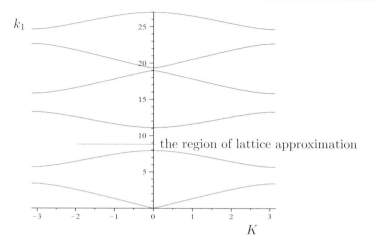

Figure 7. Dispersion diagram for Bloch waves in the high-contrast bi-material periodic system.

dian frequency ω satisfies the equation

$$\frac{\partial^2 u(x)}{\partial x^2} + \frac{\rho}{\mu}\omega^2 u(x) = 0, \tag{33}$$

where ρ and μ are the mass density and the stiffness constants, respectively. The wave form can be expressed in terms of complex amplitudes A and B

$$u(x) = A\exp(ikx) + B\exp(-ikx), \tag{34}$$

with

$$k = \omega c, \quad c = \sqrt{\frac{\rho}{\mu}}.$$

Consider a one-phase interface located at $x = x_0$ and $x = x_1 = x_0 + d$. Then displacements and tractions are related by

$$\begin{bmatrix} u(x_1) \\ \mu\dfrac{\partial u}{\partial x}(x_1) \end{bmatrix} = \begin{bmatrix} m_{11} & m_{12} \\ m_{21} & m_{22} \end{bmatrix} \begin{bmatrix} u(x_0) \\ \mu\dfrac{\partial u}{\partial x}(x_0) \end{bmatrix}, \tag{35}$$

where, according to (34), we have

$$\mathbf{M} = \begin{bmatrix} \cos \delta & \dfrac{\sin \delta}{Q} \\[2ex] -Q \sin \delta & \cos \delta \end{bmatrix}, \tag{36}$$

with $Q = \mu k$ and $\delta = k d$ being the *phase increment*. The matrix $\mathbf{M}$ in (36) is said to be the *transfer matrix* and its eigenvalues are

$$\cos \delta \pm \sqrt{\cos^2 \delta - 1} = \exp(\pm i\delta). \tag{37}$$

Along the same lines, we now analyse a *discrete interface* consisting of two springs of stiffness γ and and a mass m. The transfer matrix $\mathbf{M}^{(d)}$ of the discrete interface is defined in the form

$$\mathbf{M}^{(d)} = \begin{bmatrix} \dfrac{\gamma - m\omega^2}{\gamma} & \dfrac{2\gamma - m\omega^2}{\gamma^2} \\[2ex] -m\omega^2 & \dfrac{\gamma - m\omega^2}{\gamma} \end{bmatrix}. \tag{38}$$

The eigenvalues of the transfer matrix (38) are

$$1 - \frac{m}{\gamma}\omega^2 \pm \sqrt{\frac{m}{\gamma}\omega^2 \left(\frac{m}{\gamma}\omega^2 - 2 \right)}, \tag{39}$$

so that the dependence on ω is algebraic for the structured discrete interface, in contrast with the continuum one-phase interface (see (37)). It is also noted that the eigenvalues (39) are complex for sufficiently small ω and real when ω is sufficiently large.

If several interface regions, continuous or discrete, are placed together to form an interface stack then the transfer matrix is obtained by the appropriate multiplication of the transfer matrices of individual layers within the stack, as discussed in Brun *et al.* (2010a).

Reflected and transmitted energy

We consider an interface stack with the overall 2×2 transfer matrix $\mathbf{M} = (m_{ij})$ separating two elastic media (for the one dimensional case, we have two semi-infinite elastic rods located along the intervals $x < x_0$ and $x > x_1$), with the density ρ_- and stiffness μ_- on the left from the interface and the density ρ_+ and stiffness μ_+ on the right from the interface. The *incident* and *reflected* waves on the left of the interface admit the representation

$$u(x) = A_I \exp(ik_-x) + A_R \exp(-ik_-x), \tag{40}$$

where A_I and A_R are the *incident* and *reflection amplitudes*, while the transmitted wave on the right of the interface is given by

$$u(x) = \sqrt{\frac{\mu_-}{\mu_+}} A_T \exp(ik_+x), \tag{41}$$

with A_T denoting the *transmission amplitude*, and $k_\pm$ standing for the corresponding wave numbers in the regions outside the interface.

According the definition of the transfer matrix outlined in the text above, the coefficients A_R, A_T and A_I satisfy the equation

$$\sqrt{\frac{\mu_-}{\mu_+}} \begin{bmatrix} A_T e^{ik_+x_1} \\ iQ_+ A_T e^{ik_+x_1} \end{bmatrix} = \mathbf{M} \begin{bmatrix} A_I e^{ik_-x_0} + A_R e^{-ik_-x_0} \\ iQ_-(A_I e^{ik_-x_0} - A_R e^{-ik_-x_0}) \end{bmatrix}, \tag{42}$$

where $Q_\pm = \mu_\pm k_\pm$. In turn, the amplitudes A_R and A_T of the reflected and transmitted waves can be written (see Brun *et al.* (2010a)) in the form

$$\bar{A}_R = \frac{Q_-Q_+m_{12} + m_{21} + i(Q_-m_{22} - Q_+m_{11})}{Q_-Q_+m_{12} - m_{21} + i(Q_-m_{22} + Q_+m_{11})} e^{2ik_-x_0},$$

$$\bar{A}_T = \sqrt{\frac{\mu_+}{\mu_-}} \frac{2iQ_-}{Q_-Q_+m_{12} - m_{21} + i(Q_-m_{22} + Q_+m_{11})} e^{i(k_-x_0 - k_+x_1)},$$
$$\tag{43}$$

where we have used the normalised quantities $\bar{A}_R = A_R/A_I$, $\bar{A}_T = A_T/A_I$.

According to the standard procedure, as in Lekner (1994) and Brun *et al.* (2010a), the normalized *reflected* R and *transmitted* $T = 1 - R$ *energies* are given by

$$R = |\bar{A}_R|^2 = \frac{(Q_-Q_+m_{12} + m_{21})^2 + (Q_-m_{22} - Q_+m_{11})^2}{(Q_-Q_+m_{12} - m_{21})^2 + (Q_-m_{22} + Q_+m_{11})^2},$$

$$T = |\bar{A}_T|^2 \frac{Q_+}{Q_-} = \frac{4Q_-Q_+}{(Q_-Q_+m_{12} - m_{21})^2 + (Q_-m_{22} + Q_+m_{11})^2}, \tag{44}$$

so that the total transmission corresponds to the case of $R = 0$, $T = 1$, and total reflection corresponds to $R = 1$, $T = 0$.

Defect modes and enhanced transmission

There is a connection between the transmission properties of a finite width stack and dispersion properties of Bloch waves in infinite doubly periodic plane. In particular, if a full stop band is identified for Bloch waves, then the the finite width stack built of the same material is expected to reflect the bulk of the energy within the stop band frequency range. However, the presence of defects in the stack may alter its transmission properties. This is illustrated here for a simple one-dimensional example.

We consider the one-dimensional problem involving the spring-mass interface, as described above. In the computational example below, the discrete interface is composed of masses $m = 387.3kg$ and $M = 3m$ connected be identical springs of stiffness $\gamma = 3873N/m$. The dispersion diagram, which shows the wave number versus the radian frequency, for the two mass lattice system is given in Fig. 8. As expected, this diagram includes the band gap. If a finite sample $mM - mM - mM - mM$ of this periodic system is used as a structured interface, the bulk of the energy would be reflected within the stop band frequency range.

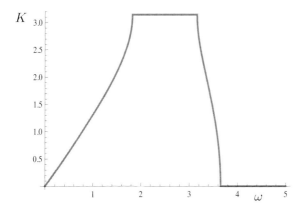

Figure 8. Dispersion diagram for discrete system of masses m and M connected by springs of stiffness γ.

The plot of the normalized reflected energy as a function of the frequency are shown (in gray colour) in Fig. 9 where the bi-atomic lattice interface $mM - mM - mM - mM$ is inserted between two different infinite continuous media. The continuous media on the left and on the right from the interface have the elastic impedance $\sqrt{\rho_-\mu_-} = 100Pa/m$ and $\sqrt{\rho_+\mu_+} = 1500Pa/m$, respectively.

On the same figure, we also show (black curve) the normalised reflected energy for the interface $mM - Mm - mM - mM$, where a defect has

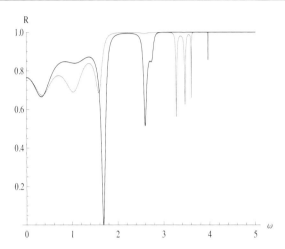

Figure 9. Normalised reflected energy for the stack consisting of 4 cells, mM-Mm-mM-mM (black curve), versus the normalised reflected energy for the stack mM-mM-mM-mM (gray curve).

been created by changing the order of masses M and m in the second cell of the stack. This change leads to an enhanced transmission (i.e. a sharp drop in the reflected energy), which is clearly visible on the diagram in Fig. 9. This effect is discussed in detail in the next section, where localisation, associated with "defect modes" within a structured interface, is shown to lead to an enhanced transmission.

0.6 Polarisation and enhanced transmission of elastic waves by a shear-type structured interface

This section reviews the results of the paper by Brun *et al.* (2010b) addressing a vector problem of two-dimensional elasticity incorporating plane pressure and shear waves interacting with a finite width structured interface of shear type. The emphasis is on the formation of the defect modes within the interface, which lead to an enhanced transmission. For this configuration, we also discuss the wave polarisation as a result of the interaction with the structured interface

Structured interface in an elastic medium

The structured interface is assumed to occupy a horizontal strip in the (x, z)-plane:

$$\Pi_D = \{(x, z) : x \in \mathbb{R}, -D \leq z \leq 0\}. \qquad (45)$$

We consider an example where the structure within the interface includes three parallel elastic infinite bars connected by transverse elastic massless links. In other words, within the interface we have an elastic frame, so that the angle between the vertical links and the horizontal elastic bars is maintained at $\pi/2$. The notations $\Omega_\pm$ are used for the half-planes above and below the structured interface. The sketch of the elastic system and relevant notations are shown in Fig. 10.

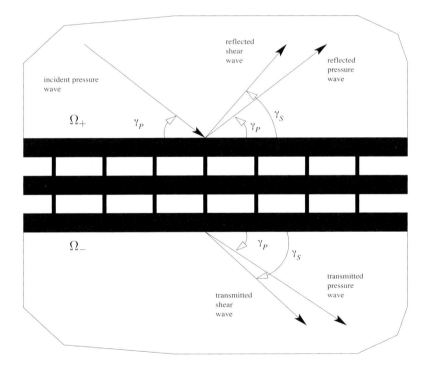

Figure 10. Structured shear-type interface, separating the elastic half-planes $\Omega_\pm$.

Each elastic bar $l_j, j = 1, 2, 3$, is assumed to be of high flexural stiffness and hence is approximated as as a one-dimensional elastic element, which moves only in the horizontal direction (along the Ox-axis), with the elastic displacement $\mathbf{u}^{(j)} \sim u_j(x,t)\mathbf{e}^{(1)}$. The equations of motion can be written separately for the elastic bar inside the interface ($j = 2$), and for the bars adjacent to Ω_+ and Ω_- ($j = 1$ and $j = 3$). The equation of motion for the interior bar is

$$E(u_2)_{xx} - \rho(u_2)_{tt} - 2\gamma u_2 + \gamma(u_3 - u_1) = 0, \tag{46}$$

where $(u_j)_{xx}$ and $(u_j)_{tt}$ stand for the second-order partial derivatives with respect to x and t, respectively.

For the bars forming the boundary of the structured interface, the equations of motion are

$$E(u_1)_{xx} - \rho(u_1)_{tt} + \gamma(u_2 - u_1) + \tau_+ = 0, \tag{47}$$

$$E(u_3)_{xx} - \rho(u_3)_{tt} + \gamma(u_2 - u_3) - \tau_- = 0. \tag{48}$$

Here τ_+ and τ_- are the shear stresses in the ambient elastic media above and below the interface.

Elastic waves in the ambient medium

The upper half-plane Ω_+ includes the incident plane pressure wave, together with reflected pressure and shear waves. The transmitted pressure and shear waves propagate in the lower half-palne Ω_-. Taking the x−axis along the interface, z−axis perpendicular to the interface and y−axis to be pointed out of the plane of Fig. 10, we use the standard representation of the displacement

$$u(x, z, t)\mathbf{e}^{(1)} + w(x, z, t)\mathbf{e}^{(3)}$$

in terms of the scalar and vector potentials $\varphi(x, z, t)$ and $\psi(x, z, t)\mathbf{e}^{(2)}$:

$$u = \frac{\partial\varphi}{\partial x} - \frac{\partial\psi}{\partial z}, \quad w = \frac{\partial\varphi}{\partial z} + \frac{\partial\psi}{\partial x}. \tag{49}$$

The notations $\{\varphi^{(I)}, \psi^{(I)}\}$, $\{\varphi^{(R)}, \psi^{(R)}\}$, $\{\varphi^{(T)}, \psi^{(T)}\}$ correspond to the incident, reflected and transmitted waves, respectively. Since the incident wave is assumed to be of pressure type, we have $\psi^{(I)} \equiv 0$.

Assume that the same elastic material occupies Ω_+ and Ω_-. If α, β are the wave speeds of the pressure and shear waves then the pressure and shear wave potentials ϕ and ψ in the elastic continuum satisfy the wave equations

$$\triangle\varphi - \frac{1}{\alpha^2}\frac{\partial^2}{\partial t^2}\varphi = 0, \quad \triangle\psi - \frac{1}{\beta^2}\frac{\partial^2}{\partial t^2}\psi = 0, \tag{50}$$

where

$$\alpha = \sqrt{\frac{\lambda + 2\mu}{\rho}}, \qquad \beta = \sqrt{\frac{\mu}{\rho}}. \qquad (51)$$

Let c be an apparent velocity of the incident pressure wave along the horizontal interface, and let $\chi_P \in [0, \pi/2]$ and $\chi_S \in [0, \pi/2]$ denote the angle between the direction of the incident or transmitted pressure wave and the horizontal interface and the angle between the direction of propagation of the reflected or transmitted shear wave and the horizontal interface. Then

$$a := \tan \chi_P = \sqrt{\frac{c^2}{\alpha^2} - 1} \quad \text{and} \quad b := \tan \chi_S = \sqrt{\frac{c^2}{\beta^2} - 1}. \qquad (52)$$

For the pressure wave, the angle of reflection is equal to the angle of incidence, and $\chi_S > \chi_P$ (see Fig. 10).

Accordingly, the pressure and shear wave potentials are

$$\begin{aligned}
\varphi_+ &= \varphi^{(I)} + \varphi^{(R)} = A_I \exp[ik(ct - x + az)] + A_R \exp[ik(ct - x - az)], \\
\psi_+ &= \psi^{(R)} = B_R \exp[ik(ct - x - bz)],
\end{aligned} \qquad (53)$$

in Ω_+, and

$$\begin{aligned}
\varphi_- &= \varphi^{(T)} = A_T \exp[ik(ct - x + az)], \\
\psi_- &= \psi^{(T)} = B_T \exp[ik(ct - x + bz)],
\end{aligned} \qquad (54)$$

in Ω_-. Here

$$k = \frac{2\pi}{\Lambda_P} \cos \chi_P = \frac{2\pi}{\Lambda_S} \cos \chi_S, \qquad (55)$$

where Λ_P and Λ_S are the wave lengths of the pressure and shear waves in the ambient elastic medium $\Omega_+ \cup \Omega_-$, and $\omega = kc$ is the radian frequency.

It is assumed that the amplitude A_I of the incident pressure wave is given, whereas A_R, B_R, A_T, B_T are evaluated with the account of the structured interface. In the text below, special attention is given to an enhanced transmission corresponding to a class of trapped modes which may occur within the interface structure.

The energy balance

Assuming that the motion is time-harmonic with the radian frequency ω, we use the notations U_j and $T_\pm$ for the amplitudes of the displacements of the bars $l_j, j = 1, 2, 3$ and for the shear stresses above and below the interface.

Then the equations of motion of elastic bars within the shear-type interface lead to the algebraic system

$$\mathcal{R}\mathbf{U} = \mathbf{T}, \tag{56}$$

where

$$\mathbf{U} = (U_1, U_2, U_3)^T, \ \mathbf{T} = (-T_+, 0, T_-)^T, \tag{57}$$

and

$$\mathcal{R} = \begin{pmatrix} -\Gamma & \gamma & 0 \\ \gamma & -\Gamma - \gamma & \gamma \\ 0 & \gamma & -\Gamma \end{pmatrix}, \tag{58}$$

with $\Gamma = k^2 E + \gamma - \rho \omega^2$.

As described in Brun *et al.* (2010b), the standard procedure involves the Betti formula applied to the displacement and its complex conjugate above and below the interface. Furthermore, using the plane wave representations (53) and (54) that include the pressure and shear wave potentials we deduce the following energy balance relation

$$E_I = E_R + E_T, \tag{59}$$

where

$$E_I = a|A_I|^2, E_R = a|A_R|^2 + b|B_R|^2, E_T = a|A_T|^2 + b|B_T|^2, \tag{60}$$

E_I, E_R and E_T represent the vertical energy fluxes of the incident, reflected and transmitted fields, respectively.

The effect of the interface on the distribution of energy can be seen by the evaluation of the coefficients A_R, B_R and A_T, B_T characterising the reflected and transmitted fields.

Trapped modes within the structured interface

Trapped vibrations within the structured interface can significantly alter its transmission properties. In the particular case involving the three-bar interface we discuss several examples in this section.

For a non-resonant regime, when $\det \mathcal{R} \neq 0$, the system (56) has a solution

$$\begin{pmatrix} U_1 \\ U_2 \\ U_3 \end{pmatrix} = \mathcal{R}^{-1} \begin{pmatrix} -T_+ \\ 0 \\ T_- \end{pmatrix}, \tag{61}$$

where

$$\mathcal{R}^{-1} = \frac{1}{(\gamma - \Gamma)(2\gamma + \Gamma)} \begin{pmatrix} \gamma - \frac{\gamma^2}{\Gamma} + \Gamma & \gamma & \frac{\gamma^2}{\Gamma} \\ \gamma & \Gamma & \gamma \\ \frac{\gamma^2}{\Gamma} & \gamma & \gamma - \frac{\gamma^2}{\Gamma} + \Gamma \end{pmatrix}. \tag{62}$$

In particular, the effective transmission relations for the shear displacements and shear stresses across the structured interface have the form:

$$\begin{aligned} U_1 &= \frac{\frac{\gamma^2}{\Gamma}(T_+ + T_-) - T_+(\gamma + \Gamma)}{(\gamma - \Gamma)(2\gamma + \Gamma)}, \\ U_3 &= \frac{T_-(\gamma + \Gamma) - \frac{\gamma^2}{\Gamma}(T_+ + T_-)}{(\gamma - \Gamma)(2\gamma + \Gamma)}. \end{aligned} \tag{63}$$

Solving the above system of algebraic equations with respect to B_R, B_T we deduce that

$$\begin{aligned} B_R &= -\frac{2a}{\Psi} \left[\Phi(\Gamma^2 + \gamma\Gamma - \gamma^2) - \gamma^2\Gamma(1 + ab) \right] A_I, \\ B_T &= \frac{2ia^2(1 + b^2)}{\Psi} k\gamma^2 \mu e^{ibDk} A_I, \end{aligned} \tag{64}$$

where $\Psi = \Gamma(1 + ab) + ik\mu a(1 + b^2)$ and $\Phi = \Psi \left[(\gamma + \Gamma)\Phi - 2\gamma^2(1 + ab) \right]$. The conditions of zero transverse displacements at the interface lead to the expressions for the coefficients A_R, A_T as follows

$$A_R = A_I - a^{-1}B_R, \qquad A_T = a^{-1}B_T e^{-ik(b-a)D}, \tag{65}$$

The full set $\{A_R, B_R, A_T, B_T\}$ determines the reflected and transmitted waves at the shear-type interface.

Furthermore, the system (56) can be written as follows

$$\mathbf{B} \begin{pmatrix} [[U]] \\ [[V]] \\ \langle U \rangle \end{pmatrix} = \begin{pmatrix} T_+ + T_- \\ \frac{1}{2}(T_- - T_+) \\ \frac{1}{3}(T_- - T_+) \end{pmatrix}, \tag{66}$$

where $[[U]] = U_3 - U_1$ represents the tangential displacement jump across the interface, $[[V]] = \frac{1}{2}(U_3 + U_1) - U_2$ is the average jump representing the difference between the average tangential displacement on the boundary and the tangential displacement of the interior bar, $\langle U \rangle = \frac{1}{3}(U_1 + U_2 + U_3)$ is the average tangential displacement over the whole structured interface, and

$$\mathbf{B} = \text{diag} \left\{ -Ek^2 + \rho\omega^2 - \gamma, -Ek^2 + \rho\omega^2 - 3\gamma, -Ek^2 + \rho\omega^2 \right\}. \tag{67}$$

The relation between $([[U]], [[V]]), \langle U \rangle)^T$ and $(U_1, U_2, U_3)^T$ has the form

$$\begin{pmatrix} [[U]] \\ [[V]] \\ \langle U \rangle \end{pmatrix} = \mathcal{Q} \begin{pmatrix} U_1 \\ U_2 \\ U_3 \end{pmatrix}, \tag{68}$$

where

$$\mathcal{Q} = \begin{pmatrix} -1 & 0 & 1 \\ \frac{1}{2} & -1 & \frac{1}{2} \\ \frac{1}{3} & \frac{1}{3} & \frac{1}{3} \end{pmatrix}, \tag{69}$$

and hence

$$\mathbf{B} = \mathcal{Q}\mathcal{R}\mathcal{Q}^{-1}. \tag{70}$$

The rows of the matrix $\mathcal{Q}$ are the eigenvectors of $\mathcal{R}$ corresponding to the eigenvalues, which coincide with the diagonal entries of the matrix $\mathbf{B}$, and

$$\det \mathbf{B} = \det \mathcal{R}. \tag{71}$$

The relation

$$\det \mathcal{R}(\omega, k) = 0 \tag{72}$$

is the dispersion equation for the elastic waves propagating horizontally along the structured interface, which is equivalent to

$$(-Ek^2 + \rho\omega^2 - \gamma)(-Ek^2 + \rho\omega^2 - 3\gamma)(-Ek^2 + \rho\omega^2) = 0. \tag{73}$$

The corresponding dispersion diagram has three branches shown in Fig. 11:

$$\begin{array}{ll} (1) & Ek^2 + 3\gamma - \rho\omega^2 = 0, \\ (2) & Ek^2 + \gamma - \rho\omega^2 = 0, \\ (3) & Ek^2 - \rho\omega^2 = 0. \end{array} \tag{74}$$

The lowest branch (3) gives a linear relation between k and ω, which corresponds to a non-dispersive wave propagating along an elastic bar. For the dispersive waves (branches (1) and (2)), there is a cut-off frequency

$$\omega^* = \sqrt{\frac{\gamma}{\rho}}, \tag{75}$$

and for $\omega < \omega^*$ no dispersive wave can propagate along the interface.

The resonance states correspond to the cases where the frequency ω of the incident wave coincides with one of the solutions of the dispersion equation (73). For the three-bar interface, such states can be classified as follows:

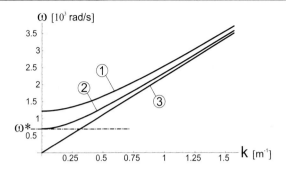

Figure 11. The dispersion curves for waves propagating horizontally along a three-bar structured interface. The material parameters for the interface are $\rho = 1000$ Kg/m, $E = 10^6$ kN, and $\gamma = 0.5$ GPa. The dispersion curves are defined by (74); the cut-off frequency for the first two branches is $\omega^* = 0.707 \times 10^3$ rad/s, as given by (75).

(a) When $Ek^2 - \rho\omega^2 = 0$, the compatibility condition for the system (66) is $T_- = T_+ = T$, which yields

$$[[V]] = 0, \qquad [[U]] = -\frac{2T}{\gamma}. \qquad (76)$$

(b) If $Ek^2 + 3\gamma - \rho\omega^2 = 0$, the equations (66) have a solution when $T_- = T_+ = T$, and hence

$$\langle U \rangle = 0, \qquad [[U]] = \frac{T}{\gamma}. \qquad (77)$$

(c) In turn, for the case when $Ek^2 + \gamma - \rho\omega^2 = 0$, the compatibility constraint is $-T_- = T_+ = T$, which leads to

$$\langle U \rangle = -\frac{2T}{3\gamma}, \qquad [[V]] = \frac{T}{2\gamma} \qquad \text{and} \qquad U_2 = -\frac{T}{\gamma}. \qquad (78)$$

The amplitudes of the reflected and transmitted shear waves B_R, B_T in the above cases (a), (b), (c) are

$(a) \quad B_R = \dfrac{a\gamma}{\gamma(1+ab)+ia(1+b^2)k\mu}A_I, \quad B_T = B_R e^{ibDk},$

$(b) \quad B_R = \dfrac{2a\gamma}{2\gamma(1+ab)-ia(1+b^2)k\mu}A_I, \quad B_T = B_R e^{ibDk},$

$(c) \quad B_R = \dfrac{2a\gamma}{2\gamma(1+ab)-ia(1+b^2)k\mu}A_I, \quad B_T = -B_R e^{ibDk}. \qquad (79)$

We note that in all above cases $|B_R| = |B_T|$. Hence we deduce that for the special resonance modes described above, the energy of the transmitted shear wave is equal to the energy of the reflected shear wave.

0.7   Concluding remarks and enhanced transmission

We finalise the article by emphasising on the important phenomena incorporating dispersive waves along the elastic bars within the structured interface as well as the enhanced transmission across the structured interface. Although the governing equations outside the interface are set for plane elastic waves, the phenomenon of the enhanced transmission is very much similar to the effect shown in the earlier section 0.5 for a simple one-dimensional example incorporating an interface with a defect.

In Fig. 12 we show the graphs of the reflected and transmitted energies as functions of c and k. In particular, the diagrams (a) and (c) show the surface plots of $E_R(k, c/\alpha)$ and $E_T(k, c/\alpha)$, whereas the parts (b) and (d) include the corresponding contour plots. We note that there is a rapid increase in the transmission energy when k is close to the value $\sqrt{2\gamma/(\rho c^2 - E)}$ for $c > \sqrt{E/\rho}$. This special case corresponds to a *defect resonance mode*, for which the upper and lower bars within the interface do not move, whereas the middle bar vibrates while being connected to the upper and lower bars via elastic links of stiffness γ. The corresponding equation of motion for such a defect mode leads to

$$(Ek^2 + 2\gamma - \rho\omega^2)U_2 = 0, \qquad (80)$$

which is fully consistent with the observed peak in transmission.

The computations in Fig. 12 are based on the algorithm of Section 0.6, and, in particular, equations (65), (64), (79) and (60). In these numerical computations, we have taken the linear mass density $\rho = 1000$ kg/m and the longitudinal stiffness $E = 5 \times 10^6$ kN for the bars, the shear stiffness $\gamma = 0.05$ GPa for the vertical elastic links and the mass density $\rho_{\text{amb}} = 1000$ kg/m^3, Poisson ratio $\nu_{\text{amb}} = 0.3$ and Young's modulus $E_{\text{amb}} = 1$ GPa for the ambient elastic medium $\Omega_+ \cup \Omega_-$; the pressure

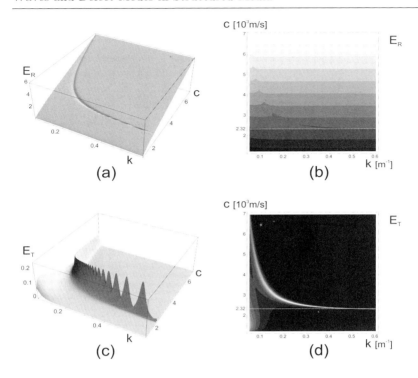

Figure 12. The reflected and transmitted energies E_R and E_T for a three-bar interface as functions of the wave number k and the apparent velocity c. Diagrams (a) and (c) show the surface plots of E_R and E_T. Diagrams (b) and (d) present the corresponding contour plots. The material parameters for the interface are $\rho = 1000$ kg/m, $E = 5 \times 10^6$ kN. The ambient medium is of the mass density $\rho_{\mathrm{amb}} = 1000$ kg/m^3, Poisson ratio $\nu_{\mathrm{amb}} = 0.3$ and Young's modulus $E_{\mathrm{amb}} = 1$ GPa; the horizontal stiffness of the vertical links is $\gamma = 0.05$ GPa/m^2.

and shear wave speeds in the ambient medium are respectively $\alpha = 1160.24$ m/s and $\beta = 620.174$ m/s.

In conclusion we emphasise on the connection between physical problems of different origins. Dispersion of waves in media with boundaries or built-in micro-structure is a fundamental phenomenon, which occurs in problems of acoustics, models of water waves, simple systems involving one-dimensional harmonic oscillators, as well as complex elastic systems leading to analysis of full vector problems of elasticity both for discrete and continuous systems. Indeed, if the system is infinite and periodic, the analysis is reduced to an elementary cell and dispersion properties of Bloch-Floquet waves can be represented via dispersion diagrams, where stop bands may exist and hence one can identify intervals of frequencies, within which the waves become evanescent and hence are confined exponentially. It is indeed natural to see the extension of such studies to structured interfaces. Despite the strong coherence between the transmission properties of structured interface built of components of periodic systems, we have highlighted an important feature of an enhanced transmission based on formation of "defect modes" within an interface. This phenomenon of a paramount importance in range of practical applications involving filtering and polarisation of waves of different physical origins.

References

Bigoni, D. and Movchan, A.B., 2002. Statics and dynamics of structural interfaces in elasticity. *International Journal of Solids and Structures* 39, 4843-4865.

Billingham, J., King, A.C., 2001. *Wave Motion.* Cambridge University Press. Cambridge.

Brillouin, L., 1953. *Wave Propagation in Periodic Structures* . Dover, NY.

Brun, M., Guenneau, S., Movchan A.B. and Bigoni, D., 2010a. Dynamics of structural interfaces: filtering and focussing effects for elastic waves. *Journal of the Mechanics and Physics of Solids* 58, Issue 9, 1212-1224.

Brun, M., Movchan A.B. and Movchan N.V., 2010b. Shear polarisation of elastic waves by a structured interface. *Continuum Mechanics and Thermodynamics*, 22, Issue 6-8, 663-677.

Cai, C.W., Liu, J.K., Yang, Y., 2005. Exact analysis of localized modes in two-dimensional bi-periodic mass-spring systems with a single disorder. *Journal of Sound and Vibration* 288, 307–320.

Jensen, J.S., 2003. Phononic band gaps and vibrations in one- and two-dimensional mass-spring structures. *Journal of Sound and Vibration* 266, Issue 5, 1053–1078.

Kittel, C. 1996. *Introduction to Solid State Physics.* 7th edition, Wiley, New York.

Kunin, I.A., 1975. *Theory of elastic media with micro-structures. Non-local theory of elasticity.* Nauka, Moscow.

Lekner, .J. 1994. Light in periodically stratified media. *Journal of Optical Society of America A* , 11 (11), 2892-2899.

Maradudin, A.A., Montroll, E.W., and Weiss, G.H., 1963. *Theory of Lattice Dynamics in the Harmonic Approximation.* Academic Press.

Martin, P.A., 2006. Discrete scattering theory: Greens function for a square lattice. *Wave Motion* 43, 619–629.

Martinsson, P.G., Movchan, A.B., 2003. Vibrations of lattice structures and phononic band gaps. *The Quarterly Journal of Mechanics and Applied Mathematics* 56, 45–64.

Movchan, A.B., Movchan, N.V., Poulton, C.G., 2002a. *Asymptotic Models of Fields in Dilute and Densely Packed Composites.* Imperial College Press. London.

Movchan, A.B., Slepyan, L.I., 2007. Band gap Green's functions and localized oscillations. *Proceedings of The Royal Soc. London A*, 463, 2709-2727.

Movchan, A.B., Zalipaev, V. and Movchan, N.V., 2002b. Photonic band gaps for fields in continuous and lattice structures. *In: Analytical and computational fracture mechanics of non-homogeneous materials. Kluwer Academic Publishers. Editor B.L. Karihaloo*, 437-451.

Ockendon, J., Howison, S., Lacey, A., Movchan, A. , 2003 *Applied Partial Differential Equations.* Oxford University Press. Oxford.

Yablonovitch, E., 1987. Inhibited spontaneous emission in solid-state physics and electronics. *Physical Review Letters* 58, 2059–2062.

Yablonovitch, E., 1993. Photonic band-gap crystals. *Journal of Physics: Condensed Matter,* 5, 2443–2460.

Piezoelectric Superlattices and Shunted Periodic Arrays as Tunable Periodic Structures and Metamaterials

Luca Airoldi, Matteo Senesi and Massimo Ruzzene

[*] D. Guggenheim School of Aerospace Engineering, Georgia Institute of Technology, Atlanta GA, USA

Abstract Two examples of internally resonating metamaterials with behavior based on multi-field coupling are illustrated. The first example consists in a 1D waveguide with a periodic array of shunted piezoelectric patches. Each patch is shunted through a passive circuit which induces resonance in the equivalent mechanical impedance of the waveguide. Analytical, numerical and experimental studies illustrate the characteristics of the system and quantify such resonant mechanical properties due to electro-mechanical coupling. Piezoelectric superlattices are presented as additional examples of *internally resonant metamaterials*. Multi-field coupling is identified as the enabler mechanism for the generation of the internal resonance. Numerical studies for 1D and 2D piezoelectric superlattices and analytical studies developed on the basis of the long wavelength approximation support the interpretation of the coupling as an internally resonant mechanism.

1 Introduction

This chapter presents the study of wave propagation in periodic systems which comprise multi-field elements. The study is based on the general observation that the presence of elements capable of energy conversion between two fields offers extensive opportunities for the achievement of unusual and novel wave propagation characteristics. In this regard, multi-field coupling in periodic materials is an excellent candidate for the design of novel *metamaterials*. One of the outstanding challenges in metamaterial development is the ability to tune their properties in response to changing operating conditions. This is particularly relevant for the case of internally resonant metamaterials (Sheng et al., 2003), which operate on the basis of strong changes in material properties around a pre-defined tuning frequency. One of the concepts illustrated in this chapter involves the use of piezoelectric

materials for the conversion of elastic into electrical energy, and the use of
shunting circuits to generate an equivalent resonant system in parallel to
the mechanical waveguide. The resonant characteristics of the shunts can be
tuned by modifying the circuit electrical impedance, which suggests the possibility for the system to achieve unusual mechanical properties at selected
frequencies. The second concept investigates acousto-electromagnetic coupling resulting from the periodic polarization of a piezoelectric waveguide.
In this case, the resonant behavior is associated with the generation and
excitation of polaritons, which resonate at frequencies defined by the periodicity and physical properties of the lattice. At these frequencies, wave motion is characterized by strong attenuation, and maximum energy transfer
between the acoustic and electromagnetic fields. The analysis of the dispersion properties of the two waveguides underlines the common characteristics
associated with internally resonating properties, and suggests potential applications such as vibration attenuation and isolation, and the development
of novel acousto-optical devices. Homogenized theories for both types of
waveguides are developed to derive expressions for their equivalent properties, which effectively illustrate their resonant characteristics, and show how
they affect the propagation of waves.

The two types of multi-field waveguides presented in this chapter provide
examples of systems where wave attenuation occurs through an internal resonance mechanism. The resonant condition is characterized by maximum
coupling between the waveguide, acting as a primary system, and a resonating secondary system. The condition of maximum coupling is identified by
the matching of the dispersion properties of the primary and secondary
system. The dispersion relation for a secondary system comprising a set of
periodically placed resonators generally appears as a flat curve, which corresponds to spatially localized modes in the resonators themselves, as defined
by a null group velocity. The intersection of this flat mode with the dispersion branch of the primary structure defines the condition of maximum
coupling between the two systems. When the primary and secondary systems belong to a multi-field domain, their coupling requires a mechanism
through which energy transfer occurs. The presence of such mechanism
leads to a new dispersion branch for the coupled system, which essentially
coincides with the branches for the primary system away from the intersection condition, and undergoes a resonance at the coupling frequency.
This behavior is illustrated in Figure 1 where the dispersion branches of
the uncoupled primary and secondary system are respectively represented
as solid black and red dashed lines. The frequency of intersection of the two
branches is the frequency of internal resonance for the uncoupled resonating
primary system. Coupling leads to the new dispersion branch (solid blue

line) which mostly follows the dispersion branch of the primary system, and is distorted by the resonant behavior at the frequency of internal resonance.

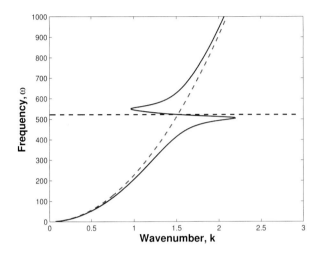

Figure 1. Typical dispersion relations for a waveguide with internal resonators: uncoupled systems (dashed lines), coupled system (solid blue line).

In the considered waveguides, the coupling mechanisms is provided by the piezoelectric effect. In the case of the periodic piezoelectric shunted network, the secondary system is effectively characterized by a flat dispersion mode corresponding to the resonant behavior of the shunting circuits, while for the case of piezoelectric superlattices, the flat dispersion modes corresponds to an elastic mode at very low wavelength. In this case, coupling is made possible by the folding of the branch caused by the periodicity of the waveguide.

The chapter is organized in two parts covering the two concepts. Part 1 presents the study of one-dimensional waveguides with periodic shunted piezo arrays. Numerical and experimental results illustrate their attenuation characteristics, and their tunable properties. Numerical investigations are performed through the application of the Transfer Matrix method, which is briefly reviewed. Experimental investigations performed on a beam structure, confirm the numerical predictions and show the attenuation characteristics of the waveguide. Part 2 is devoted to the study of piezoelectric superlattices. Numerical studies of one-dimensional and two-dimensional configurations are performed through the application of the Plane Wave

Expansion method, which is presented in some detail. Multifield coupling and internal resonant behavior of piezoelectric superlattices are illustrated through a series of numerical examples, and the evaluation of equivalent dielectric properties using a long wavelength approximation approach.

2   Periodic shunted piezoelectric arrays

Piezoelectric shunt damping is an attractive technique which offers a simple and potentially cost-effective solution for the attenuation of waves and vibrations in structures. The key element is a passive electrical network directly connected to the electrodes of the piezoelectric device. An elegant analysis of vibration damping through passive shunting was proposed by Hagood and von Flotow (1991) and is still most commonly used. The study showed how a piezoelectric material shunted through a series RL circuit, i.e., a resonant shunt, exhibits a behavior equivalent to the well known mechanical tuned vibration absorber. A resonant shunt is simple to design and offers effective damping in the vicinity of a selected mode of the underlying structure. After the initial introduction of single-mode resonant shunts, more complex shunting circuits have been investigated for suppressing multiple structural modes. Hollkamp (1994) was able to suppress two modes of a cantilever beam using a system of RLC circuits connected in parallel. The whole circuit requires as many parallel branches as there are modes to be controlled. Since no closed-form tuning solution is available for determining the component values, the method relies on the numerical optimization of a non-linear objective function fully parametrized by all of the circuit elements. As an alternative, Wu (1996) proposed the use of parallel RL shunts, each targeting an individual mode. Current-blocking LC networks are introduced in each parallel branch to reduce cross coupling and achieve multiple resonances at the desired frequencies. Specifically, each LC circuit is tuned to the frequency of an adjacent mode in order to decouple the branches. The complexity of the circuit topology greatly increases as the number of modes to be damped simultaneously increases. More recently, the current-flowing concept was presented by Behrens et al. (2003). Compared to current-blocking schemes, the current-flowing method is simpler to tune and involves less electrical components, however it appears less effective for densely spaced modes. In addition to the RL-based shunting techniques described thus far, other different strategies are available for multimode vibration reduction with piezoelectric shunts. The most common method is based on negative capacitance shunting as originally proposed by Forward (1979). In this configuration, a piezoelectric patch is shunted through a passive circuit connected to a negative impedance converter, so that the internal capacitance of the piezo is artificially cancelled and the impedance of the shunt circuit reduces to that of the passive circuit. If this circuit is frequency-independent, i.e., a resistance, a broadband damping can be achieved. Although the negative capacitance shunting strategy has been experimentally validated with success, it must be used with caution since it requires active elements that can destabilize the structure if improperly tuned. Moreover, the circuit should be tuned very close to the stability limit to achieve best performance (Preumont, 1997).

Application of the foregoing shunting methods are not limited to vibration-only studies. In the past few years, researchers have investigated the suitability and performance of piezoelectric shunt damping to increase the acoustic transmission loss in structures. Kim and Lee (2004) for example have compared the sound transmission loss performance of plates with sound-absorbing material and RL-shunted piezo patches. Multimode shunt damping with blocking circuit (Wu's solution) has been applied by Kim and Kim (2004) for noise reduction of a plate in an acoustic tunnel. Kim and Jung (2006b,a) also studied broadband reduction of noise radiated by plates with multiple resonant and negative-capacitance-converter shunt circuits, and achieved god levels of noise attenuation over a limited number of modes. A rather different approach to broadband vibration attenuation using shunts was proposed by Thorp et al. (2001). The concept involves a periodic array of simple RL-shunted piezos mounted on the structure to passively control the propagation of elastic waves and the subsequent vibration field. Periodically induced impedance-mismatch zones generate broad stop bands, i.e., frequency bands where waves are attenuated. The tunable characteristics of shunted piezo patches allow the equivalent mechanical impedance of the structure to be tuned so that stop bands are generated over desired frequency ranges. In addition, the energy dissipation mechanism of shunted piezos can be exploited to dampen the amplitude of vibration outside the stop bands. The original periodic shunting concept was numerically demonstrated on rods and fluid-loaded axisymmetric shells in Thorp et al. (2001) and Thorp et al. (2005). More recently, this strategy was extended to flat plates in Spadoni et al. (2009), where Bloch theorem was used to predict the dispersion properties of the resulting periodic assembly. The method and its effectiveness over a broad frequency range were then validated experimentally on a cantilever aluminum plate hosting a periodic layout of 4×4 RL-shunted piezo patches (Casadei et al., 2010). These latest studies illustrate how resonant piezo shunts can be utilized to affect the equivalent mechanical properties of an elastic waveguide, and therefore suggest their application for the development of a novel class of metamaterials (Huang et al., 2009)(Kushwaha et al., 1993) (Martinsson and Movchan, 2003) (Sigalas, 1998) (Sheng et al., 2003). In fact, many proposed concepts for acoustic metamaterials consider configurations that derive their unique properties from resonators contained within each unit cell. Typical designs are characterized by inclusions with a phase velocity much lower than that of the matrix (Kushwaha and Djafari-Rouhani, 1998). This allows producing attenuation bands at frequencies which are unrelated to scattering phenomena (*Bragg scattering*) which occur at wavelengths of the order of the unit cell size. The opportunity is thus given to achieve low-frequency attenuation of waves, which has relevance to vibration and noise transmission. Different types of local resonators that have been proposed include the single degree of freedom mass-in-mass lattice systems in (Huang et al., 2009). This solution was verified experimentally in (Yao et al., 2008) and is similar to those presented by (Lazarov and Jensen, 2007). More recently, the same technology has been applied to determine multiple attenuation bands through multi-resonator mass-in-mass lattice systems (Huang and Sun, 2010), or arrays of spring-mass subsystems (Pai, 2010), (Sun et al., 2010). Alternative techniques are given by oscillating coated cylinders/spheres in an epoxy matrix (Sheng et al., 2007), or Helmholtz resonators that generate negative stiffness-like behavior (Fang et al., 2006), or simultaneously negative effective mass and stiffness (Cheng et al., 2008).

In here, we illustrate how a periodic array of piezoelectric patches bonded to a one-dimensional waveguide is equivalent to mechanical resonators which generate unusual mechanical properties and attenuation bands at desired ranges of frequencies. The considered configuration reproduces the systems investigated by Thorp et al. (2001), and Spadoni et al. (2009). The concept is here re-interpreted in light of the more recent metamaterials literature, as a solution for the achievement of tunable, unusual wave properties. The metamaterial perspective is investigated through theoretical developments which derive expressions for the equivalent mechanical properties for the considered waveguides. Such properties are functions of the mechanical impedance of the waveguide, and of the electrical impedance of the shunting circuits, and feature resonant characteristics at the tuning frequencies. Specifically, we consider the wave propagation characteristics and the equivalent properties of beams with a periodic arrangement of shunted piezoelectric patches. The equations governing the longitudinal and transverse motion of the considered beam are formulated as a set of 1st order differential equations whose periodic coefficients are a function of the mechanical properties of the beam as well as of the shunting parameters for the electric circuit connected to each patch of the array. Assuming harmonic motion leads to a set of Ordinary Differential Equations (ODEs) whose solution is investigated through the application of Floquet Theorem

(Brillouin, 1946), and a Transfer Matrix approach. The resulting dispersion properties for the beam highlight the occurrence of attenuation bands at frequencies defined by the internal resonances of the shunts. Within these bands, distortion of the dispersion relations is observed as typically seen in systems featuring internal resonances, such as those illustrated in Figure 1, presented in (Fang et al., 2006), or those corresponding to the coupling of multi-field waves (Zhang et al., 2004), (Zhu et al., 2003). The internally resonant behavior of the considered system has been interpreted as the result of anomalous mass and/or stiffness variations in terms of frequency (Ding et al., 2007). Equivalent properties for the considered beams are here estimated through the analysis of the system's behavior in the long wavelength limit, which allows the calculation of the equivalent mass and stiffness properties of the system. Such analysis leads to expressions which highlight the resonant behavior of mass and stiffness of the beam, and illustrate how tuning is conveniently achieved through the selection of the shunting parameters. From this perspective, the considered beam can be regarded as an example of a tunable metamaterial.

2.1 Theoretical Background

Configuration   We consider the dynamic behavior of the beam with a periodic array of shunted piezoelectric patches (Figure 2.a), obtained as the assembly of unit cells of the kind shown in Figure 2.b. The beam behaves as a one-dimensional waveguide which supports the propagation of axial and transverse waves. In the low frequency range, the behavior of the waveguide can be conveniently described through Euler-Bernoulli theory, applied to a beam with piece-wise elastic and mass properties, as illustrated in the following section.

We refer to a coordinate system where $'1'$ and $'3'$ respectively denote the axial and thickness directions. According to Hagood and von Flotow (1991), shunting of the piezoelectric patch with electrodes across the $'3'$ direction modifies the Young's modulus of the shunted patch according to the following expression:

$$E_p^{SU}(\omega) = E_p^D \left(1 - \frac{k_{31}^2}{1 + i\omega\, C_p^\varepsilon\, Z^{SU}(\omega)} \right) \tag{1}$$

where ω is the frequency, $i = \sqrt{-1}$, C_p^ε the electrical capacitance of the piezo at constant strain, and $Z^{SU}(\omega)$ is the electrical impedance of the shunting circuit. Also in equation (1), k_{31} denotes the electro-mechanical coupling coefficient, and E_p^D is the Young's modulus of the piezoelectric material when the shunting network is in an open circuit configuration ($Z^{SU}(\omega) \rightarrow$

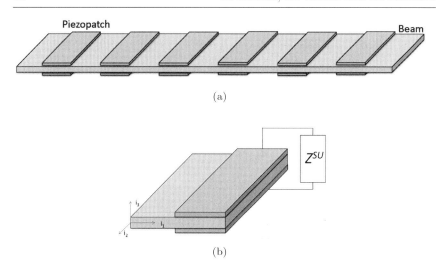

(a)

(b)

Figure 2. Periodic beam with array of shunted piezo patches (a), and unit cell with shunting through an electrical imedance Z^{SU} (b).

∞). This quantity is related to the piezo Young's modulus E_p through the expression (Hagood and von Flotow, 1991):

$$E_p^D = \frac{E_p}{1 - k_{31}^2}$$

When a resistor-inductor shunt is applied to the piezoelectric material, $Z^{SU}(\omega) = R + i\omega L$, the piezo Young's modulus features a resonant behavior at a frequency ω_T that can be selected by tuning the inductor $L = \frac{1}{\omega_T^2 C_p^\varepsilon}$ (Hagood and von Flotow, 1991), (Hollkamp, 1994), (Wu, 1996).

Governing equations   The longitudinal $u(x,t)$ and transverse $w(x,t)$ motion of the beam of Figure 2 is described by the following set of partial differential equations:

$$\frac{\partial^2}{\partial x^2}\left[D(x)\frac{\partial^2}{\partial x^2}w(x,t)\right] + m(x)\frac{\partial^2}{\partial t^2}w(x,t) = 0 \tag{2}$$

$$\frac{\partial}{\partial x}\left[K(x)\frac{\partial}{\partial x}u(x,t)\right] - m(x)\frac{\partial^2}{\partial t^2}u(x,t) = 0 \tag{3}$$

where $D(x)$, $K(x)$ respectively denote the bending and axial stiffness of the beam, while $m(x)$ is the mass per unit area. Given the beam configuration, a generic physical property of the beam $P(x)$ can be expressed as a piecewise function of period p, i.e.:

$$P(x) = P(x + p) \tag{4}$$

which over a period centred at $x = 0$ can be expressed as:

$$P(x) = \left\{ \begin{array}{ll} P_1 & -\alpha p < x < 0 \\ P_2 & 0 < x < (1-\alpha)p \end{array} \right. \tag{5}$$

For the unit cell of Figure 2, the linear mass of the beam is given by:

$$m(x) = \left\{ \begin{array}{ll} \rho_b A_b & -\alpha p < x < 0 \\ \rho_b A_b + 2\rho_p A_p & 0 < x < (1-\alpha)p \end{array} \right. \tag{6}$$

while the axial and bending stiffnesses can respectively be expressed as:

$$K(x) = \left\{ \begin{array}{ll} E_b A_b & -\alpha p < x < 0 \\ E_b A_b + 2 E_p^{SU}(\omega) A_p & 0 < x < (1-\alpha)p \end{array} \right. \tag{7}$$

and

$$D(x) = \left\{ \begin{array}{ll} \frac{1}{12} E_b b_b h_b^3 & -\alpha p < x < 0 \\ \frac{1}{12} E_b b_b h^3 + \frac{1}{6} E_p^{SU}(\omega) b_p \left[(h + 2h_p)^3 - h^3 \right] & 0 < x < (1-\alpha)p \end{array} \right. \tag{8}$$

where ρ_b, E_b are the density and the Young's modulus of the beam material, h_b, b_b define the thickness and out-of-plane width of the base beam, and $A_b = b_b h_b$. Also, ρ_p is the density of the piezoelectric material, h_p, b_p denote the thickness and out-of-plane width of the piezo patch, and $A_p = b_p h_p$.

For harmonic motion at frequency ω, equations (2,3) reduce to two ODEs in the spatial coordinate x, which can be combined in the following 1st order system:

$$A(x) \frac{d}{dx} z(x) = B(x) z(x) \tag{9}$$

where the state vector z contains displacement and stress resultants associated with axial and transverse motion, and it is defined as:

$$z(x) = [u, \ w, \ w_{,x}, \ N, \ M, \ Q]^T \tag{10}$$

In equation (10), N, Q, M are the axial stress resultant, the shear force and the bending moment at location x. Also, in the equation above and in the remainder of the chapter the notation $()_{,x}$ denotes partial derivatives with respect to the variable x, capital bold letters denote matrices, and lower case bold letters are vectors. The matrices $\boldsymbol{A}, \boldsymbol{B}$ in equation (9) are defined as:

$$\boldsymbol{A}(x) = \mathrm{diag}([K(x),\ 1,\ D(x),\ 1,\ -1,\ 1]), \tag{11}$$

and

$$\boldsymbol{B}(x) = \begin{bmatrix} 0 & 0 & 0 & 1 & 0 & 0 \\ 0 & 0 & 1 & 0 & 0 & 0 \\ 0 & 0 & 0 & 0 & 1 & 0 \\ -\omega^2 m(x) & 0 & 0 & 0 & 0 & 0 \\ 0 & 0 & 0 & 0 & 0 & 1 \\ 0 & -\omega^2 m(x) & 0 & 0 & 0 & 0 \end{bmatrix} \tag{12}$$

Equation (9) can be rewritten as:

$$\frac{d}{dx}\boldsymbol{z}(x) = \boldsymbol{C}(x)\boldsymbol{z}(x) \tag{13}$$

where $\boldsymbol{C} = \boldsymbol{A}^{-1}\boldsymbol{B}$ is a periodic matrix, i.e.:

$$\boldsymbol{C}(x) = \boldsymbol{C}(x+p)$$

which varies in a piecewise fashion over the period, i.e.:

$$\boldsymbol{C}(x) = \begin{cases} \boldsymbol{C}_1 & -\alpha p < x < 0 \\ \boldsymbol{C}_2 & 0 < x < (1-\alpha)p \end{cases} \tag{14}$$

2.2 Analysis of the Dispersion Properties

Transfer Matrix Equation (13) describes a system of ODEs with periodic coefficients. According to Floquet Theorem, its solution can be expressed as:

$$\boldsymbol{z}(x+p) = \lambda \boldsymbol{z}(x)$$

where $\lambda = e^{i\mu}$ is the Floquet multiplier, with $\mu = kp$ denoting the propagation constant, and k the wavenumber. The Floquet multipliers are obtained by relating the solution of equation (13) at a location $x+p$ and x through a transfer matrix $\boldsymbol{T}$:

$$\boldsymbol{z}(x+p) = \boldsymbol{T}\boldsymbol{z}(x) \tag{15}$$

For the case at hand, the formulation of the transfer matrix exploits the piecewise nature of the periodicity, which allows close form expressions. With reference to a unit cell, equation (15) can be expressed as:

$$\boldsymbol{z}[(1-\alpha)p] = \boldsymbol{T}\boldsymbol{z}(-\alpha p) \tag{16}$$

where T is given by:

$$T = T_2 T_1$$

In the equation above, the matrix T_1 relates the state vectors at $x = -\alpha p$ to the state vector at $x = 0$, while the matrix T_2 relates the state vectors at $x = 0$ and at $x = (1 - \alpha)p$. Matrices T_1 and T_2 can be formulated from the analytical solution of the governing equation over the first and second portion of the unit cell, i.e.:

$$\frac{d}{dx}z(x) = C_1 z(x), \quad x \in [-\alpha p, \ 0] \tag{17}$$

and

$$\frac{d}{dx}z(x) = C_2 z(x), \quad x \in [0, \ (1-\alpha)p] \tag{18}$$

which give:

$$z(0) = e^{\alpha p C_1} z(-\alpha p), \quad x \in [-\alpha p, \ 0] \tag{19}$$

$$z[(1-\alpha)p] = T_2 z(0), \quad x \in [0, \ (1-\alpha)p] \tag{20}$$

Therefore

$$T_1 = e^{\alpha p C_1}$$

and

$$T_2 = e^{(1-\alpha)p C_2}$$

Continuity of displacements and stresses at the interface of the two portions of the materials leads to:

$$\begin{aligned} z[(1-\alpha)p] &= e^{(1-\alpha)p C_2} e^{\alpha p C_1} z(-\alpha p) \\ &= T_2 T_1 z(-\alpha p) \\ &= T z(-\alpha p), \end{aligned} \tag{21}$$

Beam configuration The periodic beam is formed by the assembly of unit cells shown in Figure 3. The geometrical parameters of the unit cell are summarized in Table 1. The beam is made of aluminum (Young's Modulus $E_b = 69$ GPa and density $\rho_b = 2700$ kg/m^3), while the piezoelectric patch has the properties listed in Table 2.

Numerical Results The dispersion relations associated with the longitudinal and transverse motion of the beam are evaluated through the TM approach. Results for two different tuning frequencies are here reported. The internal resonance of the circuit is selected at 5000 Hz, and 11000

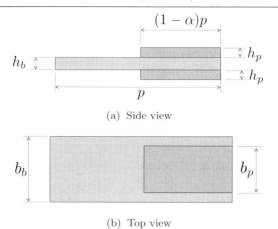

(a) Side view

(b) Top view

Figure 3. Sketch of the unit cell for the beam.

Table 1. Unit cell geometry.

p	5.4×10^{-2} m
α	0.35
h_b	3.5×10^{-3} m
b_b	3.2×10^{-2} m
h_p	0.8×10^{-3} m
b_p	1.8×10^{-2} m

Table 2. Properties of piezoelectric material.

E_p	6.3×10^{10} Pa
ρ_p	7800 kg/m^3
k_{31}	0.35
$\varepsilon_{33}{}^\sigma / \varepsilon_0$	2500
ε_0	8.854×10^{-12} F/m
C_p^ε	15 nF

Hz. Such tuning frequencies are obtained with inductance values for the shunting circuit respectively equal to $L = 33.4$ mH, and $L = 6.7$ mH. Results for different values of shunting resistance are also reported. Figure 4 presents the dispersion relations for the said beam when the electric circuit resonates at 5000 Hz (blue lines) and at 11000 Hz (red line). The latter case is analyzed for shunting resistances of 25 and 50 Ω. Specifically, Figure 4.(a) presents the branch associated with longitudinal motion, while Figure 4.(b) shows results for the transverse mode. The two modes can be studied separately as completely decoupled by the beam model employed for the analysis. Both plots feature the expected resonant behavior at the resonant (tuning) frequency of the shunting circuit, which also defines the center of an attenuation band. Such attenuation band is identified by nonzero values of the imaginary part of the wavenumber, which characterizes conditions of propagation with attenuation. Of note is the fact that the resistive component in the shunting circuit acts as a dissipation term, which has the effect of affecting the resonant behavior of the circuit. Low dissipation with low resistance leads to a sharp resonant peak, and a correspondingly large attenuation, which however occurs over a narrower frequency band. In contrast, high resistance values lead to a smoother resonant behavior, with lower attenuation occurring over a broader frequency band. Hence, if the objective is the achievement of high attenuation over a frequency band which is as large as possible, a compromise in terms of resistance must be struck between bandwidth and attenuation, as defined by the magnitude of the imaginary part of the wavenumber (Spadoni et al., 2009).

Additional analysis of the dispersion branches for axial and transverse motion reveals some interesting features. As expected, the wavenumber range identified by the TM analysis is different for the two branches, and for the transverse mode it reaches the boundary of the first Brillouin zone which is located at $k = \pi/p \approx 58.2$ 1/m. At this value, the branch appears as folding back, which is the result of the periodicity of the domain and the dual periodicity in physical and wavenumber space (Figure 4.(b)). The wavenumber range for the axial mode does not reach such bound in the considered frequency range, and therefore no branch folding is observed (Figure 4.(a)). Branch folding is a feature which is characteristics of all periodic domains, so that their dispersion properties can be solely characterized through the analysis of the First Brillouin zone. In general, however, and specifically when experimental studies are conducted, it is important to keep track of the entire frequency/wavenumber spectrum, since the branches beyond the First Brillouin zone may be measured and or need to be properly tracked. In fact, the spatial resolution of the measurements goes typically

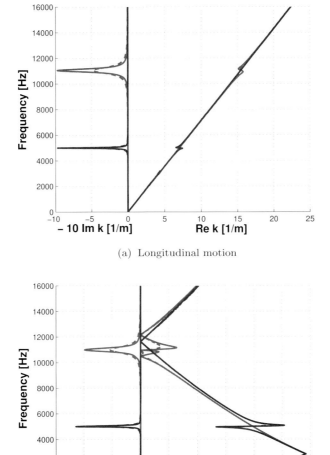

(a) Longitudinal motion

(b) Bending motion

Figure 4. Real and Imaginary part of wavenumber for axial (a) and bending (b) mode for different tuning of the piezo shunts ($f_T = 5$ kHz - blue line, $f_T = 11$ kHz - red line; $R = 25\Omega$ - solid lines, $R = 50\Omega$ - dashed lines).

beyond the spatial sampling corresponding to the limit of the First Brillouin zone, so that shorter wavelengths and correspondingly higher frequencies can be detected. This discussion is of particular relevance in light of the experimental results presented in the following sections. Figure 5 shows the complete dispersion branch for the transverse mode of the beam, and illustrates its periodic features, the First Brillouin zone highlighted by the shaded area, and the occurrence of several branches.

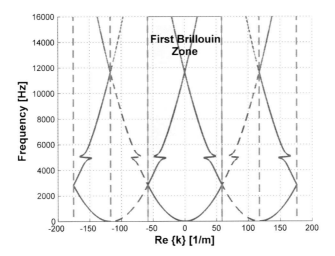

Figure 5. Real part of wavenumber k for transverse motion, tuning at 5000 Hz, $R = 33\,\Omega$, obtained through computations extended beyond the First Brillouin Zone. This region is highlighted in yellow.

2.3   Experimental Results

Experimental investigations are performed on a beam with a periodic array of piezoelectric patches. The beam is excited by a piezopatch, which induces a transient wave propagating along the beam. The motion of the beam is recorded by a Scanning Laser Doppler Vibrometer (SLDV), which maps the response over a fine grid of points. The recorded data have fine temporal and spatial resolutions sufficient for the estimation of the frequency/wavenumber content of the beam response, and the evaluation of the dispersion properties of the beam. This is performed through the application of one-dimensional and two dimensional Fourier Transforms of the recorded data. Tests are conducted for various tuning of the shunt-ing circuits to illustrate the occurrence of attenuation bands over selected frequency ranges.

The use of a SLDV as a sensor limits the investigations to the transverse motion of the beam, since the sensitivity of the Laser allows only the detection of the component of motion perpendicular to the Laser beam, which is normally incident to the beam surface.

Setup Tests were conducted on the beam shown in Figure 6. The beam is made of aluminum, it is 1.6 m long, and features an array of 11 piezo-electric patches equally spaced over a portion of the length. The unit cell configuration replicates the schematic of Figure 3 and has the dimensions listed in Table 1.

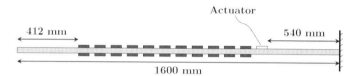

Figure 6. Sketch of the experimental beam.

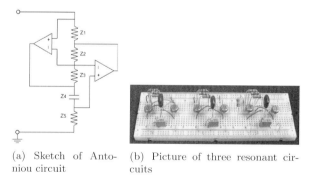

(a) Sketch of Anto-niou circuit

(b) Picture of three resonant circuits

Figure 7. Experimental resonant circuit.

The resonant shunted circuits are implemented through the application of a synthetic inductor (Antoniou's circuit) because of the high value of inductance needed for the desired frequency tunings. Also, this implementation provides the flexibility to easily tune the circuit at different frequencies. The Antoniou's circuit (Figure 7) consists in two op-amps, four resistors and one capacitor (Z_4), which altogether simulate the behavior of an inductor (Riordan, 1967), (Casadei et al., 2010). The resulting inductance is related to the value of the impedance of all the components involved, and it is given by:

$$L = \frac{Z_1 \, Z_3 \, Z_4 \, Z_5}{Z_2} \tag{22}$$

In the configuration tested, each pair of co-located piezoelectric patches are connected in series to a resonant shunting circuit. The electric networks were tuned at the frequencies of 5000 and 11000 Hz. The nominal values of the electric components required for these tunings are reported in Table 3.

Table 3. Tuning settings for the experimental circuit, nominal values.

f_T, Hz	Z_1, Ω	Z_2, Ω	Z_3, Ω	Z_4, nF	Z_5, Ω	$R\,\Omega$
5000	3300	2200	100	100	2229	10, 33, 120
11000	680	1000	100	100	988	10, 33, 100

The beam is excited by the piezo actuator shown Figure 6, which is fed a pulse signal of $40\mu s$ duration, and 50 V of amplitude. The time duration is selected for broadband excitation of the beam motion. The velocity of the beam is measured in 645 equally spaced points over the entire length by the SLDV (Polytec Model PSV-400M2). At each measurement point x_i $(i = 1, .., 645)$, the recorded time record contains 1024 samples, acquired at a sampling frequency of 256 kHz. Measurement noise is reduced through a low-pass digital filtering at 16 kHz, and by taking 10 averages at each point with a repetition rate of 10 Hz.

Response in the space/time domain   The result of the each experiment is the beam response, which is stored in a two-dimensional array $w(x,t)$ which contains the time variation the transverse velocity of the beam at location x. Figure 8.a shows an example recorded when all shunts are connected to open circuits. This represents a baseline configuration against which the performance of various shunted strategies can be compared. The space/time response clearly shows the propagation of two waves, which emanate from the excitation location at approximately $x = 1$ m, and subsequently reach the ends of the beam where they are reflected. The recorded wavefield appears, as expected, highly dispersive, as the applied pulse gets increasingly distorted as it propagates along the length of the beam. The space/time response recorded with the shunts tuned at 5000 Hz and with a resistance value of $R = 33\ \Omega$ displayed in Figure 8.b shows a similar behavior. Careful observation of the plot however unveils slightly stronger dispersion in comparison to the case with the open shunts. The evaluation of the effects of the shunts on wave propagation requires further analysis which leads to the estimation of the dispersion relations for the beam from the recorded response.

Signal Processing For the Evaluation of Dispersion   The estimation of the dispersion properties for the beam is performed by transforming the recorded response $w(x,t)$ into the frequency/wavenumber through a two-dimensional Fourier Transform (2D FT):

$$W(k,\omega) = \int_{-\infty}^{+\infty} \int_{-\infty}^{+\infty} w(x,t)e^{-i(kx+\omega t)}dtdx \qquad (23)$$

This operation, which can be simply performed through built-in FFT routines, however requires preliminary post-processing of the data in order to obtain clear dispersion representations. First, only the left-propagating wave is analyzed by windowing the response in space in the $x \in [0,\ 1]$ m interval. Time windowing is also performed for $t = [0,\ 1.2]$ msec. The next important step consists in the removal of the boundary reflections. Reflections lead to the occurrence of dominant harmonic terms, which correspond to the resonant frequencies of the beam and their associated wavenumber. The presence of such harmonic terms cause the frequency/wavenumber domain representation to be discontinuous, with energies concentrated at the frequency/wavenumber values corresponding to frequency and wavenumber of the resonant modes. As the objective is here to determine the occurrence of attenuation zones in the frequency/wavenumber relations, a discontinuous frequency/wavenumber spectrum would make the identification of such zones difficult and possibly ambiguous. Removal of the boundary reflection simulates the response of an infinite domain, which is characterized by continuous dispersion characteristics, where the presence of a frequency gap corresponding to an attenuation zone can be immediately identified.

The removal of the reflection from the boundary can be performed through proper time windowing of the signal, but is significantly complicated by the dispersive nature of the wave. In order to properly identify and then eliminate the reflection from the left boundary, dispersion is here compensated through transformation into the space-warped time domain by applying the Warped Frequency Transform proposed by De Marchi et al. (2008). The WFT is a linear transformation, which is based on the *warping* of the frequency domain according to the dispersion properties of the medium of interest. This allows the representation of the signal in warped time axis where it appears non-dispersive. The result of this operation is

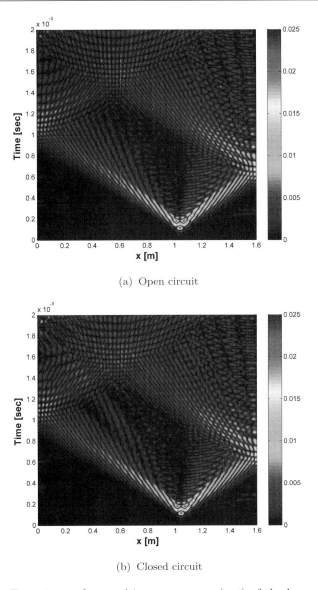

(a) Open circuit

(b) Closed circuit

Figure 8. Experimental space/time response $w(x,t)$ of the beam in space-time domain, tuning at $5000\,Hz$ and $R = 33\,\Omega$.

illustrated in Figure 9.a, where the warped-time axis is limited to the range preceding the left boundary reflection through proper windowing. The application of the inverse WFT returns the signal to the space/time domain where the boundary reflection from the left appears as completely removed (see Figure 9.b).

The windowed response is finally analyzed through a 2DFT, whose amplitude $|W(k,\omega)|$ can be represented as a surface in the frequency/wavenumber domain. The contour plot of the 2DFT surface for a beam with open circuits is shown in Figure 10 which shows how the maximuma of the contour plots outline the dispersion relation for the beam. The plot also shows that the beam is excited in a frequency range spanning approximately an interval from 3000 Hz up to the selected cut-off frequency of 16000 Hz. Within this range, a continuous frequency/wavenumber relation can be observed. Results for shunts tuned at 5000 Hz presented in Figure 11 clearly show the presence of a gap in the frequency/wavenumber spectrum at the tuning frequency. Similar behavior is observed for 11000 Hz as shown in Figure 12.

These experimental results can be used for the analysis and validation of the numerical predictions obtained through the TM approach. The dispersion relations computed numerically can in fact be overlapped to the experimental frequency/wavenumber contours for a direct comparison. The results are presented in Figure 13, where plots for the two tunings considered are presented. Numerical results presented as red lines extend to a wavenumber range that exceeds the bounds of the First Brillouin zone, in order to match the range of the experimental results. Several branches need to be included in order to follow the experimental observations, which however are very well predicted by the numerical predictions. Of note is the fact that the resonant properties of the beam as predicted by the TM correspond to the frequency gap observed experimentally at the tuning frequency. For comparison purposes, it is also interesting to plot the dispersion branch for transverse wave in a homogeneous beam, here presented as a black solid line. This dispersion branch follows very well the maximum ridge of the contour, which suggests that in the absence of shunting effects, and away from the tuning frequency, the beam essentially behaves as a homogeneous, non-periodic medium. Only the presence of the resonant piezos tuned at the selected frequencies affects the wave propagation properties of the waveguide. In essence, the physically periodicity introduced by the periodic placement of the piezos along the beam and the periodic modulation in the mechanical impedance due to the added mass and stiffness of the bonded piezos is not sufficient to create any attenuation zone through Bragg scattering although the considered wavenumber range far exceeds the limit of the First Brillouin zone. This behavior partially justifies the use of a long

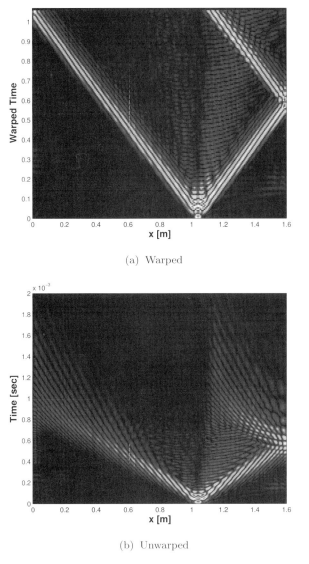

(a) Warped

(b) Unwarped

Figure 9. Warped and unwarped response upon removal of the reflection at left boundary.

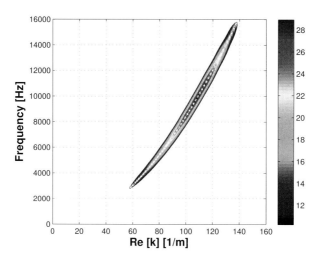

Figure 10. Contour of the of the amplitude of the 2DFT $|W(k, x)|$ outlining the dispersion properties for the beam with open shunts.

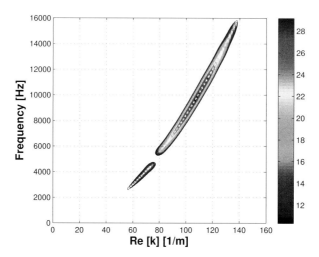

Figure 11. Contour of the of the amplitude of the 2DFT $|W(k, x)|$ outlining the dispersion properties for the beam with shunts tuned at 5000 Hz and $R = 33\ \Omega$.

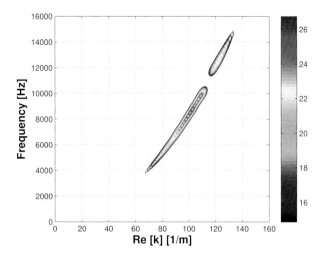

Figure 12. Contour of the of the amplitude of the 2DFT $|W(k,x)|$ outlining the dispersion properties for the beam with shunts tuned at 11000 Hz and $R = 47\ \Omega$.

wavelength approximation for the development of equivalent properties of the shunted waveguide as attempted in the next section.

Wavenumber Estimation from Experimental Measurements   The spatial resolution provided by the considered experimental set-up allows the quantitative estimation of the wavenumbers from experimental data. Both real and imaginary part of the wavenumber can be estimated as a function of frequency, so that attenuation frequency bands can be quantified. The approach is based on the estimation of the spatial variation of the Fourier Transform of the response evaluated at each frequency of interest. The procedure can be outlined by considering the measured response as the superposition of harmonic waves of the kind:

$$w(x,t) \approx \sum \hat{w}_i(x,t) \tag{24}$$

where

$$\hat{w}(x,t)_i = \hat{w}_0(\omega_i)e^{i[k(\omega_i)x - \omega_i t]} \tag{25}$$

is the i-th harmonic component, where $k = k(\omega)$ due to the dispersive nature of the medium. The Fourier Transform of $\hat{w}(x,t)_i$ at ω_i is evaluated

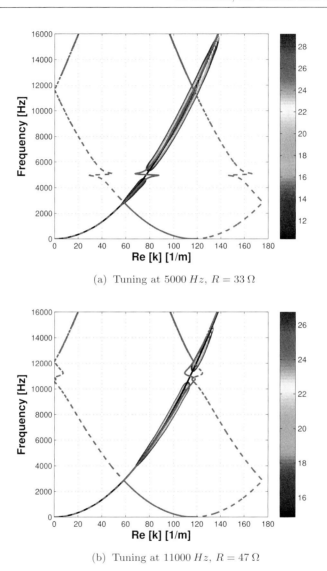

(a) Tuning at $5000\,Hz$, $R = 33\,\Omega$

(b) Tuning at $11000\,Hz$, $R = 47\,\Omega$

Figure 13. Comparison of the dispersion relations of the experimental beam and numerical results obtained with the Transfer Matrix formulation (red lines). The black solid line corresponds to the analytical dispersion relation for a homogeneous beam, with no periodic array of patches.

as follows:

$$W(x,\omega) = \int_{-\infty}^{+\infty} \hat{w}(x,t)_i e^{-i\omega t} dt \tag{26}$$

which evaluated at $\omega = \omega_i$ is approximately given by:

$$W(x,\omega_i) \approx \hat{w}_0(\omega_i)e^{ik_i x}, \tag{27}$$

where $k_i = k(\omega_i)$. The wavenumber $k_i = k_{i_R} + ik_{i_I}$ is generally a complex number, so that equation (27) can be rewritten as:

$$\hat{W}_i(x,\omega_i) \approx \hat{w}_0(\omega_i)e^{-k_{i_I} x}e^{ik_{i_R} x}, \tag{28}$$

Analysis of equation (28) reveals that the estimation of the FT of the recorded data, and the evaluation of the variation of the resulting expression along the spatial coordinate x allows for the estimation of the spatial decay of the response amplitude and the phase evolution in space at a given frequency. This in turn leads to estimation of the attenuation constant as defined by the imaginary component of the wavenumber k_{i_I}, and of its real component, which are respectively given by:

$$k_{i_I}(x_f - x_i) = \log[|W(x,\omega_i)|], \tag{29}$$

and

$$k_{i_R}(x_f - x_i) = \arg[W(x,\omega_i)|], \tag{30}$$

where $x \in [x_i \ x_f]$, with x_i, x_f denoting the initial coordinate and final coordinate over which the spatial decay and phase linear modulation are interpolated. Application of equations (29) and (30) to the experimental results allows the quantification of the wavenumber variation over the frequency range of interest, and particularly the evaluation of the attenuation constant, which was not possible through the application of the 2DFT previously illustrated.

Examples of the estimated wavenumber components (real and imaginary parts) are presented in Figures 14 and 15, where the red dashed lines correspond to the case of open circuits, while the solid blue lines are results for shunting at the tuning frequencies considered. Both results clearly illustrate how shunting of the circuits at a given frequency creates an attenuation frequency band centered at the tuning frequency. Such band is defined by large nonzero values of the imaginary part of the wavenumber, also known as *attenuation constant*. In the same frequency range, the real part of the wavenumber defining the propagation component, undergoes a resonant behavior which is consistent with the one predicted numerically (see Figures 13). The behavior of the dispersion properties around the frequency of internal resonance is typical of periodic systems with internal resonating properties as discussed in the introduction to this chapter. Of note is the fact that the case of open circuits does not lead to an absolute zero for the attenuation constant, which may be affected by other sources of dissipation which are inevitably present in an experimental setup.

2.4   Equivalent properties: a metamaterial perspective

The analysis presented in the previous sections can be further elaborated by seeking for analytical expressions which provide insight in the behavior of the system when undergoing internal resonance through the shunting circuits. The investigations developed herein aim at developing equivalent models for the considered class of waveguides which include the effects of the shunting circuit as part of a set of equivalent mechanical properties.

The theory is developed for the case of the beam undergoing axial and transverse motion. The results of the study are compared with the experimental measurements recorded for the beam in bending. The developments require the assumption that the scale of periodicity is much smaller than the wavelength considered. An interesting behavior is however observed whereby a good agreement between the predictions of the equivalent model and experiments is found in spite of the fact that the wavelengths corresponding to the frequency tunings are beyond the First Brillouin zone. This confirms that the periodicity introduced by the spacing of the piezoelectric patches does not affect the dynamic behavior of the system, and therefore that the beam in the absence of shunting behaves as a homogeneous, non-periodic system. Based on this observation, one can interpret the considered system as the embodiment of a *metamaterial* concept, whereby unusual wave mechanics is achieved in the considered waveguide through the coupling between the primary structure, and a resonating secondary system.

Long Wavelength Approximation   The general formulation of equation (13) provides the basis for the analytical evaluation of the effects of the shunting circuit parameters on the wave characteristics of the considered waveguide. The study evaluates the resonant characteristics of the shunting circuit and their effect on the equivalent mechanical properties of the beam. Such behavior can be analytically investigated in the long wavelength limit or $k \to 0$. For $k \to 0$, it is convenient to introduce two scales to describe the problem, as it is customary in homogenization problems for systems with periodically varying properties (Hassani, 1998), (Oleinik, 1985). A

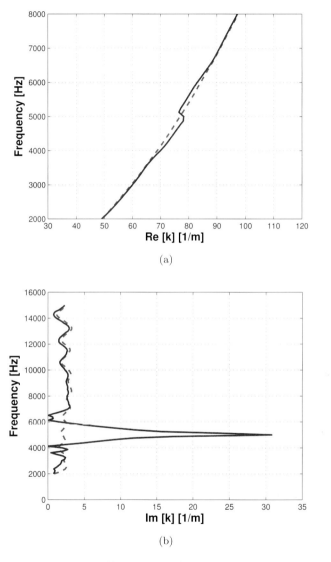

(a)

(b)

Figure 14. Experimentally estimated wavenumbers: open circuit (red dashed line), shunted circuits with tuning at 5000 Hz and resistor $R = 2.2\,\Omega$ (Blue solid line).

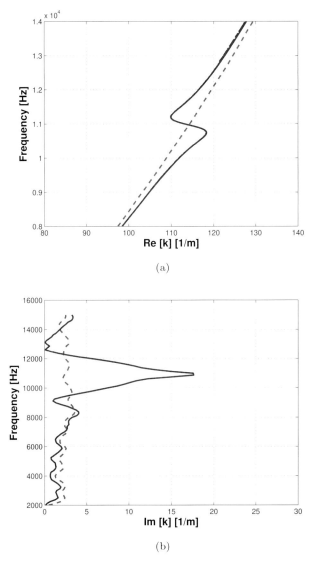

(a)

(b)

Figure 15. Experimentally estimated wavenumbers: open circuit (red dashed line), shunted circuits with tuning at 11000 Hz and resistor $R = 47\Omega$ (Blue solid line).

second scale $y = x/\epsilon$ describes the periodicity of the domain, in addition to the large scale coordinate x which governs the long wavelength behavior of the system. Assuming that $\epsilon << 1$, so that $y << x$, leads to a two-scale expansion. Accordingly, equation (13) is expressed as:

$$\frac{d}{dx}z(x,y) = C(x,y)z(x,y) \tag{31}$$

In the long wavelength, however, the properties of the beam can be considered as homogenous and therefore:

$$C(x,y) = C(x,x/\epsilon) \approx C(y)$$

Next, the state vector is expanded according to the two-scale expansion:

$$z(x) = z^{(0)}(x,y) + \epsilon z^{(1)}(x,y) + \epsilon^2 z^{(2)}(x,y) + ... \tag{32}$$

and the spatial derivative in equation (31) is rewritten as:

$$\frac{d}{dx} = \frac{\partial}{\partial x} + \frac{1}{\epsilon}\frac{\partial}{\partial y} \tag{33}$$

Substituting equations (32), (33) in equation (31) gives:

$$\frac{\partial}{\partial x}z^{(0)} + \frac{1}{\epsilon}\frac{\partial}{\partial y}z^{(0)} + \epsilon\frac{\partial}{\partial x}z^{(1)} + \frac{\partial}{\partial y}z^{(1)} + ... = C(y)(z^{(0)} + \epsilon z^{(1)} + ...) \tag{34}$$

which leads to the following set of ordered equations:

$$\epsilon^{-1}: \qquad \frac{\partial}{\partial y}z^{(0)} = 0, \tag{35}$$

$$\epsilon^{0}: \qquad \frac{\partial}{\partial x}z^{(0)} + \frac{\partial}{\partial y}z^{(1)} = C(y)z^{(0)}, \tag{36}$$

$$\epsilon^{1}: \qquad \frac{\partial}{\partial x}z^{(1)} = C(y)z^{(1)}, \tag{37}$$

Equation (35) implies that:

$$z^{(0)}(x,y) = z^{(0)}(x) \tag{38}$$

while equation (36) can be simplified by integrating both sides over a period p, which gives:

$$p\frac{\partial}{\partial x}z^{(0)}(x) = \int_{-\alpha p}^{(1-\alpha)p} C(y)dy\, z^{(0)}(x) \tag{39}$$

The result from equation (38) can be exploited along with the well known fact that the integral of the derivative of a periodic function over its period is equal to zero, i.e.:

$$\int_{-\alpha p}^{(1-\alpha)p} \frac{\partial}{\partial y} z^{(1)} dy = 0$$

Equation (39) can be rewritten as:

$$\frac{d}{dx} z^{(0)}(x) = \boldsymbol{C}_{eq} z^{(0)}(x) \tag{40}$$

which represents the governing equation for a beam of the kind considered whose equivalent homogeneous properties are given by:

$$\boldsymbol{C}_{eq} = \frac{1}{p} \int_{-\alpha p}^{(1-\alpha)p} \boldsymbol{C}(y) dy$$

Given the step-wise nature of the considered configuration, $\boldsymbol{C}_{eq}$ is given by:

$$\boldsymbol{C}_{eq} = \alpha \boldsymbol{C}_1 + (1-\alpha)\boldsymbol{C}_2 \tag{41}$$

Equivalent Mechanical Properties   For the waveguide under study, the extended expression for $\boldsymbol{C}_{eq}$ is :

$$\boldsymbol{C}_{eq} = \begin{bmatrix} 0 & 0 & 0 & \frac{1}{K_{eq}} & 0 & 0 \\ 0 & 0 & 1 & 0 & 0 & 0 \\ 0 & 0 & 0 & 0 & \frac{1}{D_{eq}} & 0 \\ -\omega^2 m_{eq} & 0 & 0 & 0 & 0 & 0 \\ 0 & 0 & 0 & 0 & 0 & -1 \\ 0 & -\omega^2 m_{eq} & 0 & 0 & 0 & 0 \end{bmatrix} \tag{42}$$

where m_{eq}, K_{eq} and D_{eq} are the equivalent linear mass, axial and bending stiffness of the beam, which are respectively given by:

$$\begin{aligned} m_{eq} &= \alpha m_1 + (1-\alpha)m_2 \\ K_{eq} &= \frac{K_1 K_2}{\alpha K_1 + (1-\alpha)K_2} \\ D_{eq} &= \frac{D_1 D_2}{\alpha D_1 + (1-\alpha)D_2} \end{aligned}$$

It is interesting to note how the results obtained within the considered long wavelength approximation correspond to the well-known relations obtained through the application of the rule of mixture for composite materials. Of note is the fact that the variability of the piezoelectric elastic modulus, equation (1), leads to a frequency-dependent axial stiffness $K_2(\omega)$ and in turn to an equivalent stiffness $K_{eq}(\omega)$ which is also frequency-dependent. Given the expressions for the elastic modulus of the shunted piezo patch (equation (1)), $K_{eq}(\omega)$ is given by:

$$K_{eq}(\omega) = \frac{E_b \, A_b \left[(1 + \bar{g}) \left(1 + \omega^2 \, L \, C_p^\varepsilon + i\omega \, R \, C_p^\varepsilon \right) - \bar{g} \, k_{31}^2 \right]}{\left[\alpha + (1 - \alpha)(1 + \bar{g}) \right] \left(1 + \omega^2 \, L \, C_p^\varepsilon + i\omega \, R \, C_p^\varepsilon \right) - (1 - \alpha) \, \bar{g} \, k_{31}^2} \tag{43}$$

where:

$$g \, E_b \, A_b = E_p^{SU} \, A_p$$

and

$$\bar{g} = g/(1 - k_{31}^2).$$

Similarly, the equivalent bending stiffness for the beam is given by:

$$D_{eq}(\omega) = \frac{E_b \, I_b \left[(1 + \bar{\gamma}) \left(1 + \omega^2 \, L \, C_p^\varepsilon + i\omega \, R \, C_p^\varepsilon \right) - \bar{\gamma} \, k_{31}^2 \right]}{\left[\alpha + (1 - \alpha)(1 + \bar{\gamma}) \right] \left(1 + \omega^2 \, L \, C_p^\varepsilon + i\omega \, R \, C_p^\varepsilon \right) - (1 - \alpha) \, \bar{\gamma} \, k_{31}^2} \tag{44}$$

where:

$$I_b = \frac{1}{12} b_b h_b^3$$

$$I_p = \frac{1}{12} b_p [2(\frac{1}{2} h_b + h_p)^3 - h_b^3]$$

$$\gamma \, E_b \, I_b = E_p^{SU} \, I_p$$

and

$$\bar{\gamma} = \gamma/(1 - k_{31}^2)$$

Frequency variations of equivalent bending and axial stiffnesses are presented in Figure 16 for different tunings of the shunt circuits. The resonant behavior of the shunts is reflected in the equivalent mechanical behavior of the waveguide, which is a behavior observed for other internally resonating metamaterials, and has been interpreted as the result of an apparent negative stiffness at the internal resonance (Huang et al., 2009). The results of Figure 16 are obtained for tuning at 5000 Hz and 11000 Hz, and illustrate how the resonant characteristics of the waveguide can be tuned at different frequencies without introducing any physical changes to the structure. Tuning is obtained by selecting different values of the inductors L in the electrical impedance of the shunts. The value of resistance R affects the magnitude of the resonance and its frequency bandwidth. High values of resistance correspond to broader ranges of frequencies, and lower amplifications at resonance for the equivalent elastic properties of the beam.

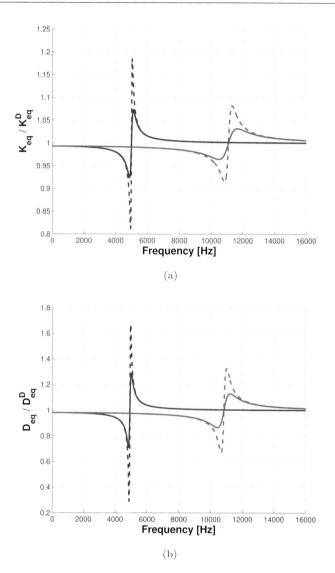

(a)

(b)

Figure 16. Equivalent axial (a) and bending (b) stiffness for different tunings of the piezo shunts (5000 Hz - blue lines, 11000 Hz - red lines; $R = 25\Omega$ - solid lines, $R = 50\Omega$ - dashed lines). Equivalent stiffnesses are normalized with respect to the open circuits values(K_{eq}^{D} and D_{eq}^{D}).

Dispersion relations   The equivalent properties found through the derivations above are used for the estimation of the dispersion properties of the waveguide. For a beam of equivalent properties given in equation (43), the dispersion relation relating frequency and wavenumber of axial waves is given by the well-known expression (Graff, 1975):

$$k_u^{(0)} = \omega \sqrt{\frac{m_{eq}}{K_{eq}}} \tag{45}$$

where $k_u^{(0)}$ denotes the approximation for $\lambda >> p$ for the wavenumber of longitudinal waves in the beam. The behavior of such wavenumber is illustrated in Figure 17 and 18, which highlight the resonant behavior of the structure at the tuning frequency and the associated attenuation frequency band. Similarly, the dispersion relation for transverse waves is also given by a well-known expression (Graff, 1975):

$$k_w^{(0)} = \left(\omega^2 \frac{m_{eq}}{D_{eq}} \right)^{\frac{1}{4}} \tag{46}$$

where $k_w^{(0)}$ denotes the approximation for $\lambda >> p$ for the wavenumber of transverse waves in the beam. The corresponding dispersion relations are also shown in Figure 17 and 18.

The dispersion relations predicted through the equivalent properties of the beam are compared with the experimental ones visualized through the contour of the magnitude of the 2DFT. Results for the two values of frequency tuning are shown in Figure 19. It is interesting to note how the equivalent properties formulation is able to identify the attenuation bands in the approximate ranges of frequency $4850 - 5150$ Hz and $10500 - 11500$ Hz. These accurate predictions are obtained in spite of the apparent inconsistence of analytical developments formulated on the basis of the long wavelength approximation, and the fact that the periodicity of the beam in this case is such that this assumption does not appear to be valid. However, the considered periodic addition of the piezoelectric patches does not affect the behavior of the beam in terms of added mass and stiffness, at least in the range of frequency here considered, and therefore the selection of a Brillouin zone on the basis of this periodicity appears as arbitrary and inappropriate. This suggests that for the case at hand, the extension of the equivalent properties estimation for the approximation of the dispersion relations can provide accurate evaluation of the equivalent properties of the beam, and a good model for the effect of piezoelectric shunting on the equivalent mechanical behavior of the beam.

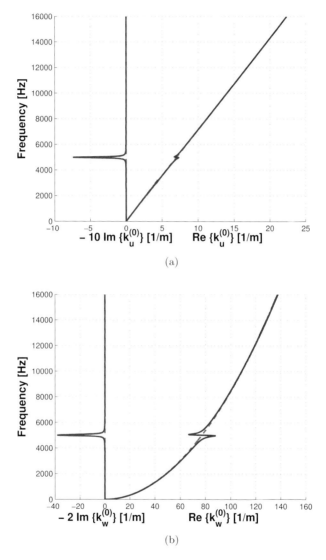

(a)

(b)

Figure 17. Dispersion relations for longitudinal (a) and transverse (b) motion: open circuit (red dashed line), tuning at 5000 Hz with $R = 33\ \Omega$ (blue solid line).

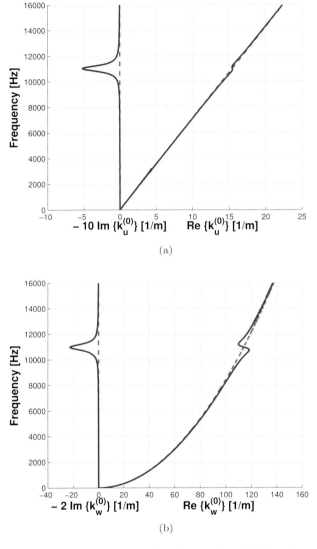

Figure 18. Dispersion relations for longitudinal (a) and transverse (b) motion: open circuit (red dashed line), tuning at 11000 Hz with $R = 47\,\Omega$ (blue solid line).

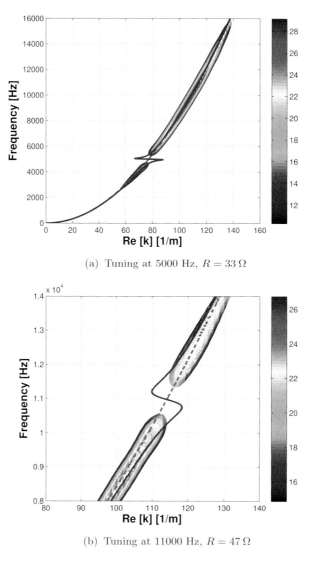

(a) Tuning at 5000 Hz, $R = 33\,\Omega$

(b) Tuning at 11000 Hz, $R = 47\,\Omega$

Figure 19. Experimental dispersion relations compared to the analytical predictions from the equivalent properties of the beam: open-circuit - thin black solid line, closed-circuit - red solid line.

2.5 Summary

This section describes the analysis of wave propagation in a periodic beam with shunted piezoelectric patches. The beam is as an example of a one-dimensional waveguide connected to a secondary system of periodic resonators. The resonating properties are due to the properties of the shunting circuits, whose impedance can be easily tuned to selected frequency values.

The wave propagation characteristics of the piezoelectric waveguide are first predicted through the application of the Transfer Matrix approach, which is conveniently derived for a structure with piecewise coefficients. The dispersion analysis highlights the occurrence of an internal resonance and the associated generation of an attenuation band at the tuning frequency of the shunts. Such a behavior is found for both axial and bending wave motion. Experimental evidence of the internal resonant behavior of the waveguide is provided through measurements performed on a beam with a periodic array of 11 patches. The experimental results, analyzed through the application of one-dimensional and two-dimensional Fourier Transforms, effectively confirm the numerical predictions, and illustrate the internal resonant characteristics of the waveguide. This behavior is achieved through the multifield coupling between the structural beam and the electrical circuits shunting the piezo patches. Further insight into the wave mechanics of the waveguide is gained through the development of analytical models of its equivalent mechanical properties. The dispersion relations predicted using this approach illustrate once again the internal resonant behavior of the beam, and capture with good accuracy the trends measured experimentally. The analytical results also suggest that the physical periodicity corresponding to the spacing between the piezo patches may not be an appropriate measure of the periodicity of the waveguide, which effectively behaves as a homogeneous structure with frequency-dependent, resonating mechanical properties.

3   Piezoelectric superlattices

As part of the investigation of multifield periodic structures, of interest is the case of piezoelectric solids with periodically modulated piezoelectric coefficients. These configurations, denoted in the literature as piezoelectric superlattices (PSL) (Chou and Yang, 2007; Zhu et al., 2003; Zhang et al., 2004), are characterized by strong coupling of acoustic and electromagnetic waves, which results from phonon-photon coupling, and the corresponding generation of *polaritons*. The coupling occurs at specified frequencies which are function of the periodicity of the medium, and is associated with the internal resonant behavior of the coupled fields. In this regard, the investigations presented in this section are very similar to those presented in the previous part of this chapter, as they also describe the formation of a bandgap due to the internal resonance of the system. The bandgaps occur at wavelengths which are much larger than the period of the domain, and therefore are not associated with Bragg scattering. At the bandgap frequencies, strong coupling between acoustics and electromagnetic waves may be associated with efficient energy transfer between the two fields. Such strong coupling, and the bandgap behavior of PSLs can be exploited for the design of novel acousto-optical devices for the manipulation of acoustic and/or electromagnetic waves for sensing and signal processing applications.

This section illustrates the wave mechanics of PSLs, which is investigated through the plane wave expansion (PWE) method (Yang et al., 2008), a technique commonly employed in the analysis of phononic and photonic crystals. The investigation of acousto-electromagnetic wave propagation in periodic superlattices illustrates their bandgaps characteristics which characterizes both one-dimensional (1D) and two-dimensional (2D) superlattices. The influence of parameters such as periodicity, piezoelectric coupling coefficients, and lattice topology is illustrated through several examples. The internal resonant behavior of the considered configurations is also interpreted through a long wavelength approximation, which allows the analysis of the bandgap behavior as the result of resonating equivalent properties of an homogenized acousto-electromagnetic medium.

3.1 Constitutive Equations for a Periodic Piezoelectric Medium

We consider wave motion in a piezoelectric material with periodic modulation of the coupling coefficients. The analysis is based on the constitutive equations for a linear piezoelectric material which have the well-known form:

$$
\begin{aligned}
\boldsymbol{S} &= \boldsymbol{d}^T \cdot \boldsymbol{E} + \boldsymbol{s}^E : \boldsymbol{T} \\
\boldsymbol{D} &= \varepsilon^S \cdot \boldsymbol{E} + \boldsymbol{d} : \boldsymbol{T},
\end{aligned}
\tag{47}
$$

where $\boldsymbol{S}$ is the strain field, $\boldsymbol{T}$ is the stress field, $\boldsymbol{D}$ is the vector of electric displacement, ε^S is the matrix of the dielectric constants at constant strain, and $\boldsymbol{s}^E$ is the compliance matrix at constant electric field. Also, $\boldsymbol{d}$ is the matrix of piezoelectric strain constants, whose ij component is the piezoelectric coefficient d_{ij}. This coefficient defines the ratio of the strain along the j-axis corresponding to an electric field applied along the i-axis, when all external stresses are held constant:

$$
d_{ij} = \frac{\varepsilon_j}{E_i},
\tag{48}
$$

Here, indexes $i = 1, 2, 3$ and $j = 1, ..., 6$ refer to different directions within the material coordinate system and the strain components. Equations (47) are known as *piezoelectric strain equations*. Sometimes it is useful to use stress rather than strain as an independent variable. To do this, we introduce the piezoelectric stress constants, e_{ij}:

$$e_{ji}^T = C_{jm}^E d_{mi}^T \tag{49}$$

where C_{jm}^E are the elastic constants, and $i = 1, 2, 3$ and $j, m = 1, 2, ..., 6$. Accordingly, (47) can be rewritten in the terms of stress to read:

$$
\begin{aligned}
\boldsymbol{T} &= -\boldsymbol{e}^T \cdot \boldsymbol{E} + \boldsymbol{C}^E : \boldsymbol{S} \\
\boldsymbol{D} &= \varepsilon^S \cdot \boldsymbol{E} + \boldsymbol{e} : \boldsymbol{S}
\end{aligned}
\tag{50}
$$

Through equations (50), the governing equations of the piezoelectric medium can be expressed as:

$$
\begin{aligned}
-\nabla \times \nabla \times \boldsymbol{E} &= \mu \varepsilon^S \frac{\partial^2 \boldsymbol{E}}{\partial t^2} + \mu \boldsymbol{e} : (\nabla_s \cdot \frac{\partial^2 \boldsymbol{u}}{\partial t^2}) \\
\nabla \cdot \boldsymbol{C}^E : (\nabla_s \cdot \boldsymbol{u}) &= \rho \frac{\partial^2 \boldsymbol{u}}{\partial t^2} + \nabla \cdot (\boldsymbol{e}^T \cdot \boldsymbol{E}),
\end{aligned}
\tag{51}
$$

which correspond to Maxwell's and Newton's equations, coupled by the piezoelectric coupling effect as defined by matrix $\boldsymbol{e}$. In equation (51), $\boldsymbol{u}$ is the displacement field, while ρ and μ respectively denote the density and the magnetic permeability of the considered medium.

In what follows, we consider the case of a periodic medium, where the spatial periodicity is introduced in the piezoelectric coupling matrix $\boldsymbol{e}$:

$$\boldsymbol{e} = \boldsymbol{e}(\boldsymbol{r}) = \boldsymbol{e}(\boldsymbol{r} + \boldsymbol{p}) \tag{52}$$

This defines a piezoelectric solid with *periodically polarized* domains, whereby polarization changes periodically with a spatial period $\boldsymbol{p}$. Given the coupled nature of the problem, this spatial periodicity affects both governing equations (51), along with the acoustic and electromagnetic dispersion properties of the medium. Such properties need to be estimated through the analysis of the coupled system. This analysis can be conveniently performed through various techniques, such as the Transfer Matrix method presented in the previous section, or the Plane Wave Expansion (PWE) method, which is briefly illustrated in the next section.

3.2   The Plane Wave Expansion Method

Dispersion estimation through the PWE method is introduced for the case of uncoupled governing equations, which is obtained simply by letting $e = \mathbf{0}$ in equations (51). The study can be therefore limited to the separate analysis of Newton's equation, or Maxwell's equation. The latter is selected with the understanding that derivations of the acoustic dispersion properties follow an identical procedure.

For simplicity, the case of a 1D system is presented, with the periodicity introduced as a 1D modulation in the dielectric properties of the considered medium, which represents a 1D example of a *photonic* crystal. The study is then extended to 2D photonic crystals in preparation for the analysis of the coupled system of PSLs.

One dimensional structures   Consider a photonic crystal featuring a one-dimensional array of air slabs penetrating a dielectric background (Figure 20). The propagation of the x_1-polarized electric field, E_1, along the x_3-direction is governed by the 1D version of Maxwell's equations (69):

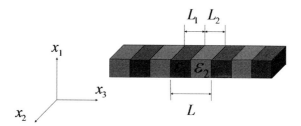

Figure 20. One-dimensional photonic crystal.

$$\mu\varepsilon(x_3)\frac{\partial^2 E_1}{\partial t^2} = \frac{\partial^2 E_1}{\partial x_3^2} \qquad (53)$$

with $\varepsilon(x_3) = \varepsilon(x_3 + L)$. Also:

$$\varepsilon(x_3) = \varepsilon_0\varepsilon_r(x_3)$$

$$\mu\varepsilon_0 = 1/c^2$$

where c is the speed of light in vacuum.

Equation (53) can be re-written as:

$$\frac{1}{c^2}\frac{\partial^2 E_1}{\partial t^2} = \frac{1}{\varepsilon_r(x_3)}\frac{\partial^2 E_1}{\partial x_3^2} \qquad (54)$$

Assuming an harmonically varying electric field, then $\frac{\partial^2}{\partial t^2} \rightarrow -\omega^2$, and equation (54) becomes:

$$-\frac{1}{\varepsilon_r(x_3)}\frac{\partial^2 E_1}{\partial x_3^2} = \frac{\omega^2}{c^2}\frac{\partial^2 E_1}{\partial t^2} \tag{55}$$

Based on Floquet-Bloch analysis for periodic media (Brillouin, 1946), the electric field can be expressed as the product of a cell-periodic term, and of a propagation term. The cell periodic part is expanded into a Fourier series representing the superposition of plane waves of wavenumber $G_n = 2n\pi/L, n \in I$, which are defined in the reciprocal lattice space of the periodic medium. Therefore the electric field is expressed as:

$$E_1(x_3) = e^{ikx_3}\sum_{G_n}\hat{E}_1(G_n)e^{iG_n x_3} = e^{ikx_3}\sum_n \hat{E}_1(G_n)e^{i\frac{2n\pi}{L}x_3} \tag{56}$$

where k is the wavenumber at the considered frequency ω, while $\hat{E}_1(G_n)$ is the n-th Fourier coefficients defining the amplitude of the corresponding plane wave. Given that the relative dielectric constant ε_r changes periodically in space, it can be also expanded in Fourier series:

$$\frac{1}{\varepsilon_r(x_3)} = \sum_{G_m}\hat{\varepsilon}_r(G_m)e^{iG_m x_3} = \sum_m \hat{\varepsilon}_r(G_m)e^{i\frac{2m\pi}{L}x_3} \tag{57}$$

where $G_m = 2m\pi/L, m \in I$ is also defined in the reciprocal lattice space, while, the $n-$th coefficient of the series is:

$$\hat{\varepsilon}_r(G_n) = \frac{1}{L}\int_{-L/2}^{L/2}\frac{1}{\varepsilon(x_3)_r}e^{-i\frac{2n\pi}{L}x_3}dx_3 \tag{58}$$

Ideally, the summations in equations (56) and (57) include an infinite number of terms, but in practice the series are truncated after a sufficient number of terms. Such number is defined on the basis of empirical considerations, previous experience and convergence studies performed on the predicted dispersion properties.

Upon Fourier expansion, equations (56) and (57) are substituted into the governing equation (55) which gives:

$$\sum_m \sum_n \left(\frac{2\pi n}{L} + k\right)^2 \hat{\varepsilon}_r(G_m)\hat{E}_1(G_n)e^{i\frac{2m\pi}{L}x_3}e^{i\frac{2n\pi}{L}x_3}e^{ikx_3} =$$

$$= \frac{\omega^2}{c^2}\sum_n \hat{E}_1(G_n)e^{i\frac{2n\pi}{L}x_3}e^{ikx_3} \tag{59}$$

Equation (59) can be considerably simplified by exploiting the orthogonality properties of harmonic functions which are bases for the Fourier
Series. Both sides of the equation are multiplied by a function $e^{-i\frac{2p\pi}{L}}$, and
then integrated over one spatial period (unit cell). This allows the elimination of one summation to give:

$$\sum_n \left(\frac{2\pi n}{L} + k\right)^2 \hat{\varepsilon}_r(G_{m-n})\hat{E}_1(G_n) = \frac{\omega^2}{c^2}\hat{E}_1(G_m) \qquad (60)$$

Equation (60) represents the m-th equation of a coupled algebraic system with n unknowns $\hat{E}_1(G_n)$. Given that the resulting system of equations
is homogeneous, the evaluation of non-trivial roots requires the solution of
an eigenvalue problem. The resulting eigenvalues are ω^2/c^2, and the eigenvectors $\boldsymbol{\hat{E}}_1$, form the expansion sought for the general solution expressed by
equation (56). The evaluation of the eigenvalue problem requires imposing
values for the integers m, n, which are defined over a range centered at zero,
i.e. $m, n = -N, ..., N$, and for the wavenumber k.

The resulting system of equations can be expressed in a compact matrix
notation as:

$$\boldsymbol{Q}\,\boldsymbol{\hat{E}}_1 = \frac{\omega^2}{c^2}\boldsymbol{\hat{E}}_1 \qquad (61)$$

where:

$$\boldsymbol{Q} \in \mathbb{R}^{(2N+1)\times(2N+1)}$$

with

$$Q_{m,n} = \left(\frac{2\pi n}{L} + k\right)^2 \hat{\varepsilon}_r(G_{m-n})$$

and

$$\boldsymbol{\hat{E}}_1 \in \mathbb{R}^{(2N+1)\times 1}$$

Given a value of k, the solution of the eigenvalue problem in equation
(61) gives the eigenvalues $\frac{\omega^2}{c^2}$ which define the dispersion relations of the
considered media, and the eigenvectors $\boldsymbol{\hat{E}}_1$ which define the wave modes
of the system at the corresponding frequency/wavenumber values. Substitution of the eigenvectors into the Fourier expansion for E_1 provides an
approximation for the field distribution at the considered frequency.

The procedure requires the evaluation of the Fourier coefficients corresponding to the periodic variation of the dielectric properties. For the
considered 1D photonic crystal, analytical expressions can be conveniently
evaluated upon the application of the definition of the Fourier coefficients:

$$\hat{\varepsilon}_r(G_n) = \frac{1}{L}\int_{-L/2}^{L/2}\frac{1}{\varepsilon_r(x_3)}e^{-i\frac{2n\pi}{L}z}dx_3 \qquad (62)$$

This integral can be properly split over the spatial range corresponding to a constant value of ε_r (Figure 21). Exploiting symmetry, the integral can be further simplified as follows:

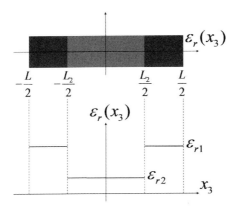

Figure 21. Unit cell of the photonic crystal and its parameters.

$$\hat{\varepsilon}_r(G_n) = 2 \left[\frac{1}{L} \int_{-L/2}^{-L_2/2} \frac{1}{\varepsilon_{r1}} e^{-i\frac{2n\pi}{L}x_3} dx_3 + \frac{1}{L} \int_{-L_2/2}^{0} \frac{1}{\varepsilon_{r2}} e^{-i\frac{2n\pi}{L}x_3} dx_3 \right] \quad (63)$$

$$\hat{\varepsilon}_r(G_n) = \frac{1}{\varepsilon_{r1}} \delta_n + \left(\frac{1}{\varepsilon_{r2}} - \frac{1}{\varepsilon_{r1}} \right) \frac{L_2}{L} \sin c \left(\frac{\pi n L_2}{L} \right) \quad (64)$$

where

$$\delta_n = \begin{cases} 1 & n = 0 \\ 0 & n \neq 0 \end{cases}$$

and

$$\sin c(nz) = \frac{\sin(nz)}{nz}.$$

The above result is used to populate the Q matrix and then solve the eigenvalue problem in equation (61). This gives the dispersion properties for the considered 1D periodic medium, which can be represented in the form of band diagrams, over a wavenumber range corresponding to the First Brioullin zone of the reciprocal lattice, i.e. $\frac{kL}{2\pi} \in [-\pi, \pi]$. The example of a band diagram for a 1D photonic crystal of period $L = 1$, $L_2/L = 0.8$, $\varepsilon_{r1} = 12.25$ and $\varepsilon_{r2} = 1$ (air slabs) is shown in Figure 22, where several band gaps are clearly visible.

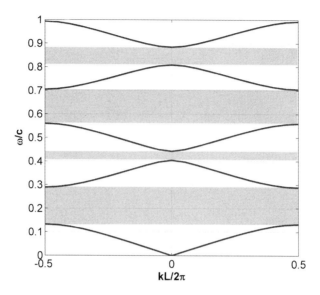

Figure 22. Dispersion diagram of the 1D photonic crystal.

Two dimensional photonic crystals   The application of the PWE is
extended to the analysis of 2D periodic structures. The case of photonic
crystals is again chosen for illustration purposes through the straightfoward
extension of the results presented in the previous paragraph to a 2D domain.

For a 2D spatially periodic domain, some attention should be devoted to
the description of the periodicity, and the definition of the topology of the
periodic structure under consideration. Five different lattice types can be
found in two-dimensions, namely the triangular lattice, the square lattice,
the rectangular lattice, the centered rectangular lattice and the oblique lattice (Kittel, 1962). A general framework describes the topology in terms of
lattice vectors, which define the directions along with the unit cell repeats to
form the periodic assembly. Based on the lattice vectors, a dual periodicity
in the wavenumber space (or reciprocal lattice space) also characterizes the
geometry of the lattice and specifically its wave propagation properties. For
periodic lattices, the reciprocal space is also periodic, and the first period
defines the First Brillouin zone (Brillouin, 1946). Knoweldge of lattice vectors, reciprocal lattice vectors and First Brillouin zone fully describes the
topology of the lattice, along with the wavenumber content of plane waves
propagating in it. Figure 23 shows for example the case of a triangular

lattice with its parameters. The lattice vectors in this case are defined as:

$$a_1 = ai$$
$$a_2 = \frac{a}{2}i + \frac{\sqrt{3}a}{2}j$$

(65)

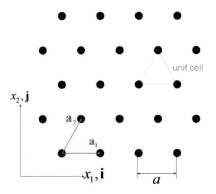

Figure 23. Triangular lattice.

The corresponding reciprocal lattice vectors are given by:

$$b_1 = \frac{2\pi}{a}\left(i - \frac{1}{\sqrt{3}}j\right)$$
$$b_2 = \frac{2\pi}{a}\frac{2}{\sqrt{3}}j$$

(66)

Figure 24 depicts the reciprocal lattice of the triangular lattice, with the light grey region indicating the first Brillouin zone, while the dark grey region defines the *irreducible* Brillouin zone which captures the minimum unit defining the characteristics of the reciprocal lattice on the basis of symmetry of the First Brillouin zone.

A further example is that of a square lattice shown in Figure 25. The lattice vectors, and reciprocal lattice vectors are respectively given by:

$$a_1 = ai$$
$$a_2 = aj$$

(67)

and

$$b_1 = \frac{2\pi}{a}i$$
$$b_2 = \frac{2\pi}{a}j$$

(68)

Reciprocal lattice space for the square lattice, and First and irreducible Brillouin zones are depicted in Figure 26.

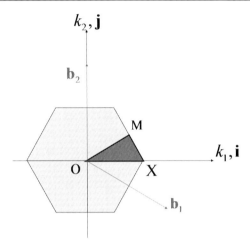

Figure 24. First Brillouin zone (light grey) and irreducible Brillouin zone (dark grey) for a triangular lattice.

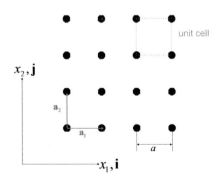

Figure 25. Square lattice.

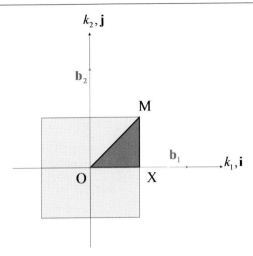

Figure 26. First Brillouin zone (light grey) and the irreducible Brillouin zone (dark grey) of the square lattice.

The PWE method is here illustrated for the case of a triangular lattice of air holes embedded in a dielectric background. The result is a two-dimensional array of air holes in a semiconductor substrate. Accordingly, the dielectric function is periodic only in the x_1, x_2 plane, and it is uniform across the thickness (x_3 direction). In the following developments, a denotes the lattice spacing, the relative dielectric constant of the medium is ε_{r1}, and the air holes have a diameter d and dielectric constant ε_{r2}.

The geometry of the lattice is depicted in Figure 27. The wave propagation dynamics of the system is described by Maxwell's equations (69):

$$-\nabla \times \nabla \times \mathbf{E} = \mu\varepsilon \frac{\partial^2 \mathbf{E}}{\partial t^2} \tag{69}$$

When restricted to the case where variables only vary along the x_1 and x_2 directions, Maxwell's equations become:

$$
\begin{aligned}
\mu\varepsilon(\mathbf{r}) \frac{\partial^2 E_1}{\partial t^2} &= \frac{\partial^2 E_1}{\partial x_2^2} - \frac{\partial^2 E_2}{\partial x_1 \partial x_2} \\
\mu\varepsilon(\mathbf{r}) \frac{\partial^2 E_2}{\partial t^2} &= \frac{\partial^2 E_2}{\partial x_1^2} - \frac{\partial^2 E_1}{\partial x_1 \partial x_2} \\
\mu\varepsilon(\mathbf{r}) \frac{\partial^2 E_3}{\partial t^2} &= \frac{\partial^2 E_3}{\partial x_1^2} + \frac{\partial^2 E_3}{\partial x_2^2} .
\end{aligned}
\tag{70}
$$

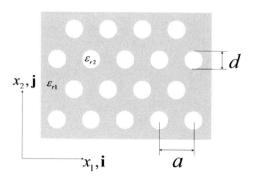

Figure 27. Geometry of the analyzed 2D photonic crystal.

Assuming a time harmonic electric field, i.e. $\frac{\partial^2}{\partial t^2} \to -\omega^2$, gives:

$$
\begin{aligned}
\frac{\omega^2}{c^2} E_1 &= \frac{1}{\varepsilon_r(\boldsymbol{r})} \left(-\frac{\partial^2 E_1}{\partial x_2^2} + \frac{\partial^2 E_2}{\partial x \partial y} \right) \\
\frac{\omega^2}{c^2} E_2 &= \frac{1}{\varepsilon_r(\boldsymbol{r})} \left(-\frac{\partial^2 E_2}{\partial x_1^2} + \frac{\partial^2 E_1}{\partial x_1 \partial x_2} \right) \\
\frac{\omega^2}{c^2} E_3 &= \frac{1}{\varepsilon_r(\boldsymbol{r})} \left(-\frac{\partial^2 E_3}{\partial x_1^2} - \frac{\partial^2 E_3}{\partial x_2^2} \right).
\end{aligned}
\tag{71}
$$

where $\varepsilon(\boldsymbol{r}) = \varepsilon_{r1}$ in the semiconductor substrate, while $\varepsilon(\boldsymbol{r}) = \varepsilon_{r2}$ in the air holes.

According to Floquet-Bloch's formalism, the electric field is expressed as the product of a cell periodic part and a wave propagation part. The periodic part is expanded in a Fourier series of plane waves whose wave vectors span the reciprocal lattice space $\boldsymbol{G} = p\boldsymbol{b}_1 + q\boldsymbol{b}_2$ with $p, q \in I$. Therefore the electric field can be expressed as:

$$
E_i(x_1, x_2) = e^{i\boldsymbol{k}\cdot\boldsymbol{r}} \sum_{\boldsymbol{G}} \hat{E}_i(\boldsymbol{G}) e^{i\boldsymbol{G}\cdot\boldsymbol{r}} = e^{i\boldsymbol{k}\cdot\boldsymbol{r}} \sum_p \sum_q \hat{E}_i(\boldsymbol{G}_{p,q}) e^{i(p\boldsymbol{b}_1 + q\boldsymbol{b}_2)\cdot\boldsymbol{r}}
\tag{72}
$$

where $\hat{E}_i(\boldsymbol{G})$ $(i = 1, 2, 3)$ are the plane wave coefficients, $\boldsymbol{k} = k_1\boldsymbol{i} + k_2\boldsymbol{j}$ is the wave vector, and $\boldsymbol{r} = x_1\boldsymbol{i} + x_2\boldsymbol{j}$ is the generic position vector.

The relative dielectric constant ε_r changes periodically and it is therefore expanded in a Fourier series:

$$\frac{1}{\varepsilon_r(\boldsymbol{r})} = \sum_{\boldsymbol{G'}} \hat{\varepsilon}_r(\boldsymbol{G'})e^{i\boldsymbol{G'}\cdot\boldsymbol{r}} = \sum_m \sum_n \hat{\varepsilon}_r(\boldsymbol{G}_{m,n})e^{i(m\boldsymbol{b}_1+n\boldsymbol{b}_2)\boldsymbol{r}} \qquad (73)$$

where $\boldsymbol{G'} = m\boldsymbol{b}_1 + n\boldsymbol{b}_2$ with $m, n \in I$ is also a reciprocal lattice vector, and $\hat{\varepsilon}_r(\boldsymbol{G'})$ are the coefficients of the series. Substituting the Fourier expansions (72) and (73) into the governing equation (71), gives:

$$\sum_{n,q,m,p} \left\{ \hat{\varepsilon}_r(\boldsymbol{G}_{m,n}) \left[\left(\tfrac{2\pi s}{a} + k_2\right)^2 \hat{E}_1(\boldsymbol{G}_{p,q}) + \right. \right.$$
$$\left. \left. - \left(\tfrac{2\pi s}{a} + k_2\right)\left(\tfrac{2\pi p}{a} + k_1\right) \hat{E}_2(\boldsymbol{G}_{p,q}) \right] e^{i\frac{2\pi(p+m)}{a}x_1} e^{i\frac{2\pi(s+r)}{a}x_2} \right\} =$$
$$= \tfrac{\omega^2}{c^2} \sum_{p,q} \left\{ \hat{E}_1(\boldsymbol{G}_{p,q})e^{i\frac{2\pi p}{a}x_1} e^{i\frac{2\pi s}{a}x_2} \right\}$$

$$\sum_{n,q,m,p} \left\{ \hat{\varepsilon}_r(\boldsymbol{G}_{m,n}) \left[-\left(\tfrac{2\pi s}{a} + k_2\right)\left(\tfrac{2\pi p}{a} + k_1\right) \hat{E}_1(\boldsymbol{G}_{p,q}) + \right. \right.$$
$$\left. \left. + \left(\tfrac{2\pi p}{a} + k_1\right)^2 \hat{E}_2(\boldsymbol{G}_{p,q}) \right] e^{i\frac{2\pi(p+m)}{a}x_1} e^{i\frac{2\pi(s+r)}{a}x_2} \right\} = \qquad (74)$$
$$= \tfrac{\omega^2}{c^2} \sum_{p,q} \left\{ \hat{E}_2(\boldsymbol{G}_{p,q})e^{i\frac{2\pi p}{a}x_1} e^{i\frac{2\pi s}{a}x_2} \right\}$$

$$\sum_{n,q,m,p} \left\{ \hat{\varepsilon}_r(\boldsymbol{G}_{m,n}) \left[\left(\tfrac{2\pi s}{a} + k_2\right)^2 + \left(\tfrac{2\pi p}{a} + k_1\right)^2 \right] \hat{E}_3(\boldsymbol{G}_{p,q}) \right.$$
$$\left. e^{i\frac{2\pi(p+m)}{a}x_1} e^{i\frac{2\pi(s+r)}{a}x_2} \right\} = \tfrac{\omega^2}{c^2} \sum_{p,q} \left\{ \hat{E}_3(\boldsymbol{G}_{p,q})e^{i\frac{2\pi p}{a}x_1} e^{i\frac{2\pi s}{a}x_2} \right\}$$

where

$$s = \frac{-p + 2q}{\sqrt{3}}, r = \frac{-m + 2n}{\sqrt{3}}$$

Following a procedure similar to that used for the 1D case, both sides of the resulting equations are multiplied by an orthogonal function $e^{-i\frac{2\pi s_1}{a}x_1}$ $e^{-i\frac{2\pi s_2}{a}x_2}$ and integrated over the unit cell, $\iint_A \dots dA$. Figure 28 shows a rectangular region as possible area of integration for the unit cell. Application of orthogonality properties upon integration, leads to the following system of algebraic equations:

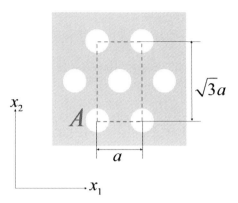

Figure 28. System unit cell.

$$\sum_{m,p}\left\{\hat{\varepsilon}_r(\boldsymbol{G}_{p-m,q-n})\left[\left(\tfrac{2\pi s}{a}+k_2\right)^2 \hat{E}_1(\boldsymbol{G}_{m,p})+\right.\right.$$
$$\left.\left.-\left(\tfrac{2\pi s}{a}+k_2\right)\left(\tfrac{2\pi m}{a}+k_1\right)\hat{E}_2(\boldsymbol{G}_{m,p})\right]\right\}=\tfrac{\omega^2}{c^2}\hat{E}_1(\boldsymbol{G}_{n,q})$$

$$\sum_{m,p}\left\{\hat{\varepsilon}_r(\boldsymbol{G}_{p-m,q-n})\left[-\left(\tfrac{2\pi s}{a}+k_2\right)\left(\tfrac{2\pi m}{a}+k_1\right)\hat{E}_1(\boldsymbol{G}_{m,p})+\right.\right.$$
$$\left.\left.+\left(\tfrac{2\pi m}{a}+k_1\right)^2 \hat{E}_2(\boldsymbol{G}_{m,p})\right]\right\}=\tfrac{\omega^2}{c^2}\hat{E}_2(\boldsymbol{G}_{n,q})$$

$$\sum_{m,p}\left\{\hat{\varepsilon}_r(\boldsymbol{G}_{p-m,q-n})\left[\left(\tfrac{2\pi s}{a}+k_2\right)^2+\left(\tfrac{2\pi m}{a}+k_1\right)^2\right]\hat{E}_3(\boldsymbol{G}_{m,p})\right\}=$$
$$=\tfrac{\omega^2}{c^2}\hat{E}_3(\boldsymbol{G}_{n,q})$$

$$(75)$$

where
$$s=\frac{-m+2p}{\sqrt{3}}$$
with $n, m, p, q = -N, ..., N$.

In compact matrix notation, the system can be rewritten as:

$$\boldsymbol{Q}\,\hat{\boldsymbol{E}}=\frac{\omega^2}{c^2}\hat{\boldsymbol{E}} \qquad (76)$$

where
$$\hat{\boldsymbol{E}}\,(\boldsymbol{G})=\begin{bmatrix} \hat{E}_1(\boldsymbol{G}) & \hat{E}_2(\boldsymbol{G}) & \hat{E}_3\,(\boldsymbol{G}) \end{bmatrix}^T$$

while $\boldsymbol{Q}\in\mathbb{R}^{3(2N+1)\times 3(2N+1)}$, is a matrix formed by 3×3 blocks $\boldsymbol{Q}(m,p)=\hat{\varepsilon}_r(\boldsymbol{G}_{n-m,q-p})\boldsymbol{M}$, with:

$$
\boldsymbol{M} = \begin{bmatrix}
\left(\frac{2\pi s}{a} + k_2\right)^2 & -\left(\frac{2\pi s}{a} + k_2\right)\left(\frac{2\pi m}{a} + k_1\right) \\
-\left(\frac{2\pi s}{a} + k_2\right)\left(\frac{2\pi m}{a} + k_1\right) & \left(\frac{2\pi m}{a} + k_1\right)^2 \\
0 & 0
\end{bmatrix}
$$

$$
\begin{bmatrix}
0 \\
0 \\
\left(\frac{2\pi s}{a} + k_2\right)^2 + \left(\frac{2\pi m}{a} + k_1\right)^2
\end{bmatrix}
$$

(77)

Equation (77) reveals that the set of equations can be decoupled in two groups. One group represents the transverse magnetic mode (TM-mode) wave, which is composed of E_3. The E_3 component implies the presence of H_1 and H_2 magnetic fields according to the Maxwell's equations. With TM-mode is thus intended the polarization of magnetic fields parallel to $x_1 - x_2$ plane. The other group represents the transverse electric mode (TE-mode) wave, which is composed of E_1 and E_2. With TE-mode wave is defined the polarization of electric fields parallel to $x_1 - x_2$ plane.

The solution of the eigenvalue problem in equation (76) is performed for an assigned pair k_1, k_2 which is chosen so that the wave vector $\boldsymbol{k}$ spans the irreducible Brillouin zone. This is sufficient to guarantee a complete analysis of the wave propagation characteristics in the periodic media (Kittel, 1962).

As in the 1D case, the process requires the evalution of the Fourier dielectric coefficients $\hat{\varepsilon}_r(\boldsymbol{G}_{m,p})$ through integration over the area of a unit cell. Analytical expressions for the coefficients is possible for simple periodic configurations and geometries of the inclusion as in this case. The general expression is given by:

$$
\hat{\varepsilon}_r(\boldsymbol{G}_{m,n}) = \frac{1}{A} \iint_A \frac{1}{\varepsilon_r(x_1, x_2)} e^{-i\frac{2\pi m}{a} x_1} e^{-i\frac{2\pi(-m+2n)}{\sqrt{3}a} x_2} dx_1 dx_2
\qquad (78)
$$

where the area of integration is represented by the rectangular region within the dashed lines in Figure 28.

For the considered geometry, cylindrical coordinates are particularly convenient. The integration is easily split in two parts, i.e. in the regions inside the air holes and in the dielectric background. The dielectric function in each region then becomes a constant and the integral thus simplifies to:

$$\hat{\varepsilon}_r(\boldsymbol{G}_{m,p}) =$$

$$= \frac{2\pi}{\sqrt{3}a^2} \left(\frac{2}{\varepsilon_{r2}} - \frac{2}{\varepsilon_{r1}} \right) \int_0^{d/2} r J_0 \left(r \sqrt{\left(\frac{2\pi m}{a} \right)^2 + \left(\frac{2\pi(-m+2p)}{\sqrt{3}a} \right)^2} \right) dr + \frac{1}{\varepsilon_{r1}} \delta_{m,p}$$

(79)

where J_n is the n-th order Bessel function. In the derivation of equation (79), the following property is used:

$$\int_{\theta=0}^{2\pi} e^{iAr\cos\theta} e^{iBr\sin\theta} r d\theta = 2\pi r J_0 \left(r \sqrt{A^2 + B^2} \right)$$

Through the identity:

$$\int_0^{d/2} r J_0 \left(Cr \right) dr = \frac{d}{2} \frac{J_1\left(C\frac{d}{2} \right)}{C}$$

equation (79) is further simplified to read:

$$\hat{\varepsilon}_r(\boldsymbol{G}_{m,p}) =$$

$$= \frac{2\pi}{\sqrt{3}a^2} \left(\frac{1}{\varepsilon_{r2}} - \frac{1}{\varepsilon_{r1}} \right) d \frac{J_1\left(\frac{d}{2} \sqrt{\left(\frac{2\pi m}{a} \right)^2 + \left(\frac{2\pi(-m+2p)}{\sqrt{3}a} \right)^2} \right)}{\sqrt{\left(\frac{2\pi m}{a} \right)^2 + \left(\frac{2\pi(-m+2p)}{\sqrt{3}a} \right)^2}} + \frac{1}{\varepsilon_{r1}} \delta_{m,p}$$

(80)

where

$$\delta_{m,p} = \begin{cases} 1 & m = p = 0 \\ 0 & m, p \neq 0 \end{cases}$$

The plane wave expansion method is applied to a case with $d/a = 0.5$, $\varepsilon_{r1} = 13.2$ and obviously $\varepsilon_{r2} = 1$ for the air holes, using 441 plane waves $(n, m, p, q = -10, ..., 10)$.

TE-modes have been separated from the results and, in particular, the modes relative to E_1 are shown in Figure 29 for values of k_1 and k_2 along the irreducible Brillouin zone. An evident band-gap is present between the first and the second branch.

3.3   One-dimensional superlattices

The PWE method is here employed to investigate the dispersion properties of PSLs. The study will first consider 1D PSLs, and will then be extended to the 2D case. The governing equations are presented in equations (51), where acousto-electromagnetic coupling is associated with periodic polarization of the piezoelectric material. The period of the alternating

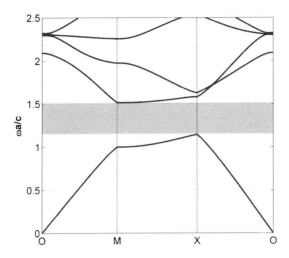

Figure 29. Dispersion diagram of the 2D photonic crystal along the irreducible Brillouin zone (TE-mode wave relative to E_1).

polarization regions defines the lattice constant of the superlattice, and provides a reference for the wavelength of wave motion in the coupled system. Based on previous derivations, the governing equations are expressed as:

$$
\begin{aligned}
-\nabla \times \nabla \times \boldsymbol{E} &= \mu \varepsilon^S \frac{\partial^2 \boldsymbol{E}}{\partial t^2} + \mu e(\boldsymbol{r}) : \left(\nabla_s \cdot \frac{\partial^2 \boldsymbol{u}}{\partial t^2}\right) \\
\nabla \cdot \boldsymbol{C}^E : (\nabla_s \cdot \boldsymbol{u}) &= \rho \frac{\partial^2 \boldsymbol{u}}{\partial t^2} + \nabla \cdot (e^T(\boldsymbol{r}) \cdot \boldsymbol{E}),
\end{aligned}
\tag{81}
$$

where

$$
e(\boldsymbol{r}) = e(\boldsymbol{r} + \boldsymbol{p})
\tag{82}
$$

We consider the 1D PSL depicted in Figure 30, which is composed of a periodic arrangement of ferroelectric domains along the x_3-axis. The polarization domains are parallel to the x_1-axis, and the considered material is LiNbO$_3$.

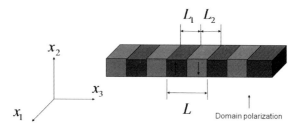

Figure 30. One-dimensional piezoelectric phononic crystal.

For LiNbO$_3$, $\boldsymbol{C}^E$, $\boldsymbol{e}$ and ε^S take the following forms:

$$\boldsymbol{C}^E = \begin{bmatrix} C_{33}^E & C_{13}^E & C_{13}^E & 0 & 0 & 0 \\ C_{13}^E & C_{11}^E & C_{12}^E & 0 & 0 & -C_{14}^E \\ C_{13}^E & C_{12}^E & C_{11}^E & 0 & 0 & C_{14}^E \\ 0 & 0 & 0 & C_{66}^E & C_{14}^E & 0 \\ 0 & 0 & 0 & C_{14}^E & C_{44}^E & 0 \\ 0 & -C_{14}^E & C_{14}^E & 0 & 0 & C_{44}^E \end{bmatrix}, \tag{83}$$

with

$$C_{66}^E = \frac{C_{11}^E - C_{12}^E}{2}$$

$$\boldsymbol{e} = \begin{bmatrix} e_{33} & e_{31} & e_{31} & 0 & 0 & 0 \\ 0 & e_{22} & -e_{22} & 0 & 0 & e_{15} \\ 0 & 0 & 0 & -e_{22} & e_{15} & 0 \end{bmatrix} \tag{84}$$

$$\boldsymbol{\varepsilon}^S = \begin{bmatrix} \varepsilon_{33}^S & 0 & 0 \\ 0 & \varepsilon_{11}^S & 0 \\ 0 & 0 & \varepsilon_{11}^S \end{bmatrix} \tag{85}$$

Substituting (83), (84) and (85) into (81), taking into consideration only the x_3-variations and the relation between the longitudinal displacement, u_3, the x_1-polarized electric field E_1, and the x_2-polarized electric field E_2, gives:

$$\frac{\partial^2 E_1(x_3,t)}{\partial x_3^2} = \mu\varepsilon_{33}\frac{\partial^2 E_1(x_3,t)}{\partial t^2} + \mu e_{31}(x_3)\frac{\partial}{\partial x_3}\frac{\partial^2 u_3(x_3,t)}{\partial t^2} \tag{86}$$

$$\frac{\partial^2 E_2(x_3,t)}{\partial x_3^2} = \mu\varepsilon_{11}\frac{\partial^2 E_2(x_3,t)}{\partial t^2} - \mu e_{22}(x_3)\frac{\partial}{\partial x_3}\frac{\partial^2 u_3(x_3,t)}{\partial t^2} \tag{87}$$

and

$$C_{11}^E \frac{\partial^2 u_3(x_3,t)}{\partial x_3^2} + \frac{\partial}{\partial x_3}[e_{22}(x_3)E_2(x_3,t)] - \frac{\partial}{\partial x_3}[e_{31}(x_3)E_2(x_3,t)] = \rho \frac{\partial^2 u_3(x_3,t)}{\partial t^2}$$
(88)

Due to spatial periodicity, the piezoelectric constants e_{ij} are expanded in Fourier series with respect to the 1D reciprocal lattice vectors, $G_m = 2m\pi/L, m \in I$:

$$e_{ij}(x_3) = \sum_{G_m} \hat{e}_{ij}(G_m)e^{iG_m x_3} = \sum_m \hat{e}_{ij}(G_m)e^{i\frac{2m\pi}{L}x_3}$$
(89)

where $\hat{e}_{ij}(G_m)$ are the Fourier coefficients defined as:

$$\hat{e}_{ij}(G_m) = \frac{1}{L}\int_{-L/2}^{L/2} e_{ij}(x_3)e^{-i\frac{2m\pi}{L}x_3}dx_3$$
(90)

The integral is evaluated analytically by exploiting the step-wise form of the function describing the periodic polarization. The integration follows a procedure identical to the one used for the 1D photonic crystal previously studied (Figure 31).

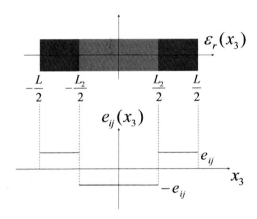

Figure 31. The unit cell of the piezoelectric phononic crystal and its parameters.

With the unit cell of the Figure 31, the integral in equation (90) can be split into:

$$\hat{e}_{ij}(G_m) = \frac{2}{L}\int_{-L/2}^{-L_2/2} e_{ij}e^{-i\frac{2m\pi}{L}x_3}dx_3 - \frac{2}{L}\int_{-L_2/2}^{0} e_{ij}e^{-i\frac{2m\pi}{L}x_3}dx_3$$
(91)

whose evaluation gives:

$$\hat{e}_{ij}(G_m) = e_{ij}\delta_m - 2e_{ij}\frac{L_2}{L}\sin c\left(\frac{\pi m L_2}{L}\right) \tag{92}$$

Similarly, expanding the displacement vector $u_3(x_3,t)$ and electric fields $E_1(x_3,t), E_2(x_3,t)$ yields:

$$u_3(x_3,t) = e^{i(kx_3-\omega t)}\sum_{G_n}\hat{u}_3(G_n)e^{iG_n x_3} = e^{i(kx_3-\omega t)}\sum_{n}\hat{u}_3(G_n)e^{i\frac{2n\pi}{L}x_3} \tag{93}$$

and

$$E_i(x_3,t) = e^{i(kx_3-\omega t)}\sum_{G_n}\hat{E}_i(G_n)e^{iG_n x_3} = e^{i(kx_3-\omega t)}\sum_{n}\hat{E}_i(G_n)e^{i\frac{2n\pi}{L}x_3} \tag{94}$$

where $i = 1, 2$, k is the wave number along x_3, and $\hat{u}_3(G_n), \hat{E}_i(G_n)$ are the coefficients of the series.

Substituting equations (92) and (93) into equations (88) and collecting terms systematically, leads to the following eigenvalue problem:

$$Q\hat{U} = \omega^2 B\hat{U} \tag{95}$$

where:

$$\hat{U} = \begin{bmatrix} \hat{u}_3(G_n) & \hat{E}_1(G_n) & \hat{E}_2(G_n) \end{bmatrix}^T$$

Also in equation (95), matrices A and B are

$$A = \begin{bmatrix} 0 & \left(\frac{2\pi n}{L}+k\right)^2\delta_{m,n} \\ 0 & 0 \\ C_{11}^E\left(\frac{2\pi m}{L}+k\right)^2\delta_{m,n} & j\left(\frac{2\pi m}{L}+k\right)\hat{e}_{31}(G_{m-n}) \end{bmatrix}$$

$$\begin{matrix} 0 \\ -\left(\frac{2\pi n}{L}+k\right)^2\delta_{m,n} \\ -j\left(\frac{2\pi m}{L}+k\right)\hat{e}_{22}(G_{m-n}) \end{matrix} \tag{96}$$

$$B = \begin{bmatrix} j\mu\left(\frac{2\pi n}{L}+k\right)\hat{e}_{31}(G_{m-n}) & \mu\varepsilon_{33}\delta_{m,n} & 0 \\ j\mu\left(\frac{2\pi n}{L}+k\right)\hat{e}_{22}(G_{m-n}) & 0 & -\mu\varepsilon_{11}\delta_{m,n} \\ \rho\delta_{m,n} & 0 & 0 \end{bmatrix} \tag{97}$$

Dispersion relations for the 1D PSL   The resulting eigenvalue problem is solved for assigned values of k in the first Brillouin zone of the lattice. The dispersion diagram is generated for the considered $LiNbO_3$ PSL, whose acoustic and electromagnetic properties are summarized in Table 3. The geometry of the crystal is defined by $L = 19\mu m$ and $L_1 = L_2 = L/2$.

Table 4. The physical parameters values of $LiNbO_3$ (according to Auld (1990))

C_{11}^E	203 GPa
ρ	4700 kg/m^3
e_{22}	2.5 Cm^{-2}
e_{31}	0.2 Cm^{-2}
ε_0	8.8542 pFm^{-1}
ε_{11}^S	$44\varepsilon_0$
ε_{33}^S	$29\varepsilon_0$

The dispersion relations are calculated by applying the PWE method using 63 wave vectors in the reciprocal space, i.e. by considering $m, n = -10, ..., 10$. For convenience, results are presented in terms of the nondimensional wavenumber $\bar{k}$,

$$\bar{k} = \frac{kL}{2\pi} \tag{98}$$

and frequency $\Omega = \omega/\omega_0$, where:

$$\omega_0 = \left(\frac{2\pi}{L}\right)\left(\frac{C_{44}^E}{\rho}\right)^{1/2} \tag{99}$$

The obtained band diagram shown in Figure 32 corresponds to the case when coupling is neglected, i.e. when $e = 0$. Two diagrams present the electromagnetic branches of the dispersion relations, which define the photon spectrum, and the acoustic branches, which describe the phonon spectrum for the considered acoustic medium. Of note is the fact that representation of the two sets of branches on the same plot would make the photon branches hardly distinguishable from the ordinates axis, as the phase velocity of the electromagnetic waves is several order of magnitudes higher than that of elastic waves. Accordingly, for a given frequency, electromagnetic wave motion occurs at wavelengths which are much longer (several orders of magnitude) than that of acoustic waves. In homogeneous media, the generation of coupled acousto-electromagnetic motion is not possible due to the lack of coincidence between the two dispersion branches. For a periodic medium as considered here however, folding of the acoustic dispersion branches provides the potential for matching of the dispersion properties of the two fields. The dispersion match enables multifield coupling, along with the coupled behavior inherent with the piezoelectric material as defined by the polarization matrix e. When such coupling is considered in the computation of the dispersion relations, coupled branches appear. Such branches define phonon-photon coupled modes of propagation and are known as *polaritons* (Kittel, 1962). Figure 33.a illustrates how the polariton branches coincide almost everywhere with phonons, as clearly illustrated by the detailed plot around the center of the Brillouin zone which is shown in Figure 32.b. This implies that the character of polaritons is extremely similar to that of phonons in an elastic material with same elastic constants.

In addition, one of the dispersion curves of polaritons coincides with that of the E_1 field photons near the very center of the first Brillouin zone as shown in Figure 33.b. These polaritons thus behave like corresponding photons in the material with the same dielectric constants. However, the polariton spectrum becomes complex and bridges the dispersion curves of phonons and E_2 field photons at specific frequencies. The corresponding frequency values are integer multiples of the fundamental frequency of the lattice ω_0.

The behavior of the polariton branch shows the typical features of an internal resonance, as the branch splits and seems to tend asymptotically to a frequency of internal resonance ω_0. This mirrors the observations made in the presence of shunted piezoelectric patches tuned at a specific frequency, as illustrated in the previous section of this chapter. As before, the internal resonance is associated with effective multi-field coupling and the resulting energy transfer between the primary system, which previously was represented by the mechanical waveguide, and a secondary system corresponding to the shunting circuit. In this case, the internal resonance appears as the condition of efficient energy transfer between the primary system corresponding to the electromagnetic waveguide, and a secondary system which is the acoustic component of the coupled system. An additional point of commonality is the occurrence of such internal resonance at very long wavelengths, which are several orders of magnitude larger than the spatial period of the medium. The internal resonance is also responsible for the generation of a bandgap centered at the coupling frequency. Figure 34 shows details of spectra in the range of $0 \leq \bar{k} \leq 3 \cdot 10^{-4}$ and $0.8 \leq \Omega \leq 1.2$. Both polariton branches and the photons and phonons curves are presented for reference.

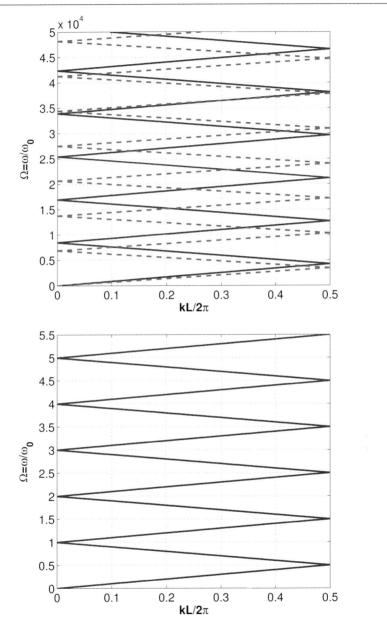

Figure 32. Photons spectrum (E_1 solid blue line, E_2 dashed red line) (a) and phonon spectrum (b) in the absence of piezoelectric coupling.

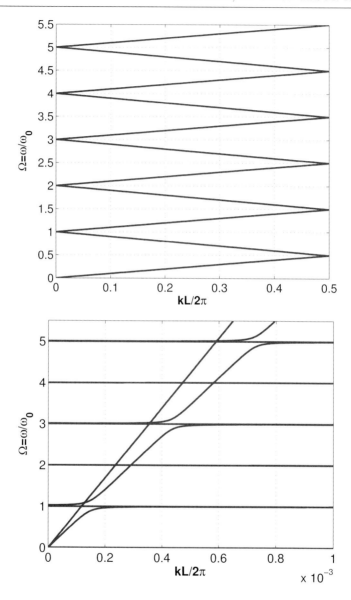

Figure 33. Polariton spectrum occurring with piezoelectric coupling (a), and detail of the coupling near the center of the first Brillouin zone (b).

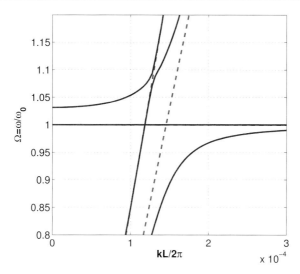

Figure 34. Detail of the dispersion diagram (zoom around $\omega/\omega_0 = 1$ in Figure 33.b). Blu solid lines, polariton branches; Red dashed lines, photons and phonons curves.

3.4 Two-dimensional PSLs

In the following, a PSL composed of a 2D periodic array (x_1x_2 plane) of material A embedded in a background material B is considered. The periodicity reproduces a square lattice topology (Figure 35), where materials A and B have identical properties, and opposite polarizations directions along the x_3-axis. Thus, the piezoelectric constants of materials A and B differ only in sign. The material used is ZnO whose properties are listed in Table 5.

Table 5. The physical parameters values of ZnO (Yang et al., 2008)

C_{11}^E	209.7 GPa
C_{12}^E	121.1 GPa
ρ	5680 kg/m^3
e_{31_0}	-0.573 Cm^{-2}
ε_0	8.8542 pFm^{-1}
ε_{33}^S	10.2ε_0

For ZnO crystal, $\boldsymbol{C}^E$, $\boldsymbol{e}$ and $\boldsymbol{\varepsilon}^S$ take the following form:

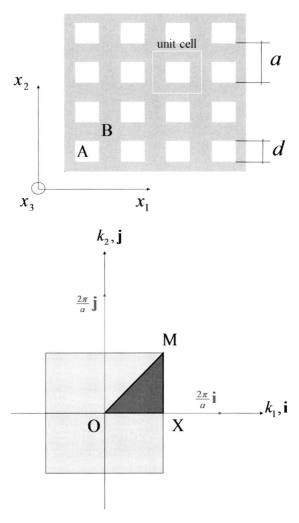

Figure 35. Schematic of PSL topology: (left) square lattice; (right) Brillouin zone relative to the square lattice.

$$\boldsymbol{C}^E = \begin{bmatrix} C_{11}^E & C_{12}^E & C_{13}^E & 0 & 0 & 0 \\ C_{12}^E & C_{11}^E & C_{13}^E & 0 & 0 & 0 \\ C_{13}^E & C_{13}^E & C_{33}^E & 0 & 0 & 0 \\ 0 & 0 & 0 & C_{44}^E & 0 & 0 \\ 0 & 0 & 0 & 0 & C_{44}^E & 0 \\ 0 & 0 & 0 & 0 & 0 & C_{66}^E \end{bmatrix} \tag{100}$$

$$\boldsymbol{e} = \begin{bmatrix} 0 & 0 & 0 & 0 & e_{15} & 0 \\ 0 & 0 & 0 & e_{15} & 0 & 0 \\ e_{31} & e_{31} & e_{33} & 0 & 0 & 0 \end{bmatrix} \tag{101}$$

$$\boldsymbol{\varepsilon}^S = \begin{bmatrix} \varepsilon_{11}^S & 0 & 0 \\ 0 & \varepsilon_{11}^S & 0 \\ 0 & 0 & \varepsilon_{33}^S \end{bmatrix} \tag{102}$$

Substituting equations (100), (101) and (125) into equation (81), and considering only x_1-x_2 variations, leads to the set of equations describing the behavior of the 2D ZnO PSL. The problem can be decoupled in two parts, one of which represents a mixed mode of the in-plane elastic wave and the TM-mode EM wave (Yang et al., 2008). The variables included in the problem therefore are u_1, u_2 and E_3. The second part corresponds to a mixed mode of the out-of-plane elastic wave, defined by u_3, and the TE-mode EM wave E_1, E_2. For brevity, only the first part (u_1, u_2 and E_3) is here considered with the understanding that treatment of the second portion of the problem follows similar steps.

Thus, the resulting set of equations is:

$$\begin{aligned} C_{11}u_{1,11} + C_{12}u_{2,21} + C_{66}\left(u_{1,22} + u_{2,12}\right) &= \rho\ddot{u}_1 + \partial_1\left[e_{31}E_3\right] \\ C_{12}u_{1,12} + C_{11}u_{2,22} + C_{66}\left(u_{1,21} + u_{2,11}\right) &= \rho\ddot{u}_2 + \partial_2\left[e_{31}E_3\right] \\ E_{3,22} + E_{3,11} &= \mu\varepsilon_{33}\ddot{E}_3 + \mu e_{31}\ddot{u}_{1,1} + \mu e_{31}\ddot{u}_{2,2} \end{aligned} \tag{103}$$

where $\partial_i = \frac{\partial}{\partial x_i}$, $g_{,j} = \frac{\partial g}{\partial x_j}$ and $g_{,ij} = \frac{\partial g}{\partial x_i \partial x_j}$.

According to the PWE method, $e_{31}(\boldsymbol{x})$ is expanded in the Fourier series with respect to the two-dimensional reciprocal lattice vectors, $\boldsymbol{G}_{mp} = \frac{2\pi}{a}[m, \ p]^T$ $(m, p = -N...N)$, which gives:

$$e_{31}(\boldsymbol{x}) = \sum_{m,p} \hat{e}_{31}(\boldsymbol{G}_{mp})e^{-i\boldsymbol{G}_{mp}\cdot\boldsymbol{x}} \tag{104}$$

The Fourier coefficients, $\hat{e}_{31}(\boldsymbol{G}_{mp})$, are defined as:

$$\bar{e}_{31}(\boldsymbol{G}_{mp}) = \frac{1}{A} \iint_A e_{31}(\boldsymbol{x})e^{i\boldsymbol{G}_{mp}\cdot\boldsymbol{x}}d\boldsymbol{x} \tag{105}$$

where $A = a^2$ is the area of the unit cell. Similarly, the displacement components u_1, u_2 and electric field component E_3 are expanded in Fourier series:

$$u_i(\boldsymbol{x}, t) = e^{i(-\boldsymbol{k} \cdot \boldsymbol{x} + \omega t)} \sum_{\boldsymbol{G}_{nq}} \hat{u}_i(\boldsymbol{G}) e^{-i \boldsymbol{G}_{nq} \cdot \boldsymbol{x}} \tag{106}$$

with $i = 1, 2$, and

$$E_3(\boldsymbol{x}, t) = e^{i(-\boldsymbol{k} \cdot \boldsymbol{x} + \omega t)} \sum_{\boldsymbol{G}_{nq}} \hat{E}_3(\boldsymbol{G}) e^{-i \boldsymbol{G}_{nq} \cdot \boldsymbol{x}} \tag{107}$$

Substituting equation (105) and equation (107) into (103) leads to the following eigenvalue problem:

$$\boldsymbol{A}\hat{U} = \omega^2 \boldsymbol{B}\hat{U} \tag{108}$$

where $\hat{U} = \begin{bmatrix} \hat{u}_1(\boldsymbol{G}_{nq}) & \hat{u}_2(\boldsymbol{G}_{nq}) & \hat{E}_3(\boldsymbol{G}_{nq}) \end{bmatrix}^T$, while $\boldsymbol{A}$ and $\boldsymbol{B}$ are:

$$\boldsymbol{A} = \begin{bmatrix} \left[C_{11}\left(\frac{2\pi n}{L} + k_1\right)^2 + C_{66}\left(\frac{2\pi q}{L} + k_2\right)^2\right]\delta_{\boldsymbol{G}_{mp}, \boldsymbol{G}_{nq}} & (C_{66} + C_{12})\left(\frac{2\pi n}{L} + k_1\right)\left(\frac{2\pi q}{L} + k_2\right)\delta_{\boldsymbol{G}_{mp}, \boldsymbol{G}_{nq}} & j\left(\frac{2\pi n}{L} + k_1\right)\hat{e}_{31}(\boldsymbol{G}_{n-m, q-p}) \\ (C_{66} + C_{12})\left(\frac{2\pi n}{L} + k_1\right)\left(\frac{2\pi q}{L} + k_2\right)\delta_{\boldsymbol{G}_{mp}, \boldsymbol{G}_{nq}} & \left[C_{11}\left(\frac{2\pi q}{L} + k_2\right)^2 + C_{66}\left(\frac{2\pi n}{L} + k_1\right)^2\right]\delta_{\boldsymbol{G}_{mp}, \boldsymbol{G}_{nq}} & j\left(\frac{2\pi q}{L} + k_2\right)\hat{e}_{31}(\boldsymbol{G}_{n-m, q-p}) \\ 0 & 0 & \left[\left(\frac{2\pi n}{L} + k_1\right)^2 + \left(\frac{2\pi q}{L} + k_2\right)^2\right]\delta_{\boldsymbol{G}_{mp}, \boldsymbol{G}_{nq}} \end{bmatrix}$$

$$\boldsymbol{B} = \begin{bmatrix} \rho\delta_{\boldsymbol{G}_{mp}, \boldsymbol{G}_{nq}} & 0 & 0 \\ 0 & \rho\delta_{\boldsymbol{G}_{mp}, \boldsymbol{G}_{nq}} & 0 \\ j\mu\left(\frac{2\pi m}{L} + k_1\right)\hat{e}_{31}(\boldsymbol{G}_{n-m, q-p}) & j\mu\left(\frac{2\pi p}{L} + k_2\right)\hat{e}_{31}(\boldsymbol{G}_{n-m, q-p}) & \mu\varepsilon_{33}\delta_{\boldsymbol{G}_{mp}, \boldsymbol{G}_{nq}} \end{bmatrix}$$

with

$$\delta_{\boldsymbol{G}_{mp},\boldsymbol{G}_{nq}} = \begin{cases} 1 & m = n \quad and \quad p = q \\ 0 & m \neq n \quad or \quad p \neq q \end{cases}$$

Dispersion relations: Band diagrams The PWE method is applied for the estimation of the dispersion properties of a ZnO superlattice of square topology defined by $d/a = 0.5$. The properties of ZnO are summarized in Table 5. The dispersion is obtained through the solution of the EVP in equation (108), for a wave vector $\boldsymbol{k}$ spanning the irreducible Brillouin zone of the lattice. The PWE method is here applied using 363 wave vectors, i.e. considering $m, n, p, q = -5, ..., 5$. Results are presented in terms of the non-dimensional wavenumber, $\bar{k}$, and non-dimensional frequency, Ω, which are respectively defined as:

$$\bar{k} = ka/2\pi$$
$$\Omega = \omega/\omega_L$$

where $\omega_L = \left(\frac{2\pi}{a}\right)\left(\frac{C_{11}^E}{\rho}\right)^{1/2}$.

Figure 36 shows the dispersion diagram of the mixed modes u_1, u_2 and E_3. The results seem two suggest that only two branches start from point O, which however is due to the fact that the photon branch is very steep and hardly distinguishable from the vertical line through O. Within the First Brillouin zone, the wavenumber range defined by the spatial periodicity of the SPL is very large for the EM wave propagation at the considered frequencies. Phonon-photon interaction can instead be observed in a limited region of the wave vector, i.e. for $\bar{k} \approx 10^{-4}$, where the coupling between phonon and photon is apparent (Figure 37). Specifically, the electromagnetic dominated dispersion curve starts from O and is strongly coupled with the folded-phonon dispersion curve near the centre of the first Brillouin zone at $\Omega \approx 1$ and $\Omega \approx \sqrt{2}$ (Zhu et al., 2003).

3.5 Equivalent Dielectric Properties Based on Long Wavelength Approximation

It would be interesting to evaluate the coupling effect predicted by the dispersion analysis through numerical simulations on finite domains. Such simulations however pose significant challenges due to the differences in time and spatial scales characterizing the dynamic behavior of the two coupled fields. Such difference is easily quantified through the estimation of the phase velocities of electromagnetic and elastic waves, which respectively are given by:

$$c_{EM} = \frac{1}{\sqrt{\mu_0 \bar{\varepsilon}}} \tag{109}$$

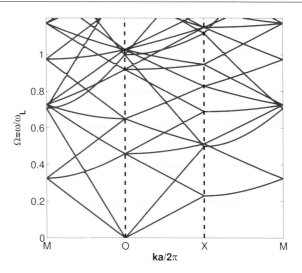

Figure 36. Band structure of mixed mode of in-plane elastic and TM electromagnetic waves along the irreducible Brillouin zone. No polariton excitation is observed at this scale.

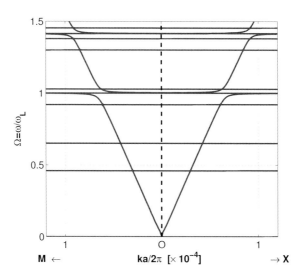

Figure 37. Band structure of mixed mode of in-plane elastic and TM electromagnetic waves near the center, O, of the first Brillouin zone. Polariton excitation is evident at $\Omega \approx 1$ and $\Omega \approx \sqrt{2}$.

and

$$c_{EL} = \sqrt{\frac{C_{11}}{\rho}} \tag{110}$$

which give $c_{EM} \approx 10^5 c_{EL}$.

Numerical simulations must be refined enough to resolve both fields, in domains which are sufficiently large to allow the development of both waves simultaneously. As an alternative, the difference in scales can be exploited to investigate the propagation and the equivalent properties of one of the two fields, and to estimate the influence of multi-field coupling on the equivalent properties of the chosen medium. In the case of PSLs, the estimation of equivalent properties reveals explicitly the resonant behavior of the coupled system, and allows the interpretation of such behavior in terms of unusual, frequency dependent properties.

The long wavelength approximation can be easily employed to estimate the equivalent dielectric properties of the considered class of PSLs. The analysis employs the governing equations for the PSL under the assumption of harmonic time variation of the displacement and electric field components. This leads to:

$$
\begin{aligned}
C_{11}u_{10,11} + C_{12}u_{20,21} + C_{66}\left(u_{10,22} + u_{20,12}\right) &= -\rho\omega^2 u_{10} + \partial_1\left[e_{31}E_{30}\right] \\
C_{12}u_{10,12} + C_{11}u_{20,22} + C_{66}\left(u_{10,21} + u_{20,11}\right) &= -\rho\omega^2 u_{20} + \partial_2\left[e_{31}E_{30}\right] \\
E_{30,22} + E_{30,11} &= -\mu\varepsilon_0\varepsilon_{33}\omega^2 E_{30} - \mu e_{31}\omega^2(u_{10,1} + u_{20,2})
\end{aligned}
\tag{111}
$$

Each of the variables in the equations above can be expressed in terms of a Fourier integral of the form:

$$u_{10}\left(\boldsymbol{x}\right) = \iint \hat{u}_{10}\left(\boldsymbol{q}\right) e^{-i\boldsymbol{q}\cdot\boldsymbol{x}} d\boldsymbol{q} \tag{112}$$

where expressions for $u_{20}(\boldsymbol{x})$ and $E_{30}(\boldsymbol{x})$ have similar formulations. Also, the relevant polarization coefficient is expanded in a Fourier series:

$$e_{31}(\boldsymbol{x}) = \sum_{m,p=-\infty}^{+\infty} e_{31mp} e^{-i\boldsymbol{G}_{mp}\cdot\boldsymbol{x}} \tag{113}$$

whose coefficients are:

$$e_{31mp} = \frac{1}{2}\frac{\sin\left(\frac{\pi}{2}m\right)}{\frac{\pi}{2}m}\frac{\sin\left(\frac{\pi}{2}p\right)}{\frac{\pi}{2}p}e_{31_0} \tag{114}$$

The first two equations in (111) thus become:

$$\iint \left[\left(\rho\omega^2 - C_{11}q_1^2 - C_{66}q_2^2 \right) \hat{u}_{10} - \left(C_{66} + C_{12}q_1q_2 \right) \hat{u}_{20} \right] e^{-i\boldsymbol{q}\cdot\boldsymbol{x}} d\boldsymbol{q} =$$

$$= \iint -i \sum_{m,p} e_{31mp} \left(G_{1m} + k_1 \right) \hat{E}_{30} e^{-i(\boldsymbol{k}+\boldsymbol{G}_{mp})\cdot\boldsymbol{x}} d\boldsymbol{k} \quad (115)$$

and

$$\iint \left[\left(\rho\omega^2 - C_{11}q_2^2 - C_{66}q_1^2 \right) \hat{u}_{20} - \left(C_{66} + C_{12}q_1q_2 \right) \hat{u}_{10} \right] e^{-i\boldsymbol{q}\cdot\boldsymbol{x}} d\boldsymbol{q} =$$

$$= \iint -i \sum_{m,p} e_{31mp} \left(G_{2p} + k_2 \right) \hat{E}_{30} e^{-i(\boldsymbol{k}+\boldsymbol{G}_{mp})\cdot\boldsymbol{x}} d\boldsymbol{k} \quad (116)$$

With reference to the photon spectrum, equivalent models for the dielectric properties of the PSL can safely adopt the long wavelength assumption, which implies that $\boldsymbol{k} \to 0$, or $\boldsymbol{k} \ll \boldsymbol{G}_{mp}$. In order for the two sides of equations (115) and (116) to be equal, and the equations to be verified, equality of the integrands must be imposed. In addition, letting $\boldsymbol{q} = \boldsymbol{k} + \boldsymbol{G}_{mp} \simeq \boldsymbol{G}_{mp}$, gives:

$$\left(\rho\omega^2 - C_{11}(\frac{2\pi}{a}m)^2 - C_{66}(\frac{2\pi}{a}p)^2 \right) \hat{u}_{10} - \left(C_{66} + C_{12}(\frac{2\pi}{a})^2 mp \right) \hat{u}_{20} =$$

$$= -i \sum_{m,p} e_{31mp} m \frac{2\pi}{a} \hat{E}_{30} (117)$$

and

$$\left(\rho\omega^2 - C_{11}(\frac{2\pi}{a}p)^2 - C_{66}(\frac{2\pi}{a}m)^2 \right) \hat{u}_{20} - \left(C_{66} + C_{12}\frac{2\pi^2}{a} mp \right) \hat{u}_{10} =$$

$$= -i \sum_{m,p} e_{31mp} \frac{2\pi}{a} p \hat{E}_{30} \quad (118)$$

Solving for $\hat{u}_{10}$ and $\hat{u}_{20}$ and transforming back to the spatial domain, leads to an expression of the displacement components as functions of the electric field:

$$u_{10}\left(\boldsymbol{x} \right) = K_1(\boldsymbol{x}) E_{30}(\boldsymbol{x}) \qquad (119)$$

$$u_{20}\left(\boldsymbol{x} \right) = K_2(\boldsymbol{x}) E_{30}(\boldsymbol{x}) \qquad (120)$$

where

$$K_1(\boldsymbol{x}) = -i \sum_{m,p} e_{31mp} \frac{Gm \, e^{-i\boldsymbol{G}_{mp}\boldsymbol{x}}}{\left(\rho\omega^2 - C_{11} \frac{2\pi}{a}^2 (m^2 + p^2) \right)} \tag{121}$$

$$K_2(\boldsymbol{x}) = -i \sum_{m,p} e_{31mp} \frac{Gp \, e^{-i\boldsymbol{G}_{mp}\boldsymbol{x}}}{\left(\rho\omega^2 - C_{11} \frac{2\pi}{a}^2 (m^2 + p^2) \right)} \tag{122}$$

Substituting equations (119) and (120) into the last equation in (111) leads to a single decoupled differential relation with E_{30} as unknown:

$$E_{30,22} + E_{30,11} = -\mu\omega^2 \left[\varepsilon_0 \varepsilon_{33} + K(\boldsymbol{x}) \right] E_{30} \tag{123}$$

where

$$K(\boldsymbol{x}) = e_{31}(\boldsymbol{x})[K_1(\boldsymbol{x}) + K_2(\boldsymbol{x})]$$

For $\boldsymbol{k} \to 0$, the periodic variation of the $K(\boldsymbol{x})$ is characterized by a period which is much smaller than the considered wavelength. Therefore, an equivalent value can be considered to simplify the form of the resulting differential equation. This value is computed as the average of $K(\boldsymbol{x})$ over one period, which gives:

$$
\begin{aligned}
K_{eq} &= \frac{1}{A} \iint_A K(\boldsymbol{x}) d\boldsymbol{x} \\
&= \sum_{m,p} \frac{16}{\rho a^2} \frac{(m^2 + p^2) \sin\left(\frac{\pi}{2}m\right) \sin\left(\frac{\pi}{2}p\right)}{(pm\pi)^2} \frac{e_{31_0}^2}{(\omega_L^2 (m^2 + p^2) - \omega^2)}
\end{aligned}
\tag{124}
$$

where $\omega_L = \frac{2\pi}{a} \left(\frac{C_{11}}{\rho} \right)^{1/2}$. This leads to an equivalent dielectric term which is frequency dependent, as described by the resulting expression:

$$\varepsilon_{33}(\omega) = \varepsilon_{33} \left(1 + \sum_{m,p} \frac{4e_{31_0}^2}{\varepsilon_0 \varepsilon_{33} C_{11}} \frac{\sin\left(\frac{\pi}{2}m\right) \sin\left(\frac{\pi}{2}p\right)}{\pi^4 p^2 m^2} \frac{\omega_{Lmp}^2}{(\omega_{Lmp}^2 - \omega^2)} \right) \tag{125}$$

where $\omega_{Lmp}^2 = \omega_L^2 (m^2 + p^2)$. The variation of ε_{33} versus frequency illustrates its resonant characteristics occurring at $\omega = \omega_{Lmp}$ (see Figure 38). Details of the behavior of ε_{33} around the resonant frequency is presented in Figure 39, whose analysis is convenient for the evaluation of the effect of the resonant behavior on the bandgap properties of the considered PSL.

The dispersion diagrams for the electromagnetic portion of the PSL, homogenized through the long wavelength approximation, can be found from the resulting governing equation, which is of the form:

$$E_{30,22} + E_{30,11} = -\mu\varepsilon_0\omega^2\varepsilon_{33}(\omega)E_{30} \tag{126}$$

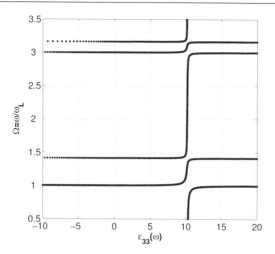

Figure 38. dielectric abnormality.

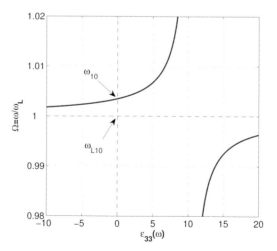

Figure 39. dielectric abnormality associated to the first band gap, which means consider $m = 1$ and $p = 0$.

The corresponding characteristic equation, obtained by imposing a plane wave solution $E_{30}(\omega, \boldsymbol{k}) = E_{30_0} e^{i\boldsymbol{k}\cdot\boldsymbol{x}} = E_{30_0} e^{i(k_1 x_1 + k_2 x_2)}$ for equation (126), is:

$$\left(k_1{}^2 + k_2{}^2\right) = \mu\varepsilon_0\omega^2\varepsilon_{33}\left(\omega\right)$$
$$k^2 = \mu\varepsilon_0\omega^2\varepsilon_{33} \tag{127}$$

The resulting dispersion branch is given by $k = \omega\sqrt{\mu\varepsilon_0\varepsilon_{33}}$ and approximates the polariton branch observed from the application of the PWE method. Results for a 1D PSL are shown in Figure 40, while the case for a 2D PSL of rectangular topology are shown in Figure 41. The polariton branch for a 2D rectangular PSL is compared to the dispersion curves obtained with the PWE method for the same case in Figure 42. A good agreement is observed between the two approaches; the green curves of the PWE method not captured by the homogenization analysis are those due to the purely elastic branches.

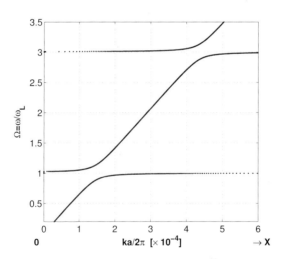

Figure 40. Polariton dispersion relation for a 1D PSL obtained with the homogenization analysis. Band gaps are located for $\Omega = 1$ and 3.

The analytical expression for the equivalent dielectric term provides the tools for the estimation of the bandgap characteristics of the considered PSL. Specifically, bandgaps correspond to the frequency intervals where $\varepsilon_{33}(\omega) < 0$, which identifies a condition of dielectric abnormality. From equation (125), we obtain the approximate upper cutoff frequencies ω_{mp} of

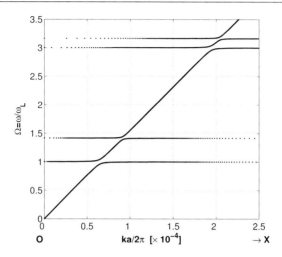

Figure 41. Polariton dispersion relation for a 2D PSL with rectangular topology obtained with the homogenization analysis. Band gaps are located for $\Omega = 1, \sqrt{2}, 3$ and $\sqrt{10}$.

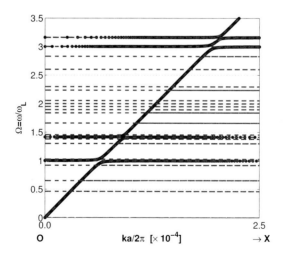

Figure 42. Comparison of the polariton dispersion relations obtained with the two alternative approaches. In blue dispersion obtained with the homogenization analysis; in green using the PWE method.

the bandgaps by setting $\varepsilon(\omega_{mp}) = 0$ (Figure 39). This gives:

$$\omega_{mp} = \omega_{Lmp} \left(1 + \frac{4e_{31_0}{}^2}{\varepsilon_0 \varepsilon_{33} C_{11}} \frac{\sin\left(\frac{\pi}{2}m\right)\sin\left(\frac{\pi}{2}p\right)}{\pi^4 p^2 m^2} \right)^{1/2} \qquad (128)$$

Since the electromechanical coupling term, $e_{31_0}^2 / \varepsilon_0 \varepsilon_{33} C_{11} \ll 1$, equation (128) can be approximated as:

$$\omega_{mp} \approx \omega_{Lmp} \left(1 + \frac{2\sin\left(\frac{\pi}{2}m\right)\sin\left(\frac{\pi}{2}p\right)}{\pi^4 p^2 m^2} \frac{e_{31_0}{}^2}{\varepsilon_0 \varepsilon_{33} C_{11}} \right) \qquad (129)$$

Therefore, the size of bandgaps $\Delta\omega_{mp}$ related to polariton excitation and dieletric abnormality can be estimated as

$$\Delta\omega_{mp} = \frac{\omega_{mp} - \omega_{Lmp}}{\omega_{Lmp}} \approx \frac{2\sin\left(\frac{\pi}{2}m\right)\sin\left(\frac{\pi}{2}p\right)}{\pi^4 p^2 m^2} \frac{e_{31_0}{}^2}{\varepsilon_0 \varepsilon_{33} C_{11}} \qquad (130)$$

Equation (130) illustrates that the width of the bandgap $\Delta\omega_{mp}$ is a function of the considered order number m, p of the resonant frequency of the dielectric constant, and depends on lattice topology and its material properties.

4   Conclusions

This chapter illustrates the wave propagation characteristics of periodic multi-field systems. The focus of the investigations presented herein is on the estimation of internal resonant mechanisms which are the result of strong coupling between the two fields. Such coupling is identified by the intersection of the dispersion properties of the two domains, where a resonant mechanism leads to unusual dispersion characteristics. Such characteristics can be interpreted as the result of unusual, frequency-dependent physical properties of the system.

Two examples are presented. The first consists of a mechanical waveguide with a periodic array of shunted piezoelectric patches. The equivalent mechanical impedance of the waveguide can be affected through the selection of the shunting circuit, which can be made resonant. This leads to an "internally resonating" metametarial, whose wave propagation properties can be tuned through the tuning of the shunting circuit. The behavior of this waveguide is investigated analytically, numerically and experimentally. The second example considers piezoelectric superlattices defined by periodic polarization of the piezoelectric medium. The resulting coupling between acoustic and electromagnetic properties of the medium leads to an internally resonant property, which occurs at wavelengths which are several order of magnitudes larger than the periodicity of polarization. Numerical results for 1D and 2D piezoelectric lattices illustrate this phenomenon and evaluate some effects of lattice topology. The long wavelength expansion is also applied to simplify the problem and to support the interpretation of the coupling behavior and the generation of polaritons as a result of an internal resonance of the multi-field system.

Bibliography

B.A. Auld. *Acoustic Fields and Waves in Solids*. Krieger Pub Co, 1990.

S. Behrens, S.O.R. Moheimani, and J. Fleming. Multiple mode current flowing passive piezoelectric shunt controller. *Journal of Sound and Vibration*, 266:929–942, 2003.

L. Brillouin. *Wave Propagation in Periodic Structures*. McGraw-Hill, 1946.

F. Casadei, M. Ruzzene, L. Dozio, and K.A. Cunefare. Broadband vibration control through periodic arrays of resonant shunts: experimental investigation on plates. *Smart Materials and Structures*, 19(1):015002, 2010.

Y. Cheng, J. Y. Xu, and X. J. Liu. One-dimensional structured ultrasonic metamaterials with simultaneously negative dynamic density and modulus. *Phys. Rev. B*, 77(4):045134, Jan 2008. doi: 10.1103/PhysRevB.77.045134.

Y. Chou and M. Yang. Energy conversion in piezoelectric superlattices. *Proc. SPIE*, 6526(65260L), 2007.

L. De Marchi, S. Caporale, N. Speciale, and A. Marzani. Ultrasonic guided-waves characterization with warped frequency transforms. In *Ultrasonics Symposium, 2008. IUS 2008. IEEE*, pages 188 –191, 2008. doi: 10.1109/ULTSYM.2008.0046.

Yiqun Ding, Zhengyou Liu, Chunyin Qiu, and Jing Shi. Metamaterial with simultaneously negative bulk modulus and mass density. *Phys. Rev. Lett.*, 99(9):093904, Aug 2007.

N. Fang, D. Xi, J. Xu, M. Ambati, W. Srituravanich, C. Sun, and X. Zhang. Ultrasonic metamaterials with negative modulus. *Nature Materials*, 5(6): 452–456, 2006.

R.L. Forward. Electronic damping of vibrations in optical structures. *Journal of applied optics*, 18(5):690–697, 1979.

K.F. Graff. *Wave Motion in Elastic Solids*. Dover, 1975.

N. Hagood and N. von Flotow. Damping of strucutal vibrations with piezoelectric materials and passive electical networks. *Journal of Sound and Vibration*, 1991.

Hassani. A review of homogenization and topology optimization iii–topology optimization using optimality criteria. *Computers and Structures*, 69(6): 739 – 756, 1998.

J. Hollkamp. Multimodal passive vibration suppression with piezoelectric materials and resonant shunt. *Journal of Intelligent Material Systems and Structures*, 1994.

G. L. Huang and C. T. Sun. Band gaps in a multiresonator acoustic metamaterial. *Journal of Vibration and Acoustics*, 132(3):031003, 2010.

H.H. Huang, C.T. Sun, and G.L. Huang. On the negative effective mass density in acoustic metamaterials. *International Journal of Engineering Science*, 47(4):610 – 617, 2009.

J. Kim and Y.-C. Jung. Broadband noise reduction of piezoelectric smart panel featuring negative-capacitive-converter shunt circuit. *Journal of the Acoustical Society of America*, 120:2017–2025, 2006a.

J. Kim and Y.-C. Jung. Piezoelectric smart panels for broadband noise reduction. *Journal of Intelligent Materials Systems and Structures*, 17 (685-690), 2006b.

J. Kim and J.-H. Kim. Multimode shunt damping of piezoelectric smart panel for noise reduction. *Journal of the Acoustical Society of America*, 116:942–948, 2004.

J. Kim and J.-K. Lee. Broadband transmission noise reduction of smart panels featuring piezoelectric shunt circuits and sound-absorbing material. *Journal of the Acoustical Society of America*, 116:9908–998, 2004.

C. Kittel. *Elementary Solid State Physics: A Short Course*. Wiley, 1962.

M. S. Kushwaha, P. Halevi, L. Dobrzynski, and B. Djafari-Rouhani. Acoustic band structure of periodic elastic composites. *Phys. Rev. Lett.*, 71 (13):2022–2025, Sep 1993.

M.S. Kushwaha and B. Djafari-Rouhani. Giant sonic stop bands in two-dimensional periodic system of fluids. *Journal of Applied Physics*, 84(9): 4677–4683, 1998.

B.S. Lazarov and J.S. Jensen. Low-frequency band gaps in chains with attached non-linear oscillators. *International Journal of Non-Linear Mechanics*, 42(10):1186 – 1193, 2007.

P. G. Martinsson and A. B. Movchan. Vibrations of Lattice Structures and Phononic Band Gaps. *The Quarterly Journal of Mechanics and Applied Mathematics*, 56(1):45–64, 2003.

O. Oleinik. On homogenisation problems. *Milan Journal of Mathematics*, 55:105, 1985.

P.F. Pai. Metamaterial-based Broadband Elastic Wave Absorber. *Journal of Intelligent Material Systems and Structures*, 21(5):517–528, 2010.

A. Preumont. *Vibration Control of Active Structures*. Kluwer, 1997.

R.H.S. Riordan. Simulated inductors using differential amplifiers. *Eletronics Letters*, 3:50–51, 1967.

Ping Sheng, X.X. Zhang, Z. Liu, and C.T. Chan. Locally resonant sonic materials. *Physica B: Condensed Matter*, 338(1-4):201 – 205, 2003.

Ping Sheng, Jun Mei, Zhengyou Liu, and Weijia Wen. Dynamic mass density and acoustic metamaterials. *Physica B: Condensed Matter*, 394(2):256 – 261, 2007.

M.M. Sigalas. Defect states of acoustic waves in a two-dimensional lattice of solid cylinders. *Journal of Applied Physics*, 84(6):3026–3030, 1998.

A. Spadoni, M. Ruzzene, and K. Cunefare. Vibration and Wave Propagation Control of Plates with Periodic Arrays of Shunted Piezoelectric Patches. *Journal of Intelligent Material Systems and Structures*, 20(8):979–990, 2009.

H. Sun, X. Du, and P.F. Pai. Theory of Metamaterial Beams for Broadband Vibration Absorption. *Journal of Intelligent Material Systems and Structures*, 21(11):1085–1101, 2010.

O. Thorp, M. Ruzzene, and A. Baz. Attenuation and localization of wave propagation in rods with periodic shunted piezoelectric patches. *Smart Materials and Structures*, 10(5):979 – 989, 2001.

O. Thorp, M. Ruzzene, and A. Baz. Attenuation and localization of waves in shells with periodic shunted piezo rings. *Smart Materials and Structures*, 14(4):594 – 604, 2005.

S.Y. Wu. Piezoelectric shunts with a parallel r-l circuit for structural damping and vibration control. In *Proceedings of SPIE - The International Society for Optical Engineering*, volume 2720, pages 259 – 269, San Diego, CA, USA, 1996.

M. Yang, L. Wu, and J. Tseng. Phonon-polariton in two-dimensional piezoelectric phononic crystals. *Physics Letters A*, 372(26):4730–4735, 2008.

S. Yao, X. Zhou, and G. Hu. Experimental study on negative effective mass in a 1d mass–spring system. *New Journal of Physics*, 10(4):043020, 2008.

X. Zhang, R. Zhu, J. Zhao, Y. Chen, and Y. Zhu. Phonon-polariton dispersion and the polariton-based photonic band gap in piezoelectric superlattices. *Phys. Rev. B*, 69(085118), 2004.

Y. Zhu, X. Zhang, Y. Lu, Y. Chen, S. Zhu, and N. Ming. New type of polariton in a piezoelectric superlattice. *Phys. Rev. Lett.*, 90(053903), 2003.

Topology optimization

Jakob S. Jensen

Department of Mechanical Engineering, Technical University if Denmark, Kgs. Lyngby, Denmark

1 Topology optimization basics.

The basic setting for the method of topology optimization will first be briefly outlined. This will include a description of the *analysis model, performance evaluation and sensitivity analysis, regularization and filtering* and finally the *optimization step*. The method will be described in general terms without specific reference to relevant problems in vibration and wave propagation. The core of this chapter follow in the subsequent chapters that include the author's contributions to different applications of topology optimization in dynamics.

First, it should be mentioned that the method of topology optimization was introduced for optimizing stiff and lightweight mechanical structures by Bendsøe and Kikuchi (1988). However, this text will not include a lengthy review of the history and applications of topology optimization. For this we will refer to the monograph by Bendsøe and Sigmund (2003).

1.1 Analysis model.

The optimization procedure in the standard form is based on a *discretized* model equation:

$$\mathbf{Su} = \mathbf{f} \qquad (1)$$

in which $\mathbf{S}$ is a *system matrix*, $\mathbf{f}$ is a *loading vector*, and $\mathbf{u}$ is a vector with the unknown discretized field values. The model equation may represent a discretized model of a static or steady-state dynamic field problem in mechanics, electromagnetics, acoustics or another physics setting. The discretized model may be obtained using one of numerous discretization techniques, such as finite elements, finite volumes, or finite differences. For a linear static mechanics application the matrix $\mathbf{S}$ will be the usual stiffness matrix, but for other applications the matrix contains the relevant physics for the problem considered. In a later chapter it will be shown how a transient dynamic problem can also be used as a basis for the optimization procedure.

Since topology optimization naturally lends itself to a model based on a *finite element discretization*, the following presentation will refer to such models exclusively.

Eq. (1) can be solved in a straightforward manner using any efficient direct or iterative solver. It should be emphasized that the choice of solver that is favorable will depend on the physical problem considered. For instance, if $\mathbf{S}$ is a dynamic stiffness matrix (as will be studied in the following chapter for steady-state vibration or wave propagation problems) then positive definiteness is not guaranteed and this will put a limitation, for instance, on the type of iterative or direct solver that can be applied.

1.2   Design parametrization.

Fig. 1 shows an example of a part of a photonic crystal waveguide (for more information on photonic crystal waveguides, see e.g. Joannopoulos et al. (1995)) and a part of the corresponding finite element model. Quadratic elements are here used for the discretization. The darkest regions in the physical structure (corresponding to the white regions in the FE model) represent air and the gray regions (black regions in the FE model) represent dielectric material. To facilitate the optimization procedure each element is assigned a material indicator variable, ξ_e, which is used to control the material properties of that element and thus indicate whether an element is of one or the other material type. Usually, the description is made so that $\xi_e = 0$ indicates one material (here air) and $\xi_e = 1$ the other material (dielectric material). The optimization problem can then be formulated as to find the optimal configuration of elements with either $\xi_e = 0$ and $\xi_e = 1$ in order to minimize (or maximize) a chosen performance criterion.

To solve the optimization problem using an exhaustive search of all combinatorial configurations is infeasible even for a small FE model. In the authors experience the same is generally true for global optimization methods based on genetic-type algorithms which are not efficient for realistic model sizes with possibly up to millions of design variables. In this presentation we will instead focus on using a *gradient-based* optimization strategy.

In order to apply a gradient-based approach, the discrete material indicator is relaxed into a *continuous* variable $\xi_e \to \rho_e$:

$$0 \le \rho_e \le 1 \tag{2}$$

so that the material properties, during the optimization procedure, are allowed also to take *intermediate* values between the two specified materials.

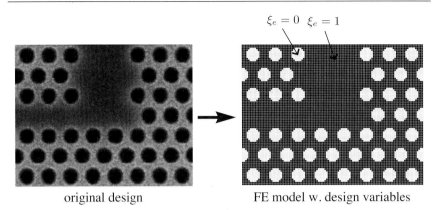

original design   FE model w. design variables

Figure 1. Illustration of an un-optimized optical component (a 90-degree bend in a photonic crystal waveguide (Jensen et al., 2005)) and a corresponding finite element (FE) model. The element material indicator ξ_e is used to control the material properties in each element with $\xi_e = 0$ (white) corresponding to the air holes and $\xi_e = 1$ (black) corresponding to the dielectric waveguide material.

The material properties of each element is then based on the value of ρ_e using a material interpolation model (Bendsøe and Sigmund, 1999). This approach is generally referred to as the density or SIMP method (Bendsøe, 1989). The final optimized design must naturally consist of design variables that are either 0 or 1, unless the intermediate values can be interpreted as specific materials. Special techniques exist for ensuring such black-white designs as will be described in section 1.5.

For a wave propagation problem in a photonic crystal the relevant material parameter is the relative dielectric constant ε. An interpolation model that interpolates between air and a dielectric material in a specific element could be:

$$\varepsilon_e = 1 + \rho_e(\varepsilon - 1) \tag{3}$$

so that for $\rho_e = 0$ the property of air ($\varepsilon_1 = 1$) is obtained and for $\rho_e = 1$ the property of the dielectric material (e.g. $\varepsilon = 11.56$ for a silicon material) is obtained. Of course, the approach could be used also to interpolate between two dielectric materials different from air.

For a static structural optimization problem where the optimal distribution of material and void is to be found, a material interpolation model for the Young's modulus E_e in an element could then be:

$$E_e = \rho_e E \tag{4}$$

so that for $\rho_e = 1$ the element modulus becomes that of the specified material: E, and for $\rho_e = 0$ the Young's modulus vanishes. When dealing with a static structural problem the definition in Eq. (2) must be modified slightly since a vanishing Young's modulus renders the stiffness matrix singular. In this case Eq. (2) is replaced by:

$$0 < \rho_{\min} \leq \rho_e \leq 1 \tag{5}$$

in which the minimum value of the design variable is set to a small value, e.g. $\rho_{\min} = 0.001$, to avoid the singularity.

1.3   Perfomance evaluation and sensitivity analysis.

A performance criterion must be specified for the given problem. For a static structural problem, a optimization goal could be to maximize the structural stiffness (or alternatively to minimize the *compliance*). The compliance is defined as:

$$\Phi = \mathbf{u}^T \mathbf{f} = \mathbf{u}^T \mathbf{S} \mathbf{u} \tag{6}$$

but naturally other choices of objective function could be made. For a static problem this could for instance be to minimize the weight of the structure. Typically, the objective is accompanied by a number of constraint functions which must be computed as well. In the case of weight minimization, a typical constraint could be a lower limit on the stiffness of the structure. In the the case of steady-state vibrations or wave propagation many options exist for defining relevant objective and constraint functions. This will be described in the following chapter.

A gradient-based optimization approach requires knowledge about the sensitivity of the performance criterion (*the objective function*) and the constraint functions with respect to changes in the design variables ρ_e. A straightforward approach for obtaining the sensitivities is to compute an approximation to the sensitivity $\frac{d\Phi}{d\rho_e}$ using finite difference calculations:

$$\frac{d\Phi}{d\rho_e} \approx \frac{\Phi(\rho_e + \Delta\rho_e) - \Phi(\rho_e)}{\Delta\rho_e} \tag{7}$$

in which $\Delta\rho_e$ is a small increment of the design variable in element e.

The advantage of this approach is that it is based solely on function evaluations, i.e. it can be obtained directly with access to the solver that was used to solve the analysis model. However, a significant drawback, which usually makes this approach unfeasible, is that the state equation needs to be solved one extra time for each design variable. As mentioned, realistic FE models may have millions of elements and design variables, which naturally renders this method unapplicable in practice for topology optimization.

Favorable is the *adjoint method* that can be used to get sensitivity information without significant computational overhead (Tortorelli and Michaleris, 1994). A detailed account for this approach will be presented in the following chapter for the case of steady-state vibrations and wave propagation. For the case of static compliance ($\mathbf{u}^T \mathbf{S} \mathbf{u}$) the adjoint approach leads to the following exact formula for the design sensitivities:

$$\frac{d\Phi}{d\rho_e} = -\mathbf{u}^T \frac{\partial \mathbf{S}}{\partial \rho_e} \mathbf{u} \qquad (8)$$

Thus, in this case no additional solutions of the state equation are necessary. One drawback, is that the method requires knowledge of $\frac{\partial \mathbf{S}}{\partial \rho_e}$ which is sometimes not directly obtainable from a commercial FE solver.

1.4 The optimization step.

With a knowledge of the system performance and computed sensitivities we use the optimization step to generate a better design. Since the sensitivities only provide information about performance change in the vicinity of the current design we must repeat the process several times in an iterative fashion. The number of iterations necessary to reach a converged design depends on the application, but for static problems usually less than 100 iterations are necessary before subsequent design changes becomes sufficiently small such that the iterations can be terminated. For other types of problem the required number of iterations may be significantly larger.

A successful optimization step is naturally of key importance for convergence and efficiency of the iterative algorithm. Many different algorithms exist, but good performance for most topology optimization problem is obtained by using the Method of Moving Asymptotes (MMA) (Svanberg, 1987, 1995).

1.5 Obtaining meaningful designs.

At this point it is worth mentioning that solving the problem exactly as specified here, will usually result in useless designs! For a static problem the optimization procedure results in a final design for which the continuous design variables in most of the design region will have values between 0 and 1, thus they will correspond to neither void nor solid material. *Penal-*

ization techniques have been developed to remedy these problems in order to ensure that the design variables in the final design are forced to either 0 or 1 (Bendsøe and Sigmund, 2003). Even, with this problem solved the design will still be unusable do to checkerboard-like patterns in the design and fine structural details that become smaller as the finite element mesh is refined. The first problem is ascribed to artificial stiffness associated with simple four-node bilinear elements and can be solved by using nodal-based or higher-order elements. The latter problem is due to the inherent physical nature of the problem that favors small structural details and requires *regularization* in order to provide unique mesh-independent designs. For the specific case of compliance minimization both problem can also be solved by the use of a proper *filtering technique* (Sigmund and Petersson, 1998).

It should be emphasized that in general each physical problem requires some special treatment. But with the right choice of objective and constraint functions in combination with suitable regularization, penalization and filtering the method of topology optimization has demonstrated to produce well defined structures with a very good performance with reasonable computational efforts. Sometimes the optimized designs have a very original appearance that is far away from previous structures reported in literature.

However, it is important to emphasize that the gradient-based approach that is the core of the efficient optimization approach can never be guaranteed to produce global optima and especially for problems in dynamics many local optima typically exist. However, with the proper choice of the tools described above we can usually ensure that the local optima corresponds to a good performance of the structure or device. Anyway, a global optimization procedure that can ensure convergence to a global optima in reasonable computational time for models with thousands or millions of design variables has not yet appeared and probably never will!

1.6   Contents of this chapter.

The organization of the following sections are as follows: following this introduction an extension of the basic topology optimization method to deal with steady-state vibration and wave propagation problems will be outlined. This includes detailed formulas for the sensitivity computation for a range of relevant objective functions. Thereafter, two recent extensions of the topology optimization method will be illustrated. First, the steady-state optimization procedure for dynamical systems is extended to a nonlinear wave propagation problem and then the topology optimization procedure is applied to transient dynamic simulations for which the optimized material distributions may vary in time.

2   Steady-state vibrations and wave propagation.

In this chapter formulas for objective functions and sensitivity formulas will be generated for the case of steady-state vibrations and wave propagation. The equations will be treated in general form and may origin in different physics settings such as structural dynamics, but also acoustic and optical wave propagation

2.1   Equation of motion.

With the use of a standard finite element procedure (Cook et al., 2002) the relevant linear wave propagation or vibration problem can be written in the following form:

$$\mathbf{M}(\boldsymbol{\rho})\ddot{\mathbf{x}} + \mathbf{C}(\boldsymbol{\rho})\dot{\mathbf{x}} + \mathbf{K}(\boldsymbol{\rho})\mathbf{x} = \mathbf{p}(\boldsymbol{\rho}, t) \tag{9}$$

where $\mathbf{x}(t)$ is a vector of time-dependent nodal displacements, $\mathbf{M}$, $\mathbf{C}$, $\mathbf{K}$ are mass, damping, and stiffness matrices, and $\mathbf{p}$ is the loading vector. The system matrices and the loading vector may all depend on the design variables, now collected in the vector $\boldsymbol{\rho}$.

It is more convenient to deal with the equation in complex form, so we introduce the complex variable transformation $\mathbf{x} \rightarrow \tilde{\mathbf{x}}$, so that the actual displacement is found as the real part of the complex displacement: $\mathbf{x} = \Re(\tilde{\mathbf{x}})$. The complex form of Equation (9) is:

$$\mathbf{M}(\boldsymbol{\rho})\ddot{\tilde{\mathbf{x}}} + \mathbf{C}(\boldsymbol{\rho})\dot{\tilde{\mathbf{x}}} + \mathbf{K}(\boldsymbol{\rho})\tilde{\mathbf{x}} = \tilde{\mathbf{p}}(\boldsymbol{\rho}, t) \tag{10}$$

where the loading vector is also transformed to complex form: $\mathbf{p} \rightarrow \tilde{\mathbf{p}}$ with $\mathbf{p} = \Re(\tilde{\mathbf{p}})$.

It is assumed that the system is loaded by *time-harmonic* excitation:

$$\tilde{\mathbf{p}}(\boldsymbol{\rho}, t) = \mathbf{f}(\boldsymbol{\rho})e^{i\Omega t} \tag{11}$$

This is inserted into Eq. (10):

$$\mathbf{M}(\boldsymbol{\rho})\ddot{\tilde{\mathbf{x}}} + \mathbf{C}(\boldsymbol{\rho})\dot{\tilde{\mathbf{x}}} + \mathbf{K}(\boldsymbol{\rho})\tilde{\mathbf{x}} = \mathbf{f}(\boldsymbol{\rho})e^{i\Omega t} \tag{12}$$

The steady-state solution to Eq. (12) is

$$\tilde{\mathbf{x}}(t) = \mathbf{u}e^{i\Omega t} \tag{13}$$

in which the *amplitude* vector $\mathbf{u}$ is generally complex. We denote the real and imaginary parts of this complex vector, $\mathbf{u}_R$ and $\mathbf{u}_I$, respectively.

Inserting the general solution form (13) into (12) yields:

$$\left(-\Omega^2\mathbf{M}(\boldsymbol{\rho}) + i\Omega\mathbf{C}(\boldsymbol{\rho}) + \mathbf{K}(\boldsymbol{\rho})\right)\mathbf{u} = \mathbf{f}(\boldsymbol{\rho}) \tag{14}$$

or alternatively we can introduce the system matrix (or dynamic stiffness matrix) $\mathbf{S}$ and write this equation as:

$$\mathbf{S}(\boldsymbol{\rho}, \Omega)\mathbf{u} = \mathbf{f}(\boldsymbol{\rho}) \tag{15}$$

After obtaining a specific solution to (15), we can obtain the instantaneous displacement, velocity, and acceleration as:

$$\mathbf{x}(t) = \Re(\mathbf{u}e^{i\Omega t}) = \mathbf{u}_R \cos(\Omega t) - \mathbf{u}_I \sin(\Omega t) \tag{16}$$

$$\dot{\mathbf{x}}(t) = -\Omega \left(\mathbf{u}_R \sin(\Omega t) + \mathbf{u}_I \cos(\Omega t)\right) \tag{17}$$

$$\ddot{\mathbf{x}}(t) = -\Omega^2 \left(\mathbf{u}_R \cos(\Omega t) - \mathbf{u}_I \sin(\Omega t)\right) \tag{18}$$

2.2 Performance evaluation and sensitivity analysis.

We will not directly specify the considered problem but rather derive the formulas for a general objective function, which we can later specialize for specific applications. We consider an objective function that can be written in the general form:

$$\Phi = \Phi_0(\boldsymbol{\rho}, \mathbf{u}_R, \mathbf{u}_I) \tag{19}$$

where it is assumed that Φ_0 is a real function.

The sensitivities will now be derived using the adjoint approach (Tortorelli and Michaleris, 1994). As usual for adjoint sensitivity analysis, we append extra terms to the objective function:

$$\Phi = \Phi_0(\boldsymbol{\rho}, \mathbf{u}_R, \mathbf{u}_I) + \boldsymbol{\lambda}^T(\mathbf{S}\mathbf{u} - \mathbf{f}) + \bar{\boldsymbol{\lambda}}^T(\bar{\mathbf{S}}\bar{\mathbf{u}} - \bar{\mathbf{f}}) \tag{20}$$

in which the two added terms vanish at equilibrium, and $\boldsymbol{\lambda}$ is a vector of complex Lagrangian multipliers. The overbars in Eq. (20) denote complex conjugates. It should be emphasized here that the choice of adding two extra terms to the objective functions, here represented by the lagrangian multiplier itself and its complex conjugate, origins in the complex solutions to the model equation. This particular form of the extra be terms proves to especially helpful for a simple derivation of the formulas.

We now compute the sensitivities of the objective function with respect to a single design variable ρ_i:

$$\begin{aligned}
\frac{d\Phi}{d\rho_e} =& \frac{\partial\Phi_0}{\partial\rho_e} + \frac{\partial\Phi_0}{\partial\mathbf{u}_R}\frac{\partial\mathbf{u}_R}{\partial\rho_e} + \frac{\partial\Phi_0}{\partial\mathbf{u}_I}\frac{\partial\mathbf{u}_I}{\partial\rho_e} \\
&+ \boldsymbol{\lambda}^T\left(\frac{\partial\mathbf{S}}{\partial\rho_e}\mathbf{u}_R + \mathbf{S}\frac{\partial\mathbf{u}_R}{\partial\rho_e} + i\frac{\partial\mathbf{S}}{\partial\rho_e}\mathbf{u}_I + i\mathbf{S}\frac{\partial\mathbf{u}_I}{\partial\rho_e} - \frac{\partial\mathbf{f}}{\partial\rho_e}\right) \\
&+ \bar{\boldsymbol{\lambda}}^T\left(\frac{\partial\bar{\mathbf{S}}}{\partial\rho_e}\mathbf{u}_R + \bar{\mathbf{S}}\frac{\partial\mathbf{u}_R}{\partial\rho_e} - i\frac{\partial\bar{\mathbf{S}}}{\partial\rho_e}\mathbf{u}_I - i\bar{\mathbf{S}}\frac{\partial\mathbf{u}_I}{\partial\rho_e} - \frac{\partial\bar{\mathbf{f}}}{\partial\rho_e}\right)
\end{aligned} \tag{21}$$

or after rearranging the terms:

$$\frac{d\Phi}{d\rho_e} = \frac{\partial \Phi_0}{\partial \rho_e} + \boldsymbol{\lambda}^T \left(\frac{\partial \mathbf{S}}{\partial \rho_e} \mathbf{u} - \frac{\partial \mathbf{f}}{\partial \rho_e} \right) + \bar{\boldsymbol{\lambda}}^T \left(\frac{\partial \bar{\mathbf{S}}}{\partial \rho_e} \bar{\mathbf{u}} - \frac{\partial \bar{\mathbf{f}}}{\partial \rho_e} \right)$$
$$+ \left(\frac{\partial \Phi_0}{\partial \mathbf{u}_R} + \boldsymbol{\lambda}^T \mathbf{S} + \bar{\boldsymbol{\lambda}}^T \bar{\mathbf{S}} \right) \frac{\partial \mathbf{u}_R}{\partial \rho_e} + \left(\frac{\partial \Phi_0}{\partial \mathbf{u}_I} + i\boldsymbol{\lambda}^T \mathbf{S} - i\bar{\boldsymbol{\lambda}}^T \bar{\mathbf{S}} \right) \frac{\partial \mathbf{u}_I}{\partial \rho_e} \tag{22}$$

The last two terms in the equation above are difficult to compute due to the quantities $\frac{\partial \mathbf{u}_R}{\partial \rho_e}$ and $\frac{\partial \mathbf{u}_I}{\partial \rho_e}$ which cannot be explicitly evaluated.

Instead we eliminate the expressions by requiring the last two parentheses to vanish:

$$\boldsymbol{\lambda}^T \mathbf{S} + \bar{\boldsymbol{\lambda}}^T \bar{\mathbf{S}} = -\frac{\partial \Phi_0}{\partial \mathbf{u}_R} \tag{23}$$

$$i\boldsymbol{\lambda}^T \mathbf{S} - i\bar{\boldsymbol{\lambda}}^T \bar{\mathbf{S}} = -\frac{\partial \Phi_0}{\partial \mathbf{u}_I} \tag{24}$$

and use these two equations to determine $\boldsymbol{\lambda}$.

Multiplying (24) by i and subtracting it from (23) yields:

$$2\boldsymbol{\lambda}^T \mathbf{S} = -\frac{\partial \Phi_0}{\partial \mathbf{u}_R} + i\frac{\partial \Phi_0}{\partial \mathbf{u}_I} \tag{25}$$

and after transposing (25) we get:

$$\mathbf{S}^T \boldsymbol{\lambda} = -\frac{1}{2} \left(\frac{\partial \Phi_0}{\partial \mathbf{u}_R} - i\frac{\partial \Phi_0}{\partial \mathbf{u}_I} \right)^T \tag{26}$$

If instead (24) is multiplied by i and added to (23) we obtain:

$$2\bar{\boldsymbol{\lambda}}^T \bar{\mathbf{S}} = -\frac{\partial \Phi_0}{\partial \mathbf{u}_R} - i\frac{\partial \Phi_0}{\partial \mathbf{u}_I} \tag{27}$$

This equation is equivalent to Eq. (25) (seen clearly by complex conjugating both sides of the equation). Thus if $\boldsymbol{\lambda}$ is found from Eq. (26) both Eqs. (23)–(24) will be fulfilled.

Thus the expression for the sensitivity becomes:

$$\frac{d\Phi}{d\rho_e} = \frac{\partial \Phi_0}{\partial \rho_e} + \boldsymbol{\lambda}^T \left(\frac{\partial \mathbf{S}}{\partial \rho_e} \mathbf{u} - \frac{\partial \mathbf{f}}{\partial \rho_e} \right) + \bar{\boldsymbol{\lambda}}^T \left(\frac{\partial \bar{\mathbf{S}}}{\partial \rho_e} \bar{\mathbf{u}} - \frac{\partial \bar{\mathbf{f}}}{\partial \rho_e} \right) \tag{28}$$

or alternatively the more simple expression:

$$\frac{d\Phi}{d\rho_e} = \frac{\partial \Phi_0}{\partial \rho_e} + 2\Re \left(\boldsymbol{\lambda}^T \left(\frac{\partial \mathbf{S}}{\partial \rho_e} \mathbf{u} - \frac{\partial \mathbf{f}}{\partial \rho_e} \right) \right) \tag{29}$$

where $\boldsymbol{\lambda}$ is the solution to

$$\mathbf{S}^T \boldsymbol{\lambda} = -\frac{1}{2}\left(\frac{\partial \Phi_0}{\partial \mathbf{u}_R} - i\frac{\partial \Phi_0}{\partial \mathbf{u}_I}\right)^T \tag{30}$$

In many cases $\mathbf{S}$ is a symmetric matrix. In this case the lhs. of Eq. (30) has the same appearance as the model Eq. (15) and if this is solved using a factorization method then the adjoint solution can be obtained virtually for free.

We can now use the derived formulas to obtain some important formulas for a number of special cases.

2.3   Static compliance.

The derived formulas can also be applied to the static case. Setting $\Omega = 0$, Eq. (14) becomes

$$\mathbf{K}\mathbf{u} = \mathbf{f} \tag{31}$$

where it is assumed that $\mathbf{f}$ does not depend on $\boldsymbol{\rho}$ (if the magnitude and direction of the applied force are independent of the design).

Eq. (31) contains no complex terms so the solution is real ($\mathbf{u}_I = 0$, $\mathbf{u}_R = \mathbf{u}$). The static compliance is found by multiplying (31) by $\mathbf{u}^T$:

$$\Phi_0 = \mathbf{u}^T \mathbf{f} \tag{32}$$

and we can then compute the necessary terms:

$$\frac{\partial \Phi_0}{\partial \mathbf{u}_R} = \mathbf{f}^T, \quad \frac{\partial \Phi_0}{\partial \mathbf{u}_I} = 0 \tag{33}$$

$$\frac{\partial \mathbf{f}}{\partial \rho_e} = 0, \quad \frac{\partial \Phi_0}{\partial \rho_e} = 0 \tag{34}$$

Inserting this into (30) we obtain:

$$\mathbf{K}\boldsymbol{\lambda} = -\frac{1}{2}\mathbf{f} \tag{35}$$

since the stiffness matrix is symmetric. When comparing Eq. (35) to Eq. (31), we can conclude that $\boldsymbol{\lambda} = -\frac{1}{2}\mathbf{u}$ and then find the sensitivities from Eq. (29):

$$\frac{d\Phi}{d\rho_e} = 2\Re\left(-\frac{1}{2}\mathbf{u}^T\frac{\partial \mathbf{K}}{\partial \rho_e}\mathbf{u}\right) = -\mathbf{u}^T\frac{\partial \mathbf{K}}{\partial \rho_e}\mathbf{u} \tag{36}$$

which is the well known result. Thus in this case there is no need to solve an adjoint equation.

2.4 Dynamic compliance I.

Now we consider the same system but including dynamic forces ($\Omega \neq 0$). The term "dynamic compliance" is not well defined and several definitions have been suggested.

One suggestion for a definition that has applied in various publications is:

$$\Phi_0 = |\mathbf{u}^T\mathbf{f}| = \sqrt{(\mathbf{f}^T\mathbf{u}_R)^2 + (\mathbf{f}^T\mathbf{u}_I)^2} \tag{37}$$

It is seen that this choice for dynamic compliance resembles the static compliance definition closely. The reason for the numerical values is that in the dynamic case, the force and the response can very well be in anti-phase such that the product $\mathbf{u}^T\mathbf{f}$ can be negative. This implies that, for instance, if the objective is minimized then adapting the objective function directly from the statical case, would not be meaningful.

Here, $\partial\Phi_0/\partial\rho_e = 0$ as in the previous case (no explicit dependence of Φ_0 on ρ_e) and we assume that $\partial\mathbf{f}/\partial\rho_e = 0$ (load independent of the design), but obtain:

$$\frac{\partial\Phi_0}{\partial\mathbf{u}_R} = \frac{\mathbf{f}^T\mathbf{u}_R\mathbf{f}^T}{\Phi_0} \tag{38}$$

$$\frac{\partial\Phi_0}{\partial\mathbf{u}_I} = \frac{\mathbf{f}^T\mathbf{u}_I\mathbf{f}^T}{\Phi_0} \tag{39}$$

Inserting this into (30) we obtain:

$$\mathbf{S}^T\boldsymbol{\lambda} = -\frac{\bar{\mathbf{u}}^T\mathbf{f}}{2\Phi_0}\mathbf{f} \tag{40}$$

The factor on the rhs. in front of $\mathbf{f}$ is a scalar and if furthermore $\mathbf{S}$ is symmetric then we can conclude:

$$\boldsymbol{\lambda} = -\frac{\bar{\mathbf{u}}^T\mathbf{f}}{2\Phi_0}\mathbf{u} = -\frac{\alpha}{2}\mathbf{u}, \quad \alpha = \frac{\bar{\mathbf{u}}^T\mathbf{f}}{\Phi_0} = \frac{\bar{\mathbf{u}}^T\mathbf{f}}{|\mathbf{u}^T\mathbf{f}|} \tag{41}$$

and the sensitivity thus becomes:

$$\frac{d\Phi}{d\rho_e} = 2\Re\left(\boldsymbol{\lambda}^T\frac{\partial\mathbf{S}}{\partial\rho_e}\mathbf{u}\right) = -\Re\left(\alpha\mathbf{u}^T\frac{\partial\mathbf{S}}{\partial\rho_e}\mathbf{u}\right) \tag{42}$$

It is seen that in a similar way as for the static case there is no need to solve an adjoint equation. Furthermore, it can be seen that in the case of an undamped system, we have $\bar{\mathbf{u}} = \mathbf{u}$ and thus $\alpha = \pm 1$ depending on if the force and corresponding displacement are in phase or anti-phase.

One drawback of this definition of the dynamic compliance is that it does not, as in the static case, directly relate to the energy level in the structure. This implies, that even if the objective is minimized a large motion of the structure can still occur away from the point of excitation.

2.5 Dynamic Compliance II.

Another possibility is to use the energy supplied to the structure during an excitation cycle as a measure of the dynamic compliance.

In order to derive an equation for the supplied energy a power flow balance for the dynamic problem is constructed by multiplying the original Eq. (9) by the velocity vector:

$$\dot{\mathbf{x}}^T \mathbf{M}\ddot{\mathbf{x}} + \dot{\mathbf{x}}^T \mathbf{C}\dot{\mathbf{x}} + \dot{\mathbf{x}}^T \mathbf{K}\mathbf{x} = \dot{\mathbf{x}}^T \mathbf{p} \tag{43}$$

where the individual contributions in form of kinetic, potential, dissipated and input power can written out as:

$$
\begin{aligned}
P_{kin} &= \dot{\mathbf{x}}^T \mathbf{M}\ddot{\mathbf{x}} = \Omega^3 (\mathbf{u}_R^T \sin \Omega t + \mathbf{u}_I^T \cos \Omega t)\mathbf{M}(\mathbf{u}_R \cos \Omega t - \mathbf{u}_I \sin \Omega t) \\
P_{pot} &= \dot{\mathbf{x}}^T \mathbf{K}\mathbf{x} = -\Omega (\mathbf{u}_R^T \sin \Omega t + \mathbf{u}_I^T \cos \Omega t)\mathbf{K}(\mathbf{u}_R \cos \Omega t - \mathbf{u}_I \sin \Omega t) \\
P_{dis} &= \dot{\mathbf{x}}^T \mathbf{C}\dot{\mathbf{x}} = \Omega^2 (\mathbf{u}_R^T \sin \Omega t + \mathbf{u}_I^T \cos \Omega t)\mathbf{C}(\mathbf{u}_R \sin \Omega t + \mathbf{u}_I \cos \Omega t) \\
P_{inp} &= \dot{\mathbf{x}}^T \mathbf{p} = -\Omega (\mathbf{u}_R^T \sin \Omega t + \mathbf{u}_I^T \cos \Omega t)(\mathbf{f}_R \cos \Omega t - \mathbf{f}_I \sin \Omega t)
\end{aligned}
\tag{44}
$$

The *energy change* over one excitation cycle can be found by integrating the power flows:

$$
\begin{aligned}
\Delta E_{kin} &= \int_0^{\frac{2\pi}{\Omega}} P_{kin} dt = 0 \\
\Delta E_{pot} &= \int_0^{\frac{2\pi}{\Omega}} P_{pot} dt = 0 \\
\Delta E_{dis} &= \int_0^{\frac{2\pi}{\Omega}} P_{dis} dt = \pi\Omega(\mathbf{u}_R^T \mathbf{C}\mathbf{u}_R + \mathbf{u}_I^T \mathbf{C}\mathbf{u}_I) \\
\Delta E_{inp} &= \int_0^{\frac{2\pi}{\Omega}} P_{inp} dt = \pi(\mathbf{u}_R^T \mathbf{f}_I - \mathbf{u}_I^T \mathbf{f}_R)
\end{aligned}
\tag{45}
$$

It is noted that in one excitation cycle, there is no net change in neither kinetic nor potential energy. Consequently, all input energy is dissipated due to the damping in the system:

$$\Delta E_{dis} = \Delta E_{inp} \Rightarrow \pi\Omega(\mathbf{u}_R^T \mathbf{C}\mathbf{u}_R + \mathbf{u}_I^T \mathbf{C}\mathbf{u}_I) = \pi(\mathbf{u}_R^T \mathbf{f}_I - \mathbf{u}_I^T \mathbf{f}_R) \tag{46}$$

Thus, if no damping is present no energy will be supplied to the system!

As in Jog (2002), we can use the input energy as objective function (omitting the factor π):

$$\Phi_0 = \mathbf{u}_R^T \mathbf{f}_I - \mathbf{u}_I^T \mathbf{f}_R \tag{47}$$

and obtain:

$$\frac{\partial \Phi_0}{\partial \mathbf{u}_R} = \mathbf{f}_I^T \tag{48}$$

$$\frac{\partial \Phi_0}{\partial \mathbf{u}_I} = -\mathbf{f}_R^T \tag{49}$$

and by inserting Eqs. (48)–(49) into Eq. (30) we get

$$\mathbf{S}\boldsymbol{\lambda} = -\frac{i}{2}\bar{\mathbf{f}} \tag{50}$$

and the sensitivity becomes:

$$\frac{d\Phi}{d\rho_e} = 2\Re\left(\boldsymbol{\lambda}^T \frac{\partial \mathbf{S}}{\partial \rho_e}\mathbf{u}\right) \tag{51}$$

Case of $\mathbf{f}_I = 0$.   If $\mathbf{f}_I = 0$ the expressions simplify into:

$$\mathbf{S}\boldsymbol{\lambda} = -\frac{i}{2}\mathbf{f} = -\alpha\mathbf{f}, \quad \alpha = \frac{i}{2} \tag{52}$$

and

$$\frac{d\Phi}{d\rho_e} = -\Re\left(i\mathbf{u}^T \frac{\partial \mathbf{S}}{\partial \rho_e}\mathbf{u}\right) \tag{53}$$

Case of $\mathbf{f}_R = 0$.   In this case we obtain

$$\mathbf{S}\boldsymbol{\lambda} = \frac{i}{2}\mathbf{f} = \alpha\mathbf{f}, \quad \alpha = \frac{i}{2} \tag{54}$$

and thus

$$\frac{d\Phi}{d\rho_e} = \Re\left(i\mathbf{u}^T \frac{\partial \mathbf{S}}{\partial \rho_e}\mathbf{u}\right) \tag{55}$$

This definition of dynamic compliance relies on the presence of damping. Therefore, the specific appearance of the optimized designs can also be expected to be dependent on the specific choice of damping model which is not particularly desirable, especially seen in light of the ad-hoc damping models which are sometimes applied.

2.6 Dynamic compliance III.

A more general form of a dynamic objective function can be written as:

$$\Phi_0 = \mathbf{u}^T \mathbf{A}(\boldsymbol{\rho})\bar{\mathbf{u}} = \mathbf{u}_R^T \mathbf{A}(\boldsymbol{\rho})\mathbf{u}_R + \mathbf{u}_I^T \mathbf{A}(\boldsymbol{\rho})\mathbf{u}_I \tag{56}$$

where $\mathbf{A}$ can be any real (symmetric) matrix that may depend on the design variables.

We can then compute the following quantities:

$$\frac{\partial \Phi_0}{\partial \mathbf{u}_R} = 2\mathbf{u}_R \mathbf{A} \tag{57}$$

$$\frac{\partial \Phi_0}{\partial \mathbf{u}_I} = 2\mathbf{u}_I \mathbf{A} \tag{58}$$

$$\frac{\partial \Phi_0}{\partial \rho_e} = \mathbf{u}^T \frac{\mathbf{A}}{\partial \rho_e} \bar{\mathbf{u}} \tag{59}$$

Inserting (57)-(58) into (30) yields

$$\mathbf{S}^T \boldsymbol{\lambda} = -\mathbf{A}^T \bar{\mathbf{u}} \tag{60}$$

and the sensitivity becomes:

$$\frac{d\Phi}{d\rho_e} = \mathbf{u}^T \frac{\partial \mathbf{A}}{\partial \rho_e} \bar{\mathbf{u}} + 2\Re\left(\boldsymbol{\lambda}^T \frac{\partial \mathbf{S}}{\partial \rho_e} \mathbf{u}\right) \tag{61}$$

In this equation $\mathbf{A}$ could, for instance, be the mass or the stiffness matrix. In this way we can consider the kinetic and/or potential energy in the structure as the objective for our optimization problem.

Case of $\mathbf{A}(\boldsymbol{\rho}) = \mathbf{I}$. This special case can be used to evaluate the "global" dynamic response, as an alternative to the previous dynamic compliance formulations:

$$\Phi_0 = \mathbf{u}^T \bar{\mathbf{u}} \tag{62}$$

which gives:

$$\mathbf{S}^T \boldsymbol{\lambda} = -\bar{\mathbf{u}} \tag{63}$$

and,

$$\frac{d\Phi}{d\rho_e} = 2\Re\left(\boldsymbol{\lambda}^T \frac{\partial \mathbf{S}}{\partial \rho_e} \mathbf{u}\right) \tag{64}$$

However, this global response measure does not directly relate to the static compliance definition. Additionally, it is seen that in this case the solution of the adjoint equation cannot be avoided. In the case of smaller

2D problems for which a direct factorization method can be used with advantage, the computational overhead is negligible, as described earlier. However, in the case of realistic-size 3D problems, efficient solution of the model equation typically necessitates the use of an iterative solver and in this case the need for solving an adjoint equation will effectively double the computational costs.

Case of $\mathbf{A}(\rho) = \mathbf{L}$. If $\mathbf{L}$ is chosen to be diagonal matrix independent on ρ, this formulation can be used to maximize or minimize the output at specific locations in the domain (corresponding to the non-zero diagonal entries in $\mathbf{L}$). We get:

$$\Phi_0 = \mathbf{u}^T \mathbf{L} \bar{\mathbf{u}} \tag{65}$$

which leads to the following expressions

$$\mathbf{S}^T \boldsymbol{\lambda} = -\mathbf{L}^T \bar{\mathbf{u}} \tag{66}$$

and

$$\frac{d\Phi}{d\rho_e} = 2\Re\left(\boldsymbol{\lambda}^T \frac{\partial \mathbf{S}}{\partial \rho_e} \mathbf{u}\right). \tag{67}$$

2.7 Energy flow.

Evaluating and optimizing the dynamic compliance is especially relevant for structural vibrations. If we consider instead the case of structures subjected to wave propagation, we may wish to consider the energy flow through some part of the structure.

In Jensen and Sigmund (2005) it is shown that energy flow through a certain part of a structure can be calculated as:

$$\Phi_0 = \Re\left(i\mathbf{u}^T \mathbf{Q}(\rho) \bar{\mathbf{u}}\right) = \mathbf{u}_R^T \mathbf{Q}(\rho) \mathbf{u}_I - \mathbf{u}_I^T \mathbf{Q}(\rho) \mathbf{u}_R \tag{68}$$

where $\mathbf{Q}$ is a non-symmetric matrix that is derived from the element shape functions. In this case it is assumed that the finite element model contains some form of absorption or transparency at the boundaries such that wave propagation can be modeled in the frequency domain. In the absence of dissipation only standing waves prevail and the energy flow vanishes.

Inserting we get:

$$\frac{\partial \Phi_0}{\partial \mathbf{u}_R} = \mathbf{u}_I^T \left(\mathbf{Q}^T - \mathbf{Q}\right) \tag{69}$$

$$\frac{\partial \Phi_0}{\partial \mathbf{u}_I} = \mathbf{u}_R^T (\mathbf{Q} - \mathbf{Q}^T) \tag{70}$$

$$\frac{\partial \Phi_0}{\partial \rho_e} = \Re \left(i \mathbf{u}^T \frac{\partial \mathbf{Q}}{\partial \rho_e} \bar{\mathbf{u}} \right) \tag{71}$$

Inserting Eqs. (69)–(70) into Eq. (30) we get

$$\mathbf{S}^T \boldsymbol{\lambda} = -\frac{i}{2} (\mathbf{Q}^T - \mathbf{Q})^T \bar{\mathbf{u}} \tag{72}$$

and the sensitivity becomes:

$$\frac{d\Phi}{d\rho_e} = \Re \left(i \mathbf{u}^T \frac{\partial \mathbf{Q}}{\partial \rho_e} \bar{\mathbf{u}} \right) + 2\Re \left(\boldsymbol{\lambda}^T \frac{\partial \mathbf{S}}{\partial \rho_e} \mathbf{u} \right) \tag{73}$$

Now, two different application examples will be considered that illustrate the use of some of the objective functions described in this section. Corresponding optimization problems are then solved with the use of the relevant sensitivity formulas and optimization algorithms.

Application example 1. In the first example we consider the optimization problem of minimizing the forced vibration response of planar elastic structures. The material presented here are based on the results from Sigmund and Jensen (2003).

In Fig. 2 we consider 3 different 2D plane structures and show the corresponding steady-state responses when subjected to harmonic loading at the left boundary. In-plane vibrations for a plane strain conditions is assumed and the response is calculated as the average amplitude along the right boundary. The structure in Fig. 2c is made from a periodic material consisting of a matrix of a soft and light material (grey) with 36 square inclusions of a stiffer and heavier material (black) and Fig. 2d shows the corresponding response of the finite structure when subjected to the external loading. A small *band gap* is identified in the response for the periodic structure, clearly appearing as a drop in response around 55 kHz, whereas for the two homogeneous structures shown in Fig. 2a,b a continuous spectrum of resonance peaks is observed.

The optimization problem is now defined as finding an optimized distribution of the two materials in order to minimize the wave magnitude on the opposite edge of the line of periodic excitation. The optimization objective can be written:

$$\min_{\rho} \quad \Phi = \mathbf{u}^T \mathbf{L} \bar{\mathbf{u}} \tag{74}$$

where $\mathbf{L}$ is a diagonal matrix with unit entries at elements corresponding to the degrees of freedom at the boundaries where the response is evaluated. According to the derivations in the previous section the sensitivities of the

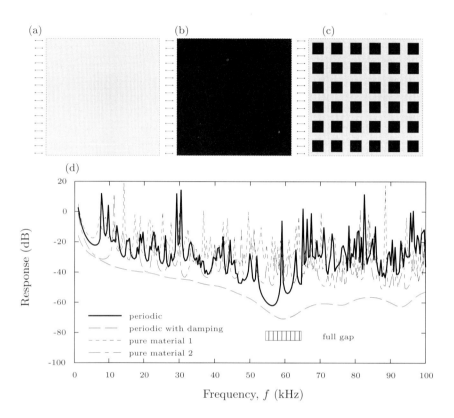

Figure 2. Response of a 12×12 cm square structure subjected to periodic loading on the left boundary, (a) Soft and light material (material 1), (b) stiff and heavy material (material 2) and (c) a periodic structure of the two materials, (d) response calculated as the average amplitude on the right boundary for the structure of pure material 1 and 2 without damping, and for the periodic structure with and without mass proportional damping. From Sigmund and Jensen (2003).

objective function can be found as:

$$\mathbf{S}\boldsymbol{\lambda} = -\mathbf{L}^T\bar{\mathbf{u}}, \tag{75}$$

and

$$\frac{d\Phi}{d\rho_e} = 2\Re\left(\boldsymbol{\lambda}^T \frac{\partial \mathbf{S}}{\partial \rho_e}\mathbf{u}\right). \tag{76}$$

where $\mathbf{S} = \mathbf{K} + i\Omega\mathbf{C} - \Omega^2\mathbf{M}$ is the symmetric system matrix.

We consider a 12×12 cm structure and wish to distribute material 1 and material 2 such that the response is minimized for $f = 63$ kHz. In order to obtain square symmetry of the resulting structure, we solve a multi-load problem. The structure is subjected to periodic excitation at the four edges independently and we then minimize the sum of responses at the opposing edges. The optimization is performed with strong damping added ($\alpha = 0, \beta = 50 \cdot 10^3$) in order to remove resonance peaks and ensure stable convergence. The effect of damping is also illustrated in Fig. 2d where the response for the periodic structure is computed with strong damping added. The smoothening effect of damping is clearly noted in this figure.

Fig.3b shows the optimized structure. The topology is periodic-like and closely resembles the periodic structure in Fig.2c. The largest difference between the two topologies is seen at the boundaries where the finite structure has inclusions directly at the boundary and also with a different periodicity. The corresponding response is seen in Fig.3c. The response with strong damping is shown with dashed line and without damping with a solid line. Although the optimization was carried out with damping included, it is seen that also when the damping is removed there is a large drop in the response due to a band gap.

Next, we try to optimize the structure in the case where the contrast between the two involved materials is smaller. The optimized topology (here obtained for $f = 52$ kHz) is no longer period-like as was seen for the high contrast case in Fig. 4b. The response shown in Fig. 4c shows only a small reduction in the response when the damping is removed. The explanation for this is that in this case the contrast is so low that a bandgap can no longer be created in the corresponding periodic material. For more details see Sigmund and Jensen (2003).

Application example 2. As the second application example we consider the design of a double 90-degree waveguide bend in a photonic crystal waveguide. The material presented here is based on results from Jensen et al. (2005).

We use the initial structure shown in Fig. 5a as a basis for the optimization procedure. Each unit cell is discretized using 14×12 four-noded

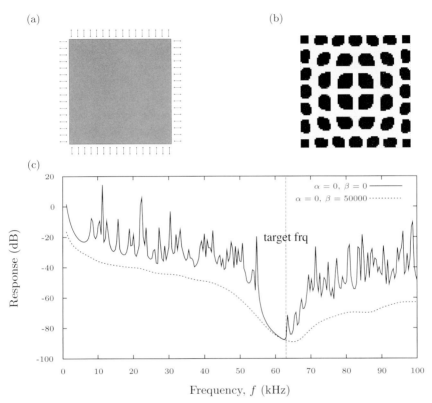

Figure 3. Optimization of structure for minimum response at $\Omega = 63$ kHz, (a) Structural domain and periodic boundary loading in the optimization procedure, (b) optimized topology, (c) average response at the right boundary when subjected to loading at the left boundary, dashed line $\alpha = 0, \beta = 50 \cdot 10^3$, solid line $\alpha = \beta = 0$. High material contrast. From Sigmund and Jensen (2003).

(a)   (b)

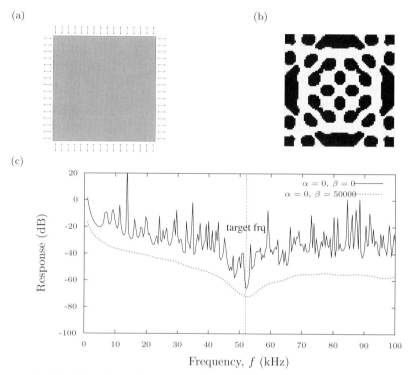

(c)

Figure 4. Optimization of structure for minimum response at $\Omega = 52$ kHz, (a) Structural domain and periodic boundary loading in the optimization procedure, (b) optimized topology, (c) average response at the right boundary when subjected to loading at the left boundary, dashed line $\alpha = 0, \beta = 50 \cdot 10^3$, solid line $\alpha = \beta = 0$. Low material contrast. From Sigmund and Jensen (2003).

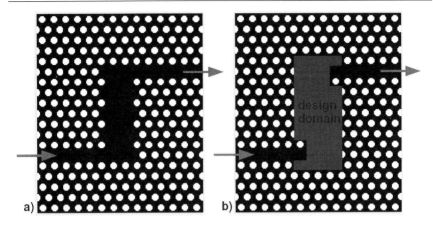

Figure 5. a) Initial configuration of the double 90-degree bend and b) indication of the design domain where the material distribution is to be modified by the optimization algorithm. From Jensen et al. (2005).

quadrilateral elements. This discretization is sufficient for describing the geometry satisfactorily, and for capturing the dynamic behavior with adequate accuracy. The full computational model consists of the domain shown in Fig. 5a as well as additional Perfectly Matching Layers (PML) (Koshiba et al., 2001) which are used to absorb waves that leave the computational domain and facilitate the simulation of wave propagation. The model comprises in total about 115000 elements of which 6720 are within the design domain.

The generic structure has a poor transmission for most frequencies as shown in Fig. 7 (here shown as wavelengths). The key advantage of the topology design method is, however, that there is no geometrical restrictions on the design so the specific performance of the initial structure is not necessarily important. In order to improve the performance sufficiently we must allow for the entire bend region to be modified, as indicated by the shaded area in Fig. 5b. This approach is slightly different than other design problems considered for photonic crystal waveguide bends (Borel et al., 2004; Frandsen et al., 2004; Borel et al., 2005), where there was only a need to modify a small part of the waveguide structure. Consequently, the design that we obtain here has a very different appearance compared to conventional PhC waveguide (PhCW) structures.

The distribution between the dielectric and air is optimized in the designated design area shown in Fig. 5b. The optimization objective is to maximize the power flow through the waveguide:

$$\max_{\rho} \quad \Phi = \Re\left(i\mathbf{u}^T \mathbf{Q}\bar{\mathbf{u}}\right) \tag{77}$$

where $\mathbf{Q}$ includes the contribution from elements in a designated energy flow evaluation area near the output waveguide port.

The adjoint variables are accordingly computed as

$$\mathbf{S}\boldsymbol{\lambda} = -\frac{i}{2}(\mathbf{Q}^T - \mathbf{Q})^T \bar{\mathbf{u}}, \tag{78}$$

in which the symmetry of the system matrix has been exploited and the sensitivity becomes:

$$\frac{d\Phi}{d\rho_e} = 2\Re\left(\boldsymbol{\lambda}^T \frac{\partial \mathbf{S}}{\partial \rho_e} \mathbf{u}\right). \tag{79}$$

in which the first term that appears in the general formula (73) vanish since the performance evaluation domain and the design domain do not overlap.

Since the design variables are continuous and not discrete, the possibility for elements in the final design with values between 0 and 1, so-called 'gray' elements, is present. This corresponds to an intermediate 'porous' material which is not feasible from a fabrication point of view. Several techniques have been developed to remedy this problem (Bendsøe and Sigmund, 2003), but here we use a method specially designed for photonic waveguides (Jensen and Sigmund, 2005), in that we artificially make design variables between 0 and 1 induce additional conduction σ_e (energy dissipation) in the following form:

$$\sigma_e \sim \alpha \rho_e (1 - \rho_e), \tag{80}$$

which enters the model equation as an additional imaginary term in the system matrix $\mathbf{S}$ and causes dissipation of energy. In this way, gray elements will be 'un-economical' and will thus be forced toward either white or black ($\rho_e = 0$ or $\rho_e = 1$), if the conduction parameter α is chosen to be sufficiently large.

The effect is seen in Fig. 6 that shows four snapshots of the optimization iteration process. After about 1000 iterations with $\alpha = 0$ the basic structure is formed (Fig. 6b) but many gray elements appear in the design. Then α is slowly increased and the gray elements gradually disappear as the final structure is reached in about 2500 iterations.

A broad bandwidth operation is ensured by maximizing the transmission through the waveguide for several frequencies simultaneously. We use a technique based on active sets (Jensen and Sigmund, 2005) in which we fix

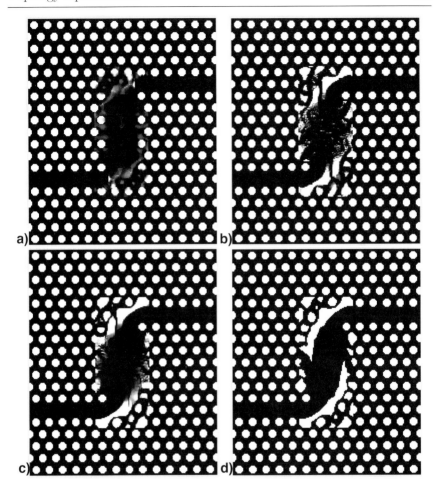

Figure 6. Snapshots of the material distribution during the optimization process at, a) 25, b) 1000, and c) 2300 iterations, and d) the final design after about 2500 iterations. From Jensen et al. (2005).

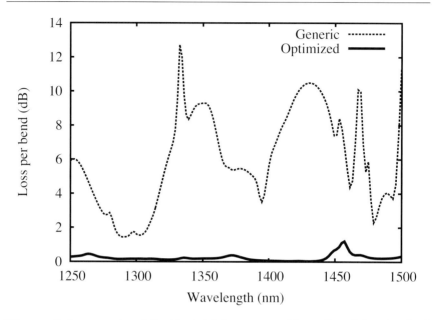

Figure 7. Loss per bend for the initial structure (dotted) and for the final optimized structure (solid) calculated with the 2D FE model. From Jensen et al. (2005).

a number of target frequencies in the desired frequency interval. During the optimization these target frequencies are repeatedly changed, according to the most critical frequencies with lowest transmission. The critical frequencies are found every 10th or 20th iteration by performing a fast frequency sweep (Jin, 2002).

The final optimized design is seen in Fig. 6d. This design is quite different from traditional photonic crystal structures but, nevertheless, shows a very good broadband performance as shown in Fig. 7, computed using 2D calculations.

The optimized waveguide has been fabricated in Silicon-on-insulator (SOI) and tested experimentally. The PhCs are defined as air holes in a triangular lattice and the PhCWs are carved out by removing a single row of holes in the nearest-neighbor direction of the crystal lattice. We use lattice pitch $\Gamma = 400$nm, diameter of the holes $D = 275$nm, and thickness of the Si/SiO$_2$ layers 340nm/1μm. The fabricated topology-optimized structure is shown in Fig. 8b and it nicely resembles the designed structure (Fig. 6d).

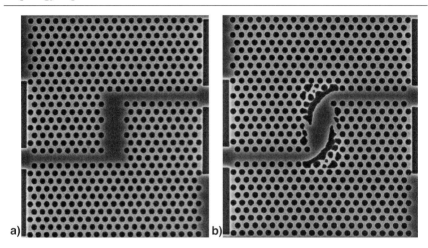

Figure 8. a) The fabricated generic structure and b) the topology optimized double 90-degree waveguide bend. From Jensen et al. (2005).

Fig. 9 shows the measured bend loss of TE-polarized light for the generic structure and the topology-optimized structure. To extract the bend loss the transmission spectra have been normalized to the transmission spectrum for a straight PhCW of similar length. A transmission loss of < 1dB/bend is obtained for the wavelength range 1250 − 1450nm.

3   A nonlinear wave propagation problem.

The previous chapter was devoted to studying linear vibration and wave propagation problems. However, fascinating dynamic behavior and novel devices may appear if we allow nonlinear behavior to be included in the optimization problem.

A nonlinear dynamic problem should generally be considered using a transient simulation of the dynamic response. However, in certain cases we may apply a simpler nonlinear model based on a non-instantaneous nonlinearity that allows us to use a steady-state formulation of the problem in a similar way as demonstrated in the previous section. An example of such a case is a nonlinear optical Kerr material for which the permittivity is a function of the electric field. In the non-instantaneous approximation we may assume that the permittivity is function of the *time-averaged* intensity of the field instead of its instantaneous value.

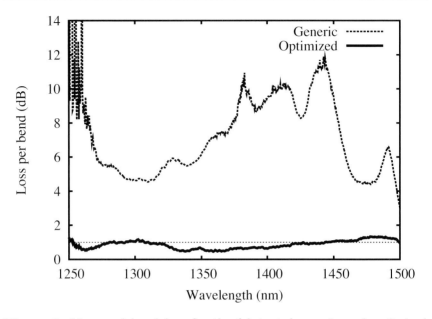

Figure 9. Measured bend loss for the fabricated generic and optimized waveguide component. From Jensen et al. (2005).

In this example we will demonstrate how we can formulate and solve the nonlinear steady-state problem and derive the design sensitivities using adjoint analysis. This will then be utilized to design a one-dimenional optical diode that allows for a greater transmission in one propagation direction than in the opposing direction. It should be emphasized that a diode in a one-dimensional structure requires a nonlinear material since with linear materials only, the transmission will always be identical in the opposing propagation directions. The material presented here is based on the results from Jensen (2011).

3.1 Nonlinear model.

Figure 10 illustrates the problem setup. A one-dimensional layered structure consisting of two different dielectric materials is to be designed. One material is linear with permittivity ε_1 and the other material is nonlinear with a permittivity that depends on the time-averaged local intensity of the electric field $\varepsilon(|E|^2)$. The objective of the optimization problem is to distribute the linear and nonlinear material so that the structure behaves as a

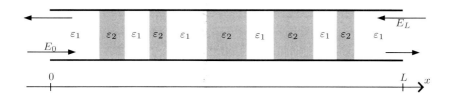

Figure 10. Optimization example for a nonlinear wave propagation problem. The difference in transmission in opposing propagation directions is to be maximized by an optimized distribution of linear (ε_1) and nonlinear material (ε_2). From Jensen (2011).

diode. This means that the difference in transmission for waves propagating in opposing directions should be maximized.

The model equation and corresponding boundary conditions for the one-dimensional wave propagation problem are:

$$\frac{\partial^2 E}{\partial x^2} + k^2 \varepsilon E = 0 \quad \text{for } 0 < x < L$$

$$-\frac{\partial E}{\partial x} + ik\sqrt{\varepsilon}E = 2ik\sqrt{\varepsilon}\bar{E}_0 \quad \text{for } x = 0 \qquad (81)$$

$$\frac{\partial E}{\partial x} + ik\sqrt{\varepsilon}E = 2ik\sqrt{\varepsilon}\bar{E}_L \quad \text{for } x = L$$

in which the two boundary conditions ensure transparent boundaries and allows for emanating waves propagating away from the boundary. Here, k is the wavenumber and $\bar{E}_0$ and $\bar{E}_L$ are the magnitudes of the propagating waves in the two directions, respectively.

A discretized version of Eq. (81) is created using a standard Galerkin FE procedure in which the computational domain is divided into N finite elements. This leads to the following set of discretized equilibrium equations:

$$\mathbf{r}(\mathbf{E}) = \mathbf{g}(\mathbf{E}) - \mathbf{f} = \mathbf{0} \qquad (82)$$

expressed in terms of the residual vector $\mathbf{r}$ that vanishes at the equilibrium electric field $\mathbf{E}$. In Eq. (82), $\mathbf{f}$ is the external excitation vector and $\mathbf{g}$ is an internal response vector given as:

$$\mathbf{g}(\mathbf{E}) = \sum_{e=1}^{N}\left(\mathbf{A}^e - k^2\varepsilon_e(|\tilde{E}|^2)\mathbf{B}^e\right)\mathbf{E}^e + ik\sqrt{\varepsilon_1}E_1\mathbf{1}_1 + ik\sqrt{\varepsilon_N}E_N\mathbf{1}_N \qquad (83)$$

$$\mathbf{f} = 2ik\sqrt{\varepsilon_1}\bar{E}_0\mathbf{1}_1 + 2ik\sqrt{\varepsilon_N}\bar{E}_L\mathbf{1}_N \tag{84}$$

in which $\mathbf{1}_i$ is a vector with a single unit entry at the i'th dof, and where $\mathbf{A}^e$, $\mathbf{B}^e$ are element matrices.:

$$\mathbf{A}^e = \int_0^L \frac{\partial \mathbf{N}^T}{\partial x}\frac{\partial \mathbf{N}}{\partial x}dx$$
$$\mathbf{B}^e = \int_0^L \mathbf{N}^T\mathbf{N}dx \tag{85}$$

in which $\mathbf{N}$ is the shape function and L the element length. In the present example simple linear two-node elements are applied.

In Eq. (83) the term $\varepsilon_e(|\tilde{E}|^2)$ has been moved outside the element integration. This is motivated by the assumption that the permittivity is taken to be a function of the intensity of the *element-averaged* electric field $\tilde{E}$.

3.2   Nonlinear solution procedure.

An incremental Newton-Raphson (NR) procedure is used to solve the nonlinear and complex Eq. (82). The NR scheme is facilitated by splitting up the governing equations into their real and imaginary parts.

The excitation vector $\mathbf{f} = \mathbf{f}_R + i\mathbf{f}_I$ is divided into M increments, where the n'th increment can be written as $\Delta\mathbf{f}^n = \Delta\mathbf{f}_R^n + i\Delta\mathbf{f}_I^n$, such that $\sum_{n=1}^N \Delta\mathbf{f}^n = \mathbf{f}$. Equilibrium at increment n dictates a vanishing residual:

$$\mathbf{r}^n = \mathbf{g}(\mathbf{E}_R^n, \mathbf{E}_I^n) - (\mathbf{f}_R^n + i\mathbf{f}_I^n) = \mathbf{0} \tag{86}$$

Now, an approximation for the residual at increment $n+1$ is obtained by applying a first order Taylor expansion of $\mathbf{r}$ in the vicinity of the equilibrium point in terms of the real and imaginary parts of $\mathbf{E}$ and $\mathbf{f}$:

$$\mathbf{r}^{n+1} \approx \mathbf{r}^n + \left(\frac{\partial \mathbf{r}}{\partial \mathbf{E}_R}\right)^n \Delta\mathbf{E}_R^n + \left(\frac{\partial \mathbf{r}}{\partial \mathbf{E}_I}\right)^n \Delta\mathbf{E}_I^n$$
$$+ \left(\frac{\partial \mathbf{r}}{\partial \mathbf{f}_R}\right)^n \Delta\mathbf{f}_R^n + \left(\frac{\partial \mathbf{r}}{\partial \mathbf{f}_I}\right)^n \Delta\mathbf{f}_I^n \tag{87}$$

The equilibrium condition in Eq. (86) is now inserted into Eq. (87) such that:

$$\mathbf{r}^{n+1} \approx \left(\frac{\partial \mathbf{g}}{\partial \mathbf{E}_R}\right)^n \Delta\mathbf{E}_R^n + \left(\frac{\partial \mathbf{g}}{\partial \mathbf{E}_I}\right)^n \Delta\mathbf{E}_I^n - \Delta\mathbf{f}_R^n - i\Delta\mathbf{f}_I^n \tag{88}$$

By requiring the rhs. of Eq. (88) to vanish, an approximation for equilibrium at increment $n+1$ is obtained. This leads to the following equation for $\Delta\mathbf{E}_R^n$ and $\Delta\mathbf{E}_I^n$:

$$\left(\frac{\partial\mathbf{g}}{\partial\mathbf{E}_R}\right)^n \Delta\mathbf{E}_R^n + \left(\frac{\partial\mathbf{g}}{\partial\mathbf{E}_I}\right)^n \Delta\mathbf{E}_I^n = \Delta\mathbf{f}_R^n + i\Delta\mathbf{f}_I^n \tag{89}$$

Eq. (89) is split up into its real and imaginary parts, leading to the following equation that can be solved for $\Delta\mathbf{E}_R^n$ and $\Delta\mathbf{E}_I^n$ at increment n:

$$\begin{bmatrix} \Re(\frac{\partial\mathbf{g}}{\partial\mathbf{E}_R})^n & \Re(\frac{\partial\mathbf{g}}{\partial\mathbf{E}_I})^n \\ \Im(\frac{\partial\mathbf{g}}{\partial\mathbf{E}_R})^n & \Im(\frac{\partial\mathbf{g}}{\partial\mathbf{E}_I})^n \end{bmatrix} \left\{ \begin{array}{c} \Delta\mathbf{E}_R^n \\ \Delta\mathbf{E}_I^n \end{array} \right\} = \left\{ \begin{array}{c} \Delta\mathbf{f}_R^n \\ \Delta\mathbf{f}_I^n \end{array} \right\} \tag{90}$$

given the specified increment components $\Delta\mathbf{f}_R^n$ and $\Delta\mathbf{f}_I^n$. On the lhs. of Eq. (90) a tangent system matrix appear in a similar way as when solving static nonlinear equilibrium equations using incremental methods.

The computed increments $\Delta\mathbf{E}_R^n$ and $\Delta\mathbf{E}_I^n$ are now used for a first approximation of $\mathbf{E}_R^{n+1}$ and $\mathbf{E}_I^{n+1}$:

$$(\mathbf{E}_R^{n+1})_1 = \mathbf{E}_R^n + \Delta\mathbf{E}_R^n \tag{91}$$

$$(\mathbf{E}_I^{n+1})_1 = \mathbf{E}_I^n + \Delta\mathbf{E}_I^n \tag{92}$$

and subsequently NR-iterations based on residual computations are used to generate successively better approximations until the residual falls below a specified tolerance.

The k'th approximations to $\mathbf{E}_R^{n+1}$ and $\mathbf{E}_I^{n+1}$ produce the residual:

$$(\mathbf{r}^{n+1})_k = \mathbf{g}((\mathbf{E}_R^{n+1})_k, (\mathbf{E}_I^{n+1})_k) - (\mathbf{f}_R^{n+1} + i\mathbf{f}_I^{n+1}) \tag{93}$$

which are used to solve an updated tangent problem:

$$\begin{bmatrix} (\Re(\frac{\partial\mathbf{g}}{\partial\mathbf{E}_R})^{n+1})_k & (\Re(\frac{\partial\mathbf{g}}{\partial\mathbf{E}_I})^{n+1})_k \\ (\Im(\frac{\partial\mathbf{g}}{\partial\mathbf{E}_R})^{n+1})_k & (\Im(\frac{\partial\mathbf{g}}{\partial\mathbf{E}_I})^{n+1})_k \end{bmatrix} \left\{ \begin{array}{c} (\Delta\mathbf{E}_R^n)_k \\ (\Delta\mathbf{E}_I^n)_k \end{array} \right\} = - \left\{ \begin{array}{c} (\mathbf{r}_R^{n+1})_k \\ (\mathbf{r}_I^{n+1})_k \end{array} \right\} \tag{94}$$

and then generate an improved approximation for $\mathbf{E}_R^{n+1}$ and $\mathbf{E}_I^{n+1}$:

$$(\mathbf{E}_R^{n+1})_{k+1} = (\mathbf{E}_R^{n+1})_k + (\Delta\mathbf{E}_R^n)_k \tag{95}$$

$$(\mathbf{E}_I^{n+1})_{k+1} = (\mathbf{E}_I^{n+1})_k + (\Delta\mathbf{E}_I^n)_k \tag{96}$$

In the examples shown later $N = 1 - 3$ increments are used which leads to $1 - 6$ NR iterations necessary for each increment in order obtain a final (relative) residual smaller than 10^{-10}.

3.3 Performance evaluation and sensitivity analysis.

As objective function we will consider the wave response in the following form

$$\Phi = \Phi_0(\mathbf{E}) \tag{97}$$

in which the response function $\Phi_0 = E_i^2$ is used to evaluate the response at specific points in the structure.

The sensitivity of the objective function with respect to a design variable is found using the adjoint method (see e.g. Tortorelli and Michaleris (1994)). First the objective function is augmented with two additional terms:

$$\Phi = \Phi_0(\mathbf{E}_R, \mathbf{E}_I) + \boldsymbol{\lambda}_R^T \mathbf{r}_R + \boldsymbol{\lambda}_I^T \mathbf{r}_I \tag{98}$$

in which $\boldsymbol{\lambda}_R$ and $\boldsymbol{\lambda}_I$ is the real and imaginary part of a new complex adjoint variable $\boldsymbol{\lambda}$, and $\mathbf{r}_R$ and $\mathbf{r}_I$ is the real and imaginary part of the residual vector, respectively.

The objective function is differentiated wrt. to a design variable in element e.

$$\frac{d\Phi}{d\rho_e} = \boldsymbol{\lambda}_R^T \mathbf{g}_R' - \boldsymbol{\lambda}_I^T \mathbf{g}_I' + \left(\frac{\partial \Phi_0}{\partial \mathbf{E}_R} + \boldsymbol{\lambda}_R^T \frac{\partial \mathbf{g}_R}{\partial \mathbf{E}_R} + \boldsymbol{\lambda}_I^T \frac{\partial \mathbf{g}_I}{\partial \mathbf{E}_R} \right) \mathbf{E}_R' \\ + \left(\frac{\partial \Phi_0}{\partial \mathbf{E}_I} + \boldsymbol{\lambda}_R^T \frac{\partial \mathbf{g}_R}{\partial \mathbf{E}_I} + \boldsymbol{\lambda}_I^T \frac{\partial \mathbf{g}_I}{\partial \mathbf{E}_I} \right) \mathbf{E}_I' \tag{99}$$

in which $()' = \partial()/\partial\rho_e$:

Following the standard approach for adjoint sensitivity analysis, the last two terms are required to vanish. This leads to the following linear equation to be solved for $\boldsymbol{\lambda}_R$ and $\boldsymbol{\lambda}_I$:

$$\begin{bmatrix} \Re(\frac{\partial \mathbf{g}}{\partial \mathbf{E}_R}) & \Re(\frac{\partial \mathbf{g}}{\partial \mathbf{E}_I}) \\ \Im(\frac{\partial \mathbf{g}}{\partial \mathbf{E}_R}) & \Im(\frac{\partial \mathbf{g}}{\partial \mathbf{E}_I}) \end{bmatrix}^T \left\{ \begin{array}{c} \boldsymbol{\lambda}_R \\ \boldsymbol{\lambda}_I \end{array} \right\} = \left\{ \begin{array}{c} -\frac{\partial \Phi_0}{\partial \mathbf{E}_R}^T \\ -\frac{\partial \Phi_0}{\partial \mathbf{E}_I}^T \end{array} \right\} \tag{100}$$

where it should be emphasized that all terms are evaluated at equilibrium for final increment M.

From the solution to Eq. (100), $\boldsymbol{\lambda} = \boldsymbol{\lambda}_R + i\boldsymbol{\lambda}_I$ can be found and the sensitivity can then be found from the simple expression:

$$\frac{d\Phi}{d\rho_e} = \Re(\bar{\boldsymbol{\lambda}}^T \mathbf{g}') \tag{101}$$

in which $\bar{\boldsymbol{\lambda}}$ is the complex conjugate adjoint vector.

The expression in Eq. (101) can be further specified by employing Eq. (83):

$$\frac{d\Phi}{d\rho_e} = -k^2 \Re((\bar{\boldsymbol{\lambda}}^e)^T \mathbf{B}^e \mathbf{E}^e) \varepsilon_e' \tag{102}$$

in which $\bar{\boldsymbol{\lambda}}^e$ and $\mathbf{E}^e$ are vectors on element level.

The position-dependent dielectric constant ε may take values corresponding to material 1 or to material 2. The predefined values for these two materials are:

$$\text{material } 1: \quad \varepsilon = 1 \tag{103}$$

$$\text{material } 2: \quad \varepsilon = \bar{\varepsilon}(1 + \gamma|\tilde{E}|^2) \tag{104}$$

in which $\bar{\varepsilon}$ is the linear permittivity of material 2 and γ is a nonlinear parameter.

The design variable ρ_e is used to interpolate linearly between the two materials:

$$\varepsilon_e = (1 + \rho_e(\bar{\varepsilon} - 1)) + \rho_e\bar{\varepsilon}\gamma|\tilde{E}|^2 \tag{105}$$

in which $|\tilde{E}|^2$ is the intensity of the element-averaged electric field. From Eq. (105) the expression ε'_e can be computed easily. The Newton-Raphson procedure and sensitivity analysis further require the computation of the following terms:

$$\frac{\partial \varepsilon_e}{\partial \mathbf{E}^e_R} = \rho_e\bar{\varepsilon}\gamma\Re(\tilde{\mathbf{E}}) \tag{106}$$

$$\frac{\partial \varepsilon_e}{\partial \mathbf{E}^e_I} = \rho_e\bar{\varepsilon}\gamma\Im(\tilde{\mathbf{E}}) \tag{107}$$

3.4   Optimization problem.

The optimization objective is written as:

$$\max_{\boldsymbol{\rho}} \quad (\Phi_0(\mathbf{E}_1))_1 - (\Phi_0(\mathbf{E}_2))_2 \tag{108}$$

in which the field $\mathbf{E}_1$ is the solution to a FE problem $\mathbf{g}(\mathbf{E}_1) - \mathbf{f}_1 = \mathbf{0}$ where $\bar{E}_0 = 1$ and $\bar{E}_L = 0$ in Eq. (81). In this situation a unit magnitude harmonic wave propagates from left to right and the response is evaluated as the wave intensity in the right end of the structure $((\Phi_0)_1 = |(E_1)_L|^2)$. The field $\mathbf{E}_2$ is the solution to the FE problem $\mathbf{g}(\mathbf{E}_2) - \mathbf{f}_2 = \mathbf{0}$ where $\bar{E}_0 = 0$ and $\bar{E}_L = 1$ so that a unit magnitude wave propagates from right to left and here the response is evaluated for $x = 0$: $(\Phi_0)_2 = |(E_2)_0|^2$.

Thus, by maximizing $(\Phi_0)_1 - (\Phi_0)_2$ the difference in transmission in the opposing propagation directions is maximized. The optimization problem in Eq. (108) is solved iteratively using the Method of Moving Asymptotes (MMA) (Svanberg, 1987).

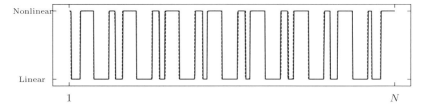

Figure 11. Optimized material distribution. The three curves obtained for different values of N and for different initial conditions all coincide. From Jensen (2011).

Optimization results.   The following set of parameters are used for this example:

   $\bar{\varepsilon} = 1.1$, $\gamma = 0.10$, $k = 1$

The nonlinear coefficient is here orders of magnitudes larger than what is normally found in dielectric materials such as silicon. This makes this example rather academic, but serves to illustrate a proof of concept of what is possible to accomplish with nonlinearities included.

   Fig. 11 shows the optimized distribution of the element-wise design variable ρ_e. The distribution is seen to have a doubly-periodic appearance. However, there is a small variation of the periodicity along the spatial position in the structure. Furthermore, the structure is seen to consist exclusively of design variables with the discrete values 0 or 1. Thus, no variables with intermediate values appear in the optimized design even though no explicit penalization of intermediate values has been carried out. Additionally, the generated distribution turns out to be robust to different initial conditions and also independent of the number of elements in the FE mesh. To illustrate this feature, the figure shows three curves of which two are for homogeneous initial structures with $\rho_e = 0.5$ and using either $N = 675$ or $N = 1125$ finite elements. The third curve is for a structure with $N = 675$, but for which all design variables have been assigned a random initial value between 0 and 1. The resulting curves are hardly distinguishable.

   Figure 12 illustrates the mechanism that allows for different transmission in opposing propagation directions. In the linear parts of the structure the permittivity is constant $\varepsilon_e = 1$, but in the nonlinear parts ε_e depends on the local intensity of the electric field. Thus, there can be a large difference in the effective permittivity in the nonlinear sections, depending on the travel direction of the wave. This consequently has a large effect on the resulting transmission.

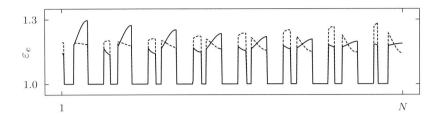

Figure 12. The effective permittivity ε_e plotted for the optimized design in Fig. 11. Solid lines are for a wave propagating from right to left and dashed lines for a wave propagating from left to right. From Jensen (2011).

The structure has been optimized for a particular value of the nonlinear parameter ($\gamma_{\mathrm{opt}} = 0.10$). The effect of changing γ_{opt} in the optimization procedure is shown together with the performance if the operating conditions are different from the optimization conditions. Figure 13 shows three curves for structures optimized for different values of the nonlinear parameter ($\gamma_{\mathrm{opt}} = 0.05, 0.10, 0.15$). The curves show the the objective $(\Phi_0)_1 - (\Phi_0)_2$ for varying values of the parameter γ for each of the three optimized structures. The objective value is seen to increase monotonously as γ is increased, except for a small section of the curve for the structure optimized for $\gamma_{\mathrm{opt}} = 0.05$. This means that the diode functionality generally increases for increasing values of γ (or alternatively the structure will also function for other wave input amplitudes than unity). It is also seen that the structure optimized for a particular value of γ is the one that performs the best for that value – as should be expected. The optimized structures for the case of $\gamma_{\mathrm{opt}} = 0.05$ and $\gamma_{\mathrm{opt}} = 0.15$ are not shown, but they both have a qualitatively similar appearance as the structure in Figure 11.

Structures optimized for this range of the nonlinear coefficient ($\gamma_{\mathrm{opt}} < \approx 0.15$) all display nice properties: mesh-independence, pure 0-1 designs, insensitivity to initial conditions and smooth behavior wrt. operating conditions (γ and wave amplitude). Although it cannot be guaranteed, they appear to be global minima. However, if γ_{opt} is increased further the situation changes. For $\gamma_{\mathrm{opt}} = 0.20$ a large value of the objective can be obtained ($(\Phi_0)_1 - (\Phi_0)_2 \approx 0.3$) but the value of the objective drops significantly for γ away from γ_{opt}, so that the structure does not perform well if the operating conditions are different from the optimization conditions.

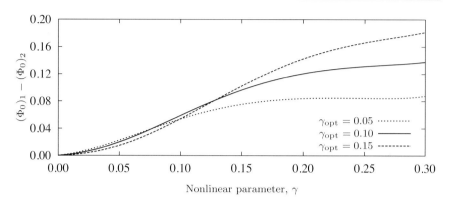

Figure 13. The value of the objective function for three optimized structures for varying values of the nonlinear parameter. The structures have been optimized for three different values of $\gamma = \gamma_{\mathrm{opt}}$. From Jensen (2011).

4 Time-space topology optimization.

The previous chapters have dealt with vibration and wave propagation problems in frequency domain and were treated using steady-state models. In this chapter a recent extension to topology optimization of transient problems is described. Whereas, transient simulation based topology optimization has been around for some time (Min et al., 1999). The case where the optimal material distribution is allowed to evolve in time is relatively new. In the following the analysis model will be described and formulas for the design sensitivities will be derived. This is followed by an application example showing the design of a dynamic bandgap structure. The material and results presented in this chapter are based on the results in Jensen (2009) and Jensen (2010).

5 Analysis model.

The starting point for the analysis and subsequent optimization study is a time-dependent FE model in which the mass matrix $\mathbf{M}(t)$ and the stiffness matrix $\mathbf{K}(t)$ may vary in time:

$$\frac{\partial}{\partial t}\big(\mathbf{M}(t)\dot{\mathbf{u}}\big) + \mathbf{C}\dot{\mathbf{u}} + \mathbf{K}(t)\mathbf{u} = \mathbf{f}(t) \tag{109}$$

Here, $\mathbf{C}$ is a constant damping matrix and $\mathbf{f}(t)$ is the transient load. The vector $\mathbf{u}(t)$ contains the unknown nodal displacements and $\dot{\mathbf{u}}(t)$ are the velocities.

The governing equation is solved in the time domain in the interval from $t = 0$ to $t = \mathcal{T}$ employing trivial initial conditions:

$$\mathbf{u}(t) = \dot{\mathbf{u}}(t) = \mathbf{0} \tag{110}$$

which imposes only limited loss of generality.

It should be noted that although the terms mass matrix/mass density and stiffness matrix/stiffness are used, the equations could be applied to an electromagnetic or an acoustic problem with proper renaming of involved parameters.

In traditional topology optimization using the density approach (Bendsøe, 1989), a single design variable is introduced for each element in the finite element model. This is illustrated in Fig. 14 for a single spatial dimension. The design variables are denoted ρ_1, ρ_2, ..., ρ_N where N is the number of elements in the model. The design variables are collected in the vector $\boldsymbol{\rho}$. The value of ρ_i governs the material properties of the corresponding element according to a specified material interpolation model (Bendsøe and Sigmund, 1999).

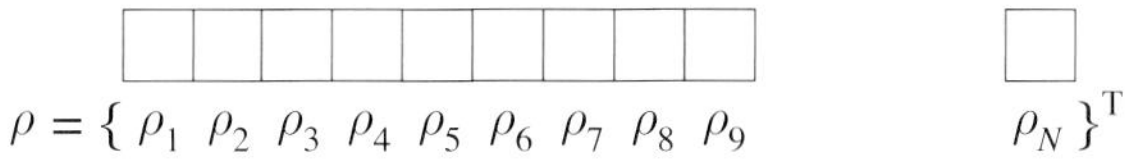

$$\rho = \{\, \rho_1 \ \ \rho_2 \ \ \rho_3 \ \ \rho_4 \ \ \rho_5 \ \ \rho_6 \ \ \rho_7 \ \ \rho_8 \ \ \rho_9 \qquad\qquad \rho_N \,\}^{\mathrm{T}}$$

Figure 14. Traditional design variable concept for topology optimization with the density approach for one spatial dimension. From Jensen (2010).

In the proposed formulation we extend the traditional approach by allowing the material properties in a spatial element to change in time. This is facilitated by introducing a two dimensional design element grid (for one spatial dimension) as can be seen in Fig. 15.

An array of design variable vectors is introduced:

$$\mathbf{X} = \{\boldsymbol{\rho}_1, \ \boldsymbol{\rho}_2, \ \ldots, \ \boldsymbol{\rho}_M\} \tag{111}$$

in which each design vector in the array contains the element-wise design variables for a specific time interval. For time interval j, the design vector components are specified as

$$\boldsymbol{\rho}_j = \{\rho_1^j, \ \rho_2^j, \ \ldots, \ \rho_N^j\}^T \tag{112}$$

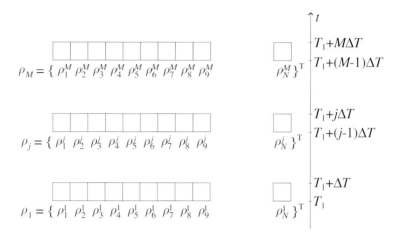

Figure 15. Extended topology optimization approach with space-time design variables for one spatial dimension. From Jensen (2010).

In Eq. (111) M is the number of time intervals in which the material properties can attain different values. The temporal design starts at $t = \mathcal{T}_1$ and continues to $t = \mathcal{T}_2 = \mathcal{T}_1 + M\Delta\mathcal{T}$. For simplicity, uniform intervals $\Delta\mathcal{T}$ are specified but non-uniform intervals can readily be used with this formulation.

The choice of $\Delta\mathcal{T}$ should depend on the specific problem considered, such as the frequency of the wave, but should also take into account the temporal discretization used in the time integration algorithm. In the present work an explicit scheme with a fixed time step Δt is applied. In this case it is necessary that $\Delta t << \Delta\mathcal{T}$ in order for the numerical results to be accurate. In the case where the mass matrix varies in time, special care should be taken in the choice of a proper time integration scheme. A further discussion of this can be found in Jensen (2010).

The value of the density variable ρ_j^e will determine the material properties of that space-time element by an interpolation between two predefined materials 1 and 2, where the variable is allowed to take any value from 0 to 1 ($\rho_j^e \in [0;1]$). By rescaling the equations with respect to the material properties of material 1, the mass and stiffness matrices can be written as:

$$\mathbf{M}_j = \sum_{e=1}^{N}(1 + \rho_j^e(\varrho - 1))\mathbf{M}^e \tag{113}$$

$$\mathbf{K}_j = \sum_{e=1}^{N}(1 + \rho_j^e(E - 1))\mathbf{K}^e \tag{114}$$

such that ϱ, E denote the *contrast* between the two materials for the mass density and stiffness, respectively. In Eqs. (113)–(114), $\mathbf{M}^e$ and $\mathbf{K}^e$ are local mass and stiffness matrices expressed in global coordinates.

5.1   Performance evaluation and sensitivity analysis.

Analytical expressions for the design sensitivities are now derived. The optimization is based on an objective that is assumed to be written as:

$$\Phi = \int_0^{\mathcal{T}} \Phi_0(\boldsymbol{\rho}, \mathbf{u})dt \tag{115}$$

in which Φ_0 is a real scalar function of the time-dependent displacement vector and $\mathcal{T}$ is the total simulation time. It should emphasized that more complicated objective functions, e.g. with evaluation time different from the total simulation time, can be treated with minor modification to the following derivation.

The derivative wrt. a single design variable in the j'th time-interval and e'th spatial variable is denoted $()' = \partial/\partial\rho_j^e$ and thus the sensitivity of Φ wrt. to ρ_j^e is:

$$\frac{d\Phi}{d\rho_j^e} = \int_0^{\mathcal{T}} (\Phi_0' + \frac{\partial\Phi_0}{\partial\mathbf{u}}\mathbf{u}')dt \tag{116}$$

Eq. (116) involves the term $\mathbf{u}'$ which cannot be explicitly evaluated. However, the adjoint method for transient computations can be used to circumvent this problem in an efficient way (Arora and Holtz, 1997). For this purpose the residual vector

$$\mathbf{R} = \frac{\partial}{\partial t}(\mathbf{M}\dot{\mathbf{u}}) + \mathbf{C}\dot{\mathbf{u}} + \mathbf{K}\mathbf{u} - \mathbf{f} \tag{117}$$

is differentiated wrt. ρ_j^e:

$$\mathbf{R}' = \frac{\partial}{\partial t}(\mathbf{M}'\dot{\mathbf{u}} + \mathbf{M}\dot{\mathbf{u}}') + \mathbf{C}\dot{\mathbf{u}}' + \mathbf{K}'\mathbf{u} + \mathbf{K}\mathbf{u}' \tag{118}$$

in which it has been used that $\mathbf{f}$ (the transient load) and $\mathbf{C}$ (the damping matrix) are both independent of the design.

With the aid of Eq. (118), Eq. (116) is reformulated as:

$$\frac{d\Phi}{d\rho_j^e} = \int_0^{\mathcal{T}} (\Phi_0' + \frac{\partial\Phi_0}{\partial\mathbf{u}}\mathbf{u}' + \boldsymbol{\lambda}^T\mathbf{R}')dt \qquad (119)$$

in which $\boldsymbol{\lambda}$ denote an unknown vector of Lagrangian multipliers to be determined in the following.

Expanding the expression in Eq. (119) yields

$$\frac{d\Phi}{d\rho_j^e} = \int_0^{\mathcal{T}} \left(\Phi_0' + \frac{\partial\Phi_0}{\partial\mathbf{u}}\mathbf{u}' + \boldsymbol{\lambda}^T(\frac{\partial}{\partial t}(\mathbf{M}'\dot{\mathbf{u}} + \mathbf{M}\dot{\mathbf{u}}') + \mathbf{C}\dot{\mathbf{u}}' + \mathbf{K}'\mathbf{u} + \mathbf{K}\mathbf{u}')\right)dt \quad (120)$$

and afterwards using integration by parts leads to the following equation:

$$\frac{d\Phi}{d\rho_j^e} = \int_0^{\mathcal{T}} (\Phi_0' + \boldsymbol{\lambda}^T\mathbf{K}'\mathbf{u} - \dot{\boldsymbol{\lambda}}^T\mathbf{M}'\dot{\mathbf{u}})dt$$

$$+ \int_0^{\mathcal{T}} (\frac{\partial\Phi_0}{\partial\mathbf{u}} + \frac{\partial}{\partial t}(\dot{\boldsymbol{\lambda}}^T\mathbf{M}) - \dot{\boldsymbol{\lambda}}^T\mathbf{C} + \boldsymbol{\lambda}^T\mathbf{K})\mathbf{u}'dt \qquad (121)$$

$$+ \left[\boldsymbol{\lambda}^T(\mathbf{M}'\dot{\mathbf{u}} + \mathbf{M}\dot{\mathbf{u}}' + \mathbf{C}\mathbf{u}') - \dot{\boldsymbol{\lambda}}^T\mathbf{M}\mathbf{u}'\right]_0^{\mathcal{T}}$$

Now the unknowns $(\boldsymbol{\lambda}, \dot{\boldsymbol{\lambda}})$ can be chosen so that the last integral in expression (121) vanishes along with the bracketed term. This leads to the following adjoint equation:

$$\frac{\partial}{\partial t}(\mathbf{M}^T\dot{\boldsymbol{\lambda}}) - \mathbf{C}^T\dot{\boldsymbol{\lambda}} + \mathbf{K}^T\boldsymbol{\lambda} = -(\frac{\partial\Phi_0}{\partial\mathbf{u}})^T \qquad (122)$$

along with the following terminal conditions:

$$\boldsymbol{\lambda}(\mathcal{T}) = \dot{\boldsymbol{\lambda}}(\mathcal{T}) = \mathbf{0} \qquad (123)$$

The terminal value problem stated above can be transformed into an initial value problem by introducing the adjoint time $\tau = \mathcal{T} - t$:

$$\frac{\partial}{\partial\tau}(\mathbf{M}(\mathcal{T} - \tau)\dot{\boldsymbol{\lambda}}) + \mathbf{C}\dot{\boldsymbol{\lambda}} + \mathbf{K}(\mathcal{T} - \tau)\boldsymbol{\lambda} = -(\frac{\partial\Phi_0}{\partial\mathbf{u}(\mathcal{T} - \tau)})^T \qquad (124)$$

$$\boldsymbol{\lambda}(0) = \dot{\boldsymbol{\lambda}}(0) = \mathbf{0} \qquad (125)$$

where also the symmetry of the involved matrices has been exploited. Additionally, it has been explicitly stated that the matrices and field should be evaluated at the physical time $t = \mathcal{T} - \tau$.

The sensitivities are computed from the remaining expression:

$$\frac{d\Phi}{d\rho_j^e} = \int_0^{\mathcal{T}} (\Phi_0' + \boldsymbol{\lambda}^T \mathbf{K}'\mathbf{u} - \dot{\boldsymbol{\lambda}}^T \mathbf{M}'\dot{\mathbf{u}})dt = \int_{\mathcal{T}_j^-}^{\mathcal{T}_j^+} (\Phi_0' + \boldsymbol{\lambda}^T \mathbf{K}'\mathbf{u} - \dot{\boldsymbol{\lambda}}^T \mathbf{M}'\dot{\mathbf{u}})dt$$
(126)

in which the integral can be reduced to the j'th time interval ranging from $\mathcal{T}_j^-$ to $\mathcal{T}_j^+$ simply because $\mathbf{K}'$ and $\mathbf{M}'$ vanish outside the interval belonging to the specific design variable.

The expression can be further reduced to element level as follows:

$$\frac{d\Phi}{d\rho_j^e} = \int_{\mathcal{T}_j^-}^{\mathcal{T}_j^+} \left(\Phi_0' + (E-1)(\boldsymbol{\lambda}^e)^T \mathbf{K}^e \mathbf{u}^e - (\varrho-1)(\dot{\boldsymbol{\lambda}}^e)^T \mathbf{M}^e \dot{\mathbf{u}}^e\right)dt$$
(127)

by using the material interpolations defined in Eqs. (113)–(114).

5.2   Application example.

It will now be illustrated how the developed theory can be used to design a dynamic bandgap structure. The design problem is illustrated in Fig. 16.

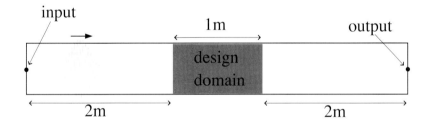

Figure 16. Design problem.   The transmission of a sinusoidal Gauss-modulated pulse is minimized. From Jensen (2011).

A sinusoidal Gauss-modulated pulse is send through a one-dimensional elastic rod and the transmitted wave is recorded at the output point. The purpose of the study is to design the structure so that the transmission is minimized. The following objective function is considered:

$$\min_{\mathbf{X}} \Phi = \int_0^{\mathcal{T}} \Phi_0(\mathbf{u})dt$$
(128)

in which $\Phi_0(\mathbf{u}) = u_{\text{out}}^2$ where u_{out} is the displacement of the output point and $\mathcal{T}$ is the total simulation time. Thus, the considered objective function is proportional to the total transmitted wave energy.

The wave pulse is generated by applying the the following force at the input point:

$$f_1(t) = -2u_0\omega_0 \sin\left(\omega_0(t-t_0)\right)e^{-\delta(t-t_0)^2} \qquad (129)$$

in which u_0 is the amplitude of the generated pulse, ω_0 is the center-frequency of excitation and spectral δ determines the width of the pulse.

The parameters used in this example are given as:
$\mathcal{T} = 10\,\text{s}$, $u_0 = 1\,\text{m}$, $t_0 = 2.5\,\text{s}$, $N = 150$, $\omega_0 = 15.7\,\text{rad/s}$, $\Delta t = 0.06\,\text{s}$, $\beta = 0.1\,(\text{kg}\,\text{m})^{-1}$, $\delta = 1.5\,\text{s}^{-2}$
and the material contrasts are chosen as $E = 1.25$ and $\varrho = 1$, so that no variation in mass is present in this example.

To be able to clearly separate input and reflected waves in the time series, inlet and outlet sections of $2\,\text{m}$ with constant material properties have been added on each side of the design domain of length $1\,\text{m}$. The material properties of the inlet and outlet sections are normalized to $E = 1$ and $\varrho = 1$ so that the incoming wave propagates with the speed $c = 1\,\text{m/s}$. In order to simulate wave propagation in this finite structure, fully absorbing boundaries are added at the input and output points by appropriate viscous dampers.

Two materials can be distributed in the design domain: the normalized material used for the the inlet and output sections, and a material with a slightly higher stiffness $E = 1.25$. A small stiffness contrast combined with added stiffness proportional damping ($\beta = 0.1$ used throughout this example) eliminate problems with numerical stability while still displaying the main qualitative features, see Jensen (2009).

Space-time bandgap structure. We aim at designing a highly wave-reflecting structure. In the "static" case, i.e. if the material parameters are not allowed to change in time, an optimal structure would appear as a layered structure with alternating sections of high and low stiffness material – a band gap structure.

However here we allow the design to change in time as well as in space. An optimized static bandgap structure is used as a starting point for the dynamic structure. This is illustrated in Fig. 17 where the space-time design domain is indicated. A temporal design interval of $\Delta\mathcal{T} = 1.5\,\text{s}$ is chosen, which is sufficiently long to significantly modify the wave motion but still keeps the total number of design variables at a manageable level. The start and finish point for the optimization is chosen as $\mathcal{T}_1 = 4.25\,\text{s}$ and $\mathcal{T}_2 = 5.75\,\text{s}$ and the number of sub-intervals is $M = 225$, thus the material

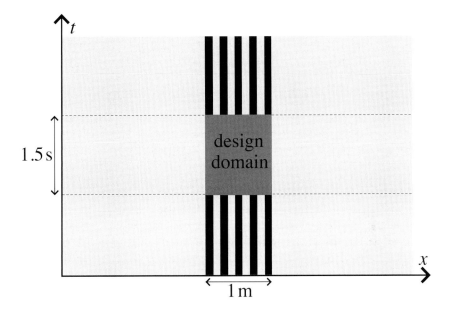

Figure 17. Design domain for the space-time optimization problem. From Jensen (2011).

properties in each element are allowed to change 150 times per second. The spatial discretization is chosen as 150 elements in the design domain. Thus, the two-dimensional design grid is composed of "square" elements with the dimensions $\frac{1}{150}$ s$\times\frac{1}{150}$ m and a total of 33750 design variables.

The initial value of all design variables is chosen as $\rho_e = 0.5$, implying that the stiffness in all design elements is initially $E = 1.125$. The optimized design is obtained after about 100 iterations and is shown in Fig. 18. The structure is seen to be a kind of spatio-temporal laminate with space-time layered inclusions. Properties of spatio-temporal laminates are discussed e.g. in Lurie (1997); Weekes (2001); Lurie (2006).

At each time instance the structure is layered in a similar way as a static bandgap structure. However, the inclusion layers move with a constant speed corresponding to the wave speed in the layered medium so that the wave peaks and valleys actually move together with the front of the inclusions. This is illustrated in Fig. 19, in which the material distribution is shown together with the wave motion at the two time instances $t = 5.0$ s

Figure 18. Optimized space-time structure for minimum transmission of the wave pulse. From Jensen (2011).

and $t = 5.4\,\text{s}$.

Fig. 20 shows the input and output point time response. The objective is reduced to 23 % relative to the undisturbed input signal – a significant reduction compared to a standard "static" bandgap structure. The large reduction is not a consequence of an increased reflection. Actually, the reflected energy is reduced to only about 1 % of the input compared to about 17 % for a corresponding static bandgap structure. Instead the main part of the energy (about 75 %) is extracted from the system via the time dependent force that is needed to create a temporal stiffness variation. Only a small fraction (about 1 %) is dissipated by the stiffness proportional damping. However, without the presence of damping the optimization procedure becomes unstable after only a few iterations.

The output time response in Fig. 20 indicates that the frequency content of the wave has changed. This is confirmed by an FFT-analysis shown in Fig. 21. In addition to the main frequency component $\omega_0 = 15.7\,\text{rad/s}$, higher order components at $\approx 3\omega_0$ and $\approx 5\omega_0$ (and more further up in the spectrum) are seen as well. This might seem surprising since the system is completely linear. However, as mentioned, the material properties can only be changed in time with a time dependent external force acting on the system, and this force generates these higher order frequency components in the time signal.

Similar structures as the one shown, can be obtained with different values of the material contrast E. Only the wave speed in the layered medium changes and thereby also the speed of the bandgap material front which leads to a different "slope" of the spatio-temporal laminates.

Realizable structures – patches & checkerboards. The optimized structure presented in the previous section is not easy to realize in practice. In order to generate structures that could more easily be realized, the number of design variables is reduced. This is done by lumping spatial elements

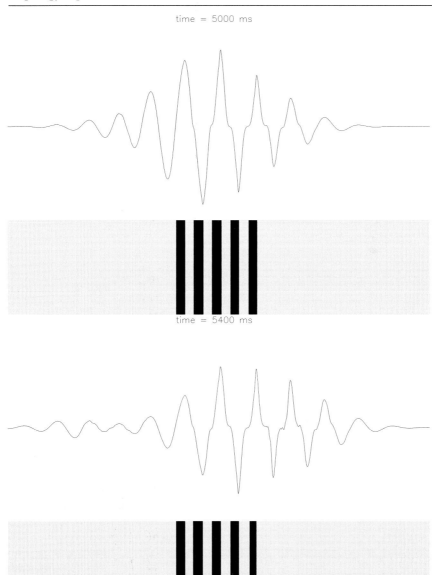

Figure 19. Instantaneous material distribution and wave motion at $t =$ 5.0 s and $t = 5.4$ s for the structure in Fig. 18. From Jensen (2011).

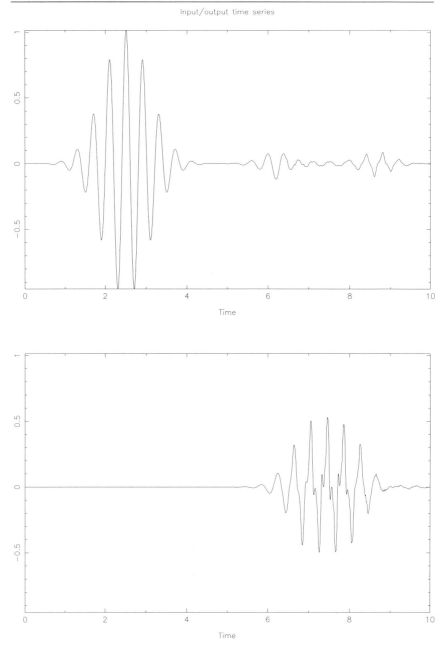

Figure 20. Response for the optimized structure shown in Fig. 18. Top: displacement of input point, bottom: displacement of output point. From Jensen (2011).

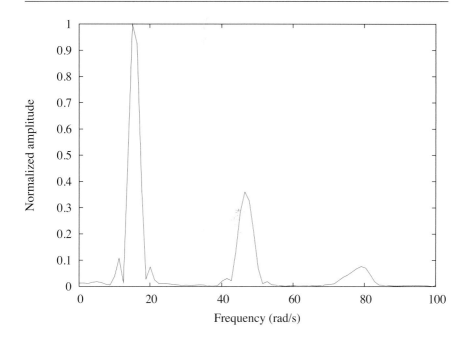

Figure 21. FFT-analysis of the output point time response shown in Fig. 20. From Jensen (2011).

together in patches and reducing the number of temporal design intervals as well.

Fig. 22 (top and middle) shows two examples of simplified optimized structures. Both structures have 30 spatial design elements (patches of 5 elements). Using fewer design variables than this, makes it impossible to resolve the layered structures adequately. The two structures have $M = 45$ and $M = 15$ time design intervals, respectively. The finer structure (1350 design variables in total) displays the same spatio-temporal laminate structure as the fully discretized structure in Fig. 18 and the corresponding objective function is only marginally higher (27 % compared to 23 %). For the coarse design with only 15 time design variables (total of 450 design variables) the laminated structure can no longer be created. Instead the structure has a checkerboard appearance. The objective is increased to 45 % which is significantly higher than for the laminated structure, but still much better than can be obtained with a static bandgap structure.

Figure 22. Simplified optimized structures with a reduced number of design variables. Top: 30×45, middle 30×15 variables. Bottom: optimized checkerboard structure with special checkerboard design variable model. From Jensen (2011).

The checkerboard structure is an example of a space-time material pattern that is easier to realize in practice than the spatio-temporal laminate. A detailed analysis of the properties of such structures can be found in Lurie (2006). To further exploit this, a design parametrization is now constructed that *ensures* a checkerboard structure as outcome. This is done by replacing the material interpolation model in Eq. (114) by:

$$E_e = 1 + (\tilde{\rho}_i - \tilde{\rho}_j)^2 (E - 1) \tag{130}$$

in which $\tilde{\rho}_i$ is a vector of *spatial* design variables and $\tilde{\rho}_j$ is a vector of *temporal* design variables. The computation of the design sensitivities should now be done directly based on Eq. (126) since the simplification in Eq. (127) no longer holds. This increases the computation time for each design variable but the number of variables is correspondingly smaller. The fully discretized optimized structure in Fig. 18 has $150 \times 225 = 33750$ variables, but with the new parametrization, the same discretization results in $150 + 225 = 375$ variables. The resulting checkerboard structure is seen in Fig. 22 (bottom) and the resulting response in Fig. 23. The structure is seen to qualitatively resemble the structure in Fig. 22 (bottom) with the same number (fifteen) of temporal inclusions. The objective is also similar (40 % compared to 45 %). Compared to the response for the fully discretized structure in Fig. 20 the response peaks are now higher but the qualitative nature of the response is unchanged.

Long wavelengths – temporal laminates.   The appearance of the optimized spatio-temporal laminates depends strongly on the frequency and wavelength contents of the wave. In the previous examples the main wavelength of the wave was $\lambda = 0.4$ m for a wave with center frequency $\omega_0 = 15.7$ rad/s propagating in the material with $c = 1$ m/s. Thus, the design domain had a spatial extent corresponding to 2.5 wavelengths.

If the design domain length is short compared to the main wavelength, a spatially layered structure is no longer efficient for minimizing the wave transmission. Fig. 24 shows two optimized structures for $\lambda = 2.5$ m and $\lambda = 10$ m, respectively. The temporal design intervals have been increased accordingly to 5 s and 10 s. As appears, longer wavelengths result in structures that are spatially more homogeneous at a given time. In the long wavelength limit the optimized structure approaches that of a pure temporal laminate with instant simultaneous switching of the material properties in the entire structure. For a discussion of temporal laminates see Weekes (2002); Lurie (2006).

Bibliography

J. S. Arora and D. Holtz. An efficient implementation of adjoint sensitivity analysis for optimal control problems. *Structural Optimization*, 13:223–229, 1997.

M. P. Bendsøe. Optimal shape design as a material distribution method. *Structural Optimization*, 10:193–202, 1989.

M. P. Bendsøe and N. Kikuchi. Generating optimal topologies in structural design using a homogenization method. *Computer Methods in Applied Mechanics and Engineering*, 71(2):197–224, 1988.

M. P. Bendsøe and O. Sigmund. *Topology Optimization - Theory, Methods and Applications*. Springer Verlag, Berlin Heidelberg, 2003.

M. P. Bendsøe and O. Sigmund. Material interpolations in topology optimization. *Archive of Applied Mechanics*, 69:635–654, 1999.

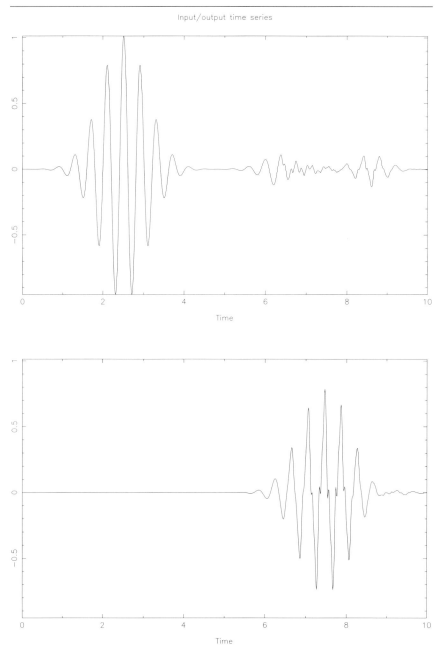

Figure 23. Response for the optimized structure shown in Fig. 22 (bottom). Top: displacement of input point, bottom: displacement of output point. From Jensen (2011).

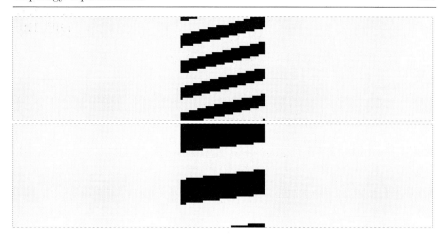

Figure 24. Optimized structures for long wavelengths. Top: $\lambda = 2.5\,\mathrm{m}$ and bottom: $\lambda = 10\,\mathrm{m}$. From Jensen (2011).

P. I. Borel, A. Harpøth, L. H. Frandsen, M. Kristensen, J. S. Jensen, P. Shi, and O. Sigmund. Topology optimization and fabrication of photonic crystal structures. *Optics Express*, 12(9):1996–2001, 2004. http: //www. opticsexpress.org/abstract.cfm?URI=OPEX-12-9-1996.

P. I. Borel, L. H. Frandsen, A. Harpøth, M. Kristensen, J. S. Jensen, and O. Sigmund. Topology optimised broadband photonic crystal Y-splitter. *Electronics Letters*, 41(2):69–71, 2005.

R. D. Cook, D. S. Malkus, M. E. Plesha, and R. J. Witt. *Concepts and Applications of Finite Element Analysis*. John Wiley and Sons, 2002.

L. H. Frandsen, A. Harpøth, P. I. Borel, M. Kristensen, J. S. Jensen, and O. Sigmund. Broadband photonic crystal waveguide 60° bend obtained utilizing topology optimization. *Optics Express*, 12(24):5915–5921, 2004. http: //www. opticsexpress.org/abstract.cfm?URI=OPEX-12-24-5916.

J. S. Jensen. Space-time topology optimization for one-dimensional wave propagation. *Methods in Applied Mechanics and Engineering*, (198):705–715, 2009.

J. S. Jensen. Optimization of space-time material layout for 1d wave propagation with varying mass and stiffness parameters. *Control and Cybernetics*, (39):599–614, 2010.

J. S. Jensen. Topology optimization of nonlinear photonic devices. *Structural and Multidisciplinary Optimization*, page to appear, 2011.

J. S. Jensen and O. Sigmund. Topology optimization of photonic crystal structures: A high-bandwidth low-loss T-junction waveguide. *Journal of the Optical Society of America B: Optical Physics*, 22(6):1191–1198, 2005.

J. S. Jensen, O. Sigmund, L. H. Frandsen, P. I. Borel, A. Harpøth, and M. Kristensen. Topology design and fabrication of an efficient double 90-degree photonic crystal waveguide bend. *IEEE Photonics Technology Letters*, 41(2):69–71, 2005.

J. Jin. *The finite element method in electromagnetics*. John Wiley & Sons, 2nd edition, 2002.

J. D. Joannopoulos, R. D. Meade, and J. N. Winn. *Photonic Crystals*. Princeton University Press, New Jersey, 1995.

C. S. Jog. Topology design of structures subjected to periodic loading. *Journal of Sound and Vibration*, 253(3):687–709, 2002.

M. Koshiba, Y. Tsuji, and S. Sasaki. High-performance absorbing boundary conditions for photonic crystal waveguide simulations. *IEEE Microwave Wireless Components Lett.*, 11:152–154, April 2001.

K. A. Lurie. *An Introduction to Mathematical Theory of Dynamic Materials*. Springer Verlag, 2006.

K. A. Lurie. Effective properties of smart elastic laminates and the screening phenomenon. *International Journal of Solids and Structures*, 34:1633–1643, 1997.

S. Min, N. Kikuchi, Y. C. Park, S. Kim, and S. Chang. Optimal topology design of structures under dynamic loads. *Structural Optimization*, 17: 208–218, 1999.

O. Sigmund and J. S. Jensen. Systematic design of phononic band gap materials and structures by topology optimization. *Philosophical Transactions of the Royal Society A: Mathematical, Physical and Engineering Sciences*, 361:1001–1019, 2003.

O. Sigmund and J. Petersson. Numerical instabilities in topology optimization: A survey on procedures dealing with checkerboards, mesh-dependencies and local minima. *Structural Optimization*, 16:68–75, 1998.

K. Svanberg. The method of moving asymptotes - a new method for structural optimization. *International Journal for Numerical Methods in Engineering*, 24:359–373, 1987.

K. Svanberg. A globally convergent version of mma without linesearch. In N. Olhoff and G. I. N. Rozvany, editors, *Proceedings of the First World Congress of Structural and Multidisciplinary Optimization WCSMO-1*, pages 9–16, Goslar, Germany, 1995.

D. A. Tortorelli and P. Michaleris. Design sensitivity analysis: overview and review. *Inverse Problems in Engineering*, 1:71–105, 1994.

S. L. Weekes. Numerical computation of wave propagation in dynamic materials. *Applied Numerical Mathematics*, 37:417–440, 2001.

S. L. Weekes. A stable scheme for the numerical computation of long wave propagation in temporal laminates. *Journal of Computational Physics*, 176:345–362, 2002.

Map-based approaches for periodic structures

Francesco Romeo

Dipartimento di Ingegneria Strutturale e Geotecnica,
SAPIENZA University of Rome, Rome, Italy

Abstract In this chapter the dynamic behavior of periodic structures is dealt with by means of maps. Continuous and discrete models of both linear and nonlinear mechanical systems are considered. The first part of the chapter is devoted to linear problems; general multi-coupled periodic systems are presented and they are dealt with by means of linear maps, namely the transfer matrices of single units. An exhaustive description of the free wave propagation patterns is given on the invariants' space where propagation domains with qualitatively different character are identified. The problem of minimizing transmitted vibrations through finitely long periodic structures as well as computational issues arising in the wave vector approach are also addressed. The second part of the chapter concerns nonlinear periodic systems, the dynamic analysis of which hinges on nonlinear maps conceived according to two different approaches. At first, a perturbation method is applied to the transfer matrix of a chain of continuous nonlinear beams. Afterwards, nonlinear maps are derived from the governing difference equations of a chain of nonlinear oscillators.

1 Linear periodic structures

Starting from the early 70's the dynamics of linear periodic structures has been extensively analysed in literature by means of the transfer matrix method (Gupta, 1970; Faulkner and Hong , 1985). This method features the computational advantage of reducing the dimension of the underlying problem to the number of degrees of freedom (d.o.f.) coupling basic periodic elements. Such dimension is thereby independent from the number of elements constituting the whole structure. The dynamic behavior of mono-coupled systems (1 d.o.f. at the interface) has been thoroughly addressed by Mead (1975); already in the late 60's it was evidenced that the disturbance propagation through mono-coupled periodic structures is governed by the frequency dependent transfer matrix eigenvalues. On the frequency axis there exist intervals or bands where disturbances propagate harmonically without attenuation (*pass-bands*), in which the eigenvalues are complex with unitary modulus, and bands where the disturbances decay (*stop-bands*), in which the eigenvalues are real and different from 1. Analytical studies have also been proposed for bi-coupled periodic structures by Mead (1975) while mostly numerical approaches have been developed for multi-coupled cases by Signorelli and von Flotow (1988) and Yong and Lin (1989). It has been found that as soon as the coupling coordinates are more than one there exist further frequency bands characterized by disturbance harmonic propagation with attenuation (*complex-bands*) where pairs of complex conjugate eigenvalues, with modulus different from 1, exist.

The description of the dynamics of multi-coupled periodic structures usually proposed in literature relies on the frequency as the sole parameter (Mead, 1970; Bouzit and Pierre, 2000; Mead, 1986; Bansal, 1997; Koo and Park, 1998). Nevertheless such description is generally not exhaustive of the propagation properties of the system, since it qualitatively depends on the values fixed for the remaining physical parameters. Indeed, when these are modified, some existing bands generally disappear and some new bands appear somewhere. Therefore, a multi-dimensional representation is necessary to completely depict the propagation properties scenario. However, not all the parameters have the same qualitative influence on such properties, so that the problem of selecting the smallest set of parameters, able to furnish a complete representation, arises. The task has some similarities with that of linear bifurcation analysis of dynamical systems (Guckenheimer and Holmes, 1983). The problem was solved by Romeo and Luongo (2002) by referring to the invariants of the transfer matrix characteristic equation and identifying in the invariants' space the domains in which the eigenvalues are of the same type thereby achieving an exhaustive geometrical representation of the propagation properties. Such description is universal, as it is system-independent. However, when dealing with a specific n-coupled system, it would be desirable to transform the invariant space into a physical space. This transformation is accomplished by expressing the invariants I_j as functions of n chosen parameters $\mathbf{p}_c$, ahead referred to as *control parameters*, namely $I_j = I_j(\mathbf{p}_c)$. These relationships map the invariant into the physical space. However, since the relations are in general of nonlinear type, one point of the former space is mapped in more points of the latter space. Thus the connected domains of the invariant representation split in the physical one, thus explaining the bands alternating pattern in the usual one-dimensional representation. Among the control parameters, the frequency is the most meaningful one and should be included in the set as *distinguished control parameter*. The choice of the remaining $n-1$

control parameters is arbitrary, whilst it is expected that they are not all equivalent. The optimum choice would consist in obtaining a representation qualitatively independent of the remaining (auxiliary) parameters, at least in some range of their values.

The material presented in the following sections aims at providing a consistent framework enabling to encompass analytical, computational and optimal design aspects pertaining to the dynamics of linear periodic structure. The sections covering the propagation regions derivation on the invariants' space and associated mechanical models, up to three-coupled cases, is based on the results published in (Romeo and Luongo, 2002; Romeo and Paolone, 2007) . Then, according to the material presented in Romeo and Luongo (2003), the optimal reduction of transmitted vibrations through finitely long periodic structures is addressed. Computational aspects of the wave vector approach conclude the topics in linear periodic structures (Luongo and Romeo, 2005).

1.1 Invariant propagation scenarios for multi-coupled periodic structures

A generic periodic structure whose elements are coupled through n degrees of freedom to the adjacent ones is considered. Let $\mathbf{z}_k = (\mathbf{d}_k , \mathbf{f}_k)^T$ be the state vector of generalized displacement $\mathbf{d}_k$ and forces $\mathbf{f}_k$ at the coupling point k; according to the transfer matrix approach, the state vector $\mathbf{z}_{k+1}$ at the coupling point $k + 1$ is related to the state vector $\mathbf{z}_k$ by

$$\mathbf{z}_{k+1} = \mathbf{T}\mathbf{z}_k \tag{1}$$

where $\mathbf{T}$ is the $2n \times 2n$ frequency dependent transfer matrix which is real in absence of damping. It follows that the matrix $\mathbf{T}$ and, more specifically, its invariants summarize all the propagation features of the periodic cell.

1.2 Propagation regions on the invariants' plane

Let's consider a generic undamped n-coupled periodic structure; due to the spectral properties of the symplectic matrix $\mathbf{T}$ (Zhong and Williams, 1992, 1993), the characteristic equation $\det[\ \mathbf{T} - \lambda\ \mathbf{I}] = 0$ reads

$$\lambda^{2n} + \sum_{k=1}^{2n-1} I_k \lambda^{2n-k} + 1 = 0 \qquad I_k \in \mathbb{R} \tag{2}$$

where the coefficients I_k are the invariants of $\mathbf{T}$ satisfying the condition $I_k = I_{2n-k}$; therefore the latter reversibility property halves the number of the transfer matrix invariants.

The meaning of the eigenvalues λ_i emerges form the Floquet's theorem: there exist free wave motions (characteristic waves) in which $\mathbf{z}_{k+1} = \lambda_i \, \mathbf{z}_k$, each associated with an eigenvalue of $\mathbf{T}$. If $|\lambda_i| < 1$ the wave amplitude decays in the positive direction (forward wave), if $|\lambda_i| > 1$ it decays in the negative direction (backward wave), if $|\lambda_i| = 1$ no attenuation exists in the two directions. Due to the reversibility of the coefficients of equation (2), if λ_i is an eigenvalue, then λ_i^{-1} is also an eigenvalue (called the *adjoint* eigenvalue). Therefore forward and backward waves always exist in pairs and have both the same propagation properties. It follows that n eigenvalues such that $|\lambda_i| \leq 1$ $(i = 1, n)$, completely define the propagation properties of a multi-coupled periodic structure; they will be here referred to as *principal* eigenvalues. When a complex eigenvalue has unitary modulus, its adjoint coincide with the complex conjugate, so that more than n eigenvalues satisfy the inequality $|\lambda_i| \leq 1$. To avoid indeterminacies, n eigenvalues with positive phase $0 \leq \vartheta \leq \pi$ will be taken as principal eigenvalues. If $\vartheta = 0$ or $\vartheta = \pi$, only one of the coincident eigenvalues must be taken. As usually done in literature, previous findings can be restated in terms of the propagation constants μ_i, instead of the eigenvalues λ_i, by defining $\lambda_i = e^{\mu_i}$. This position maps the unitary circle of the $\mathrm{Re}(\lambda_i)$-$\mathrm{Im}(\lambda_i)$ plane into the left half-space of the $\mathrm{Re}(\mu_i)$-$\mathrm{Im}(\mu_i)$ plane.

Mono-coupled case   For the mono-coupled case the 2×2 transfer matrix is characterized by the sole invariant $I_1 = Tr(\mathbf{T})$; letting $\lambda = e^{\mu}$, $\mu \in \mathbb{C}$, and multiplying by $e^{-\mu}$ equation (2) becomes

$$2 \cosh \mu + I_1 = 0 \qquad\qquad (3)$$

From equation (3) it follows that $\cosh \mu = -I_1/2 =: F(I_1)$, where $F \in \mathbb{R}$. If $F \in \mathcal{I} := [-1, 1]$, then $\mu = i\vartheta$ and $\lambda = e^{i\vartheta}$ is complex with unitary modulus; if $F \notin \mathcal{I}$, $\mu = \alpha + ij\pi$ with j integer, and $\lambda = \pm e^{\alpha}$ is real. The possible location of the eigenvalues on the complex plane are summarized in Figure 1. The wave propagation characteristics can be described through a straightforward geometric representation on the I_1 axis. When $\cosh \mu = \mp 1$, two points are obtained in the I_1 axis given, respectively, by $r := \{I_1 \mid I_1 + 2 = 0\}$, $s := \{I_1 \mid I_1 - 2 = 0\}$. Therefore the points r, s divide the invariant axis in domains (propagation zones) where the eigenvalues are of the same type. The propagation zones are labeled according to the notation commonly used in literature. The region where the pair of λ lay on the unit circle is referred to as *pass* (P); the regions where only a real pair of eigenvalues occurs are the *stop* (S) domains. It is worth noting that only

Figure 1. Allowable scenarios for λ. $\mathcal{I} = [-1, 1]$. (a): $F \in \mathcal{I}$; (b): $F \notin \mathcal{I}$.

the P region is bounded while the remaining S regions are unbounded.

Bi-coupled case   To solve equation (2) it is convenient to rewrite it in terms of the propagation constant μ. By letting $\lambda = e^{\mu}$ and multiplying by $e^{-2\mu}$, equation (2) reads

$$e^{2\mu} + I_1 e^{\mu} + I_2 + I_1 e^{-\mu} + e^{-2\mu} = 0 \qquad (4)$$

leading to the following quadratic equation in $\cosh\mu$

$$4\cosh^2\mu + 2I_1\cosh\mu + I_2 - 2 = 0 \qquad (5)$$

The roots of equation (5) are

$$(\cosh\mu)_{1,2} = \frac{1}{4}\left(-I_1 \pm \sqrt{8 + I_1^2 - 4I_2}\right) =: F_{1,2}(I_1, I_2) \qquad (6)$$

To discuss equation (6) the cases $F_{1,2} \in \mathbb{R}$ and $F_{1,2} \in \mathbb{C}$ are separately considered. As far as $F_{1,2} \in \mathbb{R}$, if $F_{1,2} \in \mathcal{I} := [-1, 1]$, then $\mu = i\vartheta$ and $\lambda = e^{i\vartheta}$ is complex with unitary modulus; if $F_{1,2} \notin \mathcal{I}$, $\mu = \alpha + ij\pi$ with j integer, and $\lambda = \pm e^{\alpha}$ is real. Differently, if $F_{1,2} \in \mathbb{C}$, $\mu = \alpha + i\vartheta$ and $\lambda = e^{\alpha + i\vartheta}$ is complex with modulus different from 1. The possible location of the eigenvalues on the complex plane are summarized in Figure 2. The wave propagation characteristics of bi-coupled periodic systems can be conveniently described through a geometric representation on the invariants' plane I_1-I_2 (Figure 3). When $\cosh\mu = \mp 1$, two curves are obtained in the I_1-I_2 plane given, respectively, by

$$r := \{(I_1, I_2) \mid 2 - 2I_1 + I_2 = 0\} \quad ; \quad s := \{(I_1, I_2) \mid 2 + 2I_1 + I_2 = 0\} \qquad (7)$$

A further curve, dividing the real roots of equation (6) from the complex ones, is given by the parabola

$$p := \{(I_1, I_2) \mid 8 + I_1^2 - 4I_2 = 0\} \qquad (8)$$

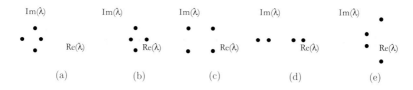

Figure 2. Allowable scenarios for λ. $\mathcal{I} = [-1, 1]$. (a)-(d): $F_{1,2} \in \mathbb{R}$; (a): $F_1 \notin \mathcal{I}$, $F_2 \in \mathcal{I}$; (b): $F_1 \in \mathcal{I}$, $F_2 \notin \mathcal{I}$; (c): $F_{1,2} \in \mathcal{I}$; (d): $F_{1,2} \notin \mathcal{I}$; (e) $F_{1,2} \in \mathbb{C}$.

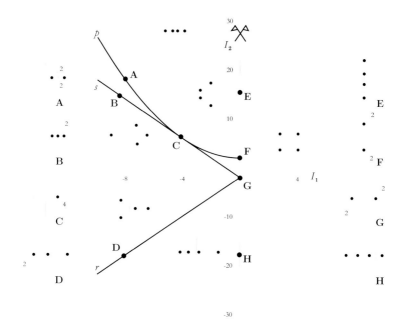

Figure 3. Position on the invariants' plane I_1-I_2 according to the location of the eigenvalues λ on the complex plane.

The three curves r, s and p divide the invariants' plane in domains (propagation zones) where the eigenvalues are of the same type.

In Figure 4 the propagation zones are labeled according to the notation commonly used in literature. The region where both the pairs of λ lay on the unit circle is referred to as *pass-pass* (PP); the regions where only one pair of λ lays on the unit circle while the other pair is real are referred to as *pass-stop* (PS); the regions where only real pairs of eigenvalues occurs are the *stop-stop* (SS) domains. Moreover, the curve p bounds the so called *complex* region (C) characterized by complex conjugate eigenvalues. It is worth noting that only the PP region is bounded while the remaining domains are unbounded.

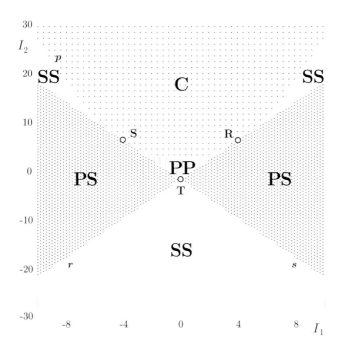

Figure 4. Propagation zones on the invariants' plane I_1-I_2 .

Three-coupled case To solve equation (2) it would be convenient to rewrite it in terms of the propagation constant μ. By letting $\lambda = e^{\mu}$ and multiplying by $e^{-3\mu}$, equation (2) reads

$$\left(e^{3\mu} + e^{-3\mu}\right) + I_1 \left(e^{2\mu} + e^{-2\mu}\right) + I_2 \left(e^{\mu} + e^{-\mu}\right) + I_3 = 0 \qquad (9)$$

which is also written as

$$2 \cosh 3\mu + 2\,I_1 \cosh 2\mu + 2\,I_2 \cosh \mu + I_3 = 0 \qquad (10)$$

After some algebraic manipulation the latter equation can be expressed as a cubic equation in $\cosh \mu$ as follows

$$\cosh^3 \mu + \frac{1}{2} I_1 \cosh^2 \mu + \frac{1}{4} \left(I_2 - 3\right) \cosh \mu + \frac{1}{8} \left(I_3 - 2\,I_1\right) = 0 \qquad (11)$$

To discuss the three roots $F_{1,2,3}$ of equation (11) for $\cosh \mu$, the cases $F_{1,2,3} \in \mathbb{R}$ and $F_{1,2,3} \in \mathbb{C}$ are separately considered. As far as $F_{1,2,3} \in \mathbb{R}$, if $F_{1,2,3} \in \mathcal{I} := [-1, 1]$, then $\mu = i\vartheta$ and $\lambda = e^{i\vartheta}$ is complex with unit modulus; if $F_{1,2,3} \notin \mathcal{I}$, $\mu = \alpha + ij\pi$ with j integer, and $\lambda = \pm e^{\alpha}$ is real. Differently, if $F_{1,2,3} \in \mathbb{C}$, $\mu = \alpha + i\vartheta$ and $\lambda = e^{\alpha + i\vartheta}$ is complex with modulus different from 1. The possible location of the eigenvalues on the complex plane are summarized in Figure 5. The wave propagation characteristics

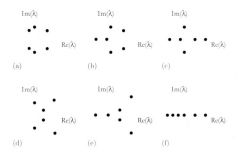

Figure 5. Transfer matrix eigenvalues scenarios: a) Pass-Pass-Pass; b) Pass-Pass-Stop; c) Pass-Stop-Stop; d) Complex-Pass; e) Complex-Stop; f) Stop-Stop-Stop.

of three-coupled periodic structures can be properly described through a geometric representation on the space of the invariants I_1-I_2-I_3 (Figure 6). When $\cosh \mu = \mp 1$, two surfaces are obtained in the I_1-I_2-I_3 space given, respectively, by

$$\begin{aligned} R: \ &= \{(I_1, I_2, I_3) \mid 2 - 2I_1 + 2I_2 - I_3 = 0\} \\ S: \ &= \{(I_1, I_2, I_3) \mid 2 + 2I_1 + 2I_2 + I_3 = 0\} \end{aligned} \qquad (12)$$

Further surfaces, dividing the real roots of equation (11) from the complex ones, are given by

$$P_{1,2} := \{(I_1, I_2, I_3) \mid 27I_1 - 27I_3 - 2I_1^3$$
$$\pm 2\sqrt{(9 + I_1^2 - 3I_2)^3} + 9I_1 I_2 = 0\} \qquad (13)$$

The four surfaces R, S and $P_{1,2}$ divide the space of the invariants in domains (propagation zones) where the eigenvalues do not differ in type. In Figure 7 the section $I_1 = 0$ of the space of the invariants is shown, where the curves r, s, $p_{1,2}$ represent the traces of the matching surfaces on the chosen plane; the propagation zones are labelled according to the notation commonly used in literature. As well known (Signorelli and von Flotow, 1988), such eigenvalues govern the stationary wave transmission properties: if the eigenvalues lie on the unit circle, then free waves propagate harmonically without attenuation (pass band, P); if the eigenvalues are real, then free waves decay without oscillations (stop band, S); furthermore, if the coupling coordinates are more than one, pairs of complex conjugate eigenvalues with modulus different from 1 can exist and harmonic propagation with attenuation of the characteristic waves takes place (complex-band, C). Accordingly, the region where the three pairs of λ lay all on the unit circle is referred to as *pass-pass-pass* (PPP); the regions where two pairs of λ lay on the unit circle while the other pair is real are referred to as *pass-pass-stop* (PPS) ; the regions where one pair lays on the unit circle while the remaining two pairs of λ are real are referred to as *pass-stop-stop* (PSS); the regions where only real pairs of eigenvalues occurs are the *stop-stop-stop* (SSS) domains. Moreover, the curves $p_{1,2}$ bound the regions characterized by a pair of complex conjugate eigenvalues and the remaining pair is either real *complex-stop* (CS) or lays on the unit circle *complex-pass* (CP). It is worth noticing that only the PPP and PPS regions, unlike the remaining domains, are bounded.

1.3   Bi-coupled mechanical models

Up to this point the analysis has been carried out on the invariants' plane where no assumptions were required on the nature of the bi-coupled periodic system. To apply it to a specific problem, two steps must be performed after deriving the transfer matrix $\mathbf{T}$; first, the invariants must be expressed in terms of the physical parameters; then, the propagation regions (and possibly the iso-attenuation curves) must be mapped in a plane of two control parameters suitably selected. The procedure is illustrated here referring to a specific class of bi-coupled periodic structures whose repetitive elements, as sketched in Figure 8, are given by Euler beams of length l,

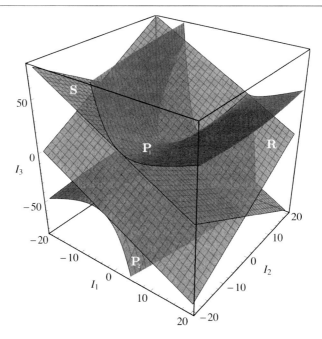

Figure 6. Propagation zones on the invariants' space.

flexural stiffness EI, resting on elastic supports with translational stiffness $k_t/2$, with distributed mass m and a lumped mass M located at x. For such periodic elements, the state vector at the coupling point k is given by $\mathbf{z}_k = (v_k, \varphi_k, V_k, M_k)^T$, where v, φ and V, M represent the generalized displacement and forces components respectively. The following four non-dimensional parameters govern the propagation properties of a cell:

$$\beta = \sqrt[4]{m\omega^2 l^4/(EI)} \quad ; \quad \kappa = k_t l^3/EI \quad ; \quad \xi = x/l \quad ; \quad \delta = M/ml \quad (14)$$

Different sub-models will be considered, namely: no lumped mass (two-parameter β,κ model); mass located at midspan (three-parameter β,κ,δ model); free mass location (four-parameter β,κ,ξ,δ model). The main steps pertaining the derivation of the transfer matrix of the beam element sketched in Figure 8 are here summarized.

Due to the presence of the lumped mass, it is convenient to divide the integration domain of the differential problem $m\,\ddot{v}+EI\,v'''' = 0$ in two intervals, to the left and to the right of the lumped mass located at the abscissa x. By letting $v_i = \phi_i(s)\,e^{i\omega t}$ $(i = 1, 2)$, two space dependent ordinary differential

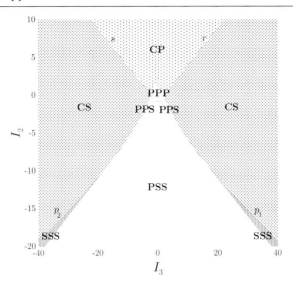

Figure 7. Propagation zones on the invariants' plane $I_3 - I_2$ ($I_1 = 0$).

equations follow, namely $\phi_i'''' - a^4\phi_i = 0$ ($i = 1, 2$), where $a^4 = m\omega^2/EI$. The solutions of such equations are expressed as

$$\phi_i(s) = C_{i1}\sin as + C_{i2}\cos as + C_{i3}\sinh as + C_{i4}\cosh as \quad i = 1, 2 \quad (15)$$

The eight constants C_{ij} are found by by solving an 8×8 algebraic problem obtained by imposing the nodal displacements and rotations at the left (v_L, φ_L) and right (v_R, φ_R) ends of the beam segment and enforcing the compatibility and equilibrium at the mass location x. Afterwards, the end forces are evaluated as functions of the end displacements through:

$$\begin{cases} f_L = EI\phi_1'''(0) \\ m_L = -EI\phi_1''(0) \end{cases} \qquad \begin{cases} f_R = -EI\phi_2'''(l) \\ m_R = EI\phi_2''(l) \end{cases} \quad (16)$$

from which the dynamic stiffness matrix $\mathbf{S}$ is derived. The elastic supports are taken into account by adding the spring translational stiffness to the diagonal terms S_{ii}, $i = 1, 3$. After partitioning the matrix $\mathbf{S}$ in 2×2 submatrices, the transfer matrix $\mathbf{T}$ is eventually obtained by using the well-known relations Zhong and Williams (1992): $\mathbf{T}_{LL} = -\mathbf{S}_{LR}^{-1}\mathbf{S}_{LL}$, $\mathbf{T}_{LR} = \mathbf{S}_{LR}^{-1}$, $\mathbf{T}_{RL} = -\mathbf{S}_{RL} + \mathbf{S}_{RR}\mathbf{S}_{LR}^{-1}\mathbf{S}_{LL}$, $\mathbf{T}_{RR} = -\mathbf{S}_{RR}\mathbf{S}_{LR}^{-1}$.

The first class of beams considered in this section refers to systems with distributed mass only. The model physical parameters are β and κ. From

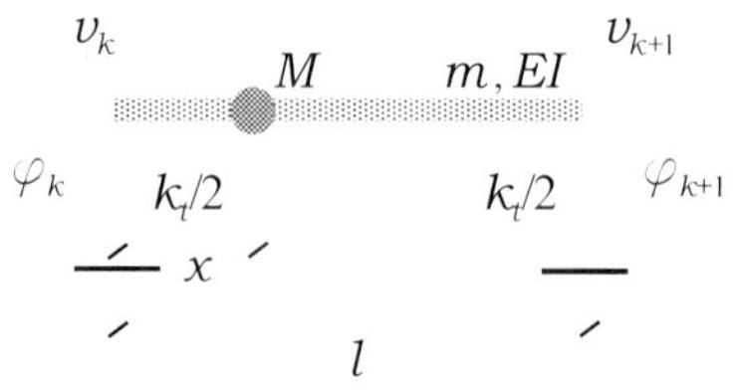

Figure 8. Bi-coupled beam element.

the expression of the transfer matrix (see e.g. Pestel and Leckie (1963)), the following invariants are drawn:

$$
\begin{aligned}
I_1 &= -2\cosh\beta - 2\cos\beta - \tfrac{\kappa}{2\beta^3}(\sin\beta - \sinh\beta) \\
I_2 &= 2 + 4\cos\beta\cosh\beta + \tfrac{\kappa}{\beta^3}(\cosh\beta\sin\beta - \cos\beta\sinh\beta)
\end{aligned}
\tag{17}
$$

Equations (17) represent a nonlinear mapping from points on the invariant plane I_1-I_2 to points on the physical parameters plane β-κ. Thus, the curves r, s and p of Figure 4 become the branches r_i, s_i and p_i of Figure 9 , obtained by substituting (17) in the equations (7,8); in particular, the parabola (8) is mapped into the closed curves p_i. Such branches define the propagation regions of the periodic structure. In Figure 9 closer views of the first three Pass-Pass zones are also presented. The points T, S and R of Figure 4 are mapped into the points T_i, S_i and R_i, the index i representing the i-th Pass-Pass region.

The curves r_i and s_i are given by

$$
r_i, s_i := \left\{ (\beta,\kappa) \mid
\begin{cases}
1 \pm \cos\beta = 0 & \text{for } i \text{ odd} \\
\kappa = \dfrac{\mp 4\beta^3(1\pm\cosh\beta)}{\frac{\sin\beta}{1\pm\cos\beta}(1\pm\cosh\beta)-\sinh\beta} & \text{for } i \text{ even}
\end{cases}
\right\}
$$

while the curves p_i are made up of two branches, p_{iu} (upper) and p_{il} (lower), defined in the intervals $\beta \in [n\pi, (n+1)\pi]$ (n even), having equations

$$
p_{iu,l} := \left\{ (\beta,\kappa) \mid \kappa = \frac{4\beta^3(\cosh\beta - \cos\beta)}{(\sqrt{\sin\beta} \pm \sqrt{\sinh\beta})^2} \right\}
\tag{18}
$$

As shown in Figure 10, the two branches p_{iu} and p_{il} in equation (18) coincide for $\beta = n\pi$ and they tend to coalesce for large values of β; indeed, for $\beta \to \infty$, $p_{iu} \approx p_{il} \approx 4\beta^3/\tanh\beta \approx 4\beta^3$. Such monotonically increasing trend implies that, for each value of the springs translational stiffness, only one complex region can be crossed. Furthermore, it can be inferred that, for

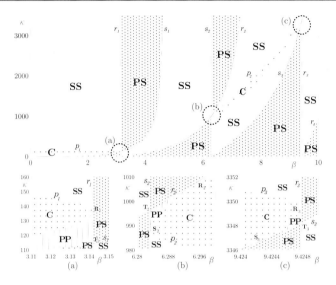

Figure 9. Propagation zones and closer views around the PP zones on the β-κ plane for the two-parameter model.

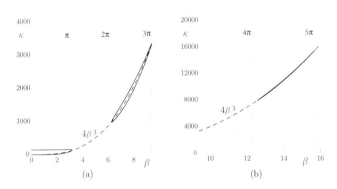

Figure 10. Complex zones on the β-κ plane; (a) $0 \le \beta \le 9.42$, (b) $9.42 \le \beta \le 16$

this class of periodic structures, complex regions disappear as the frequency increases. Such result is consistent with the asymptotic analysis of the preceding section. Indeed, from the expression of I_1 in (17) it follows that as $\beta \to \infty$, $I_1 \to -\infty$ so that at least one pair of real roots λ will always exist, preventing the eigenvalues from complex zones crossings.

The propagation zones shown in Figure 9 provide a complete description of the dependence of the wave propagation characteristics on the spring translational stiffness. The usual representation (Mead, 1975) of the real and imaginary parts of the propagation constants (Figures 11a,b) can be interpreted as a sections of the three-dimensional graphs of the functions $\mathrm{Re}\,\mu = f(\beta, \kappa)$ and $\mathrm{Im}\,\mu = g(\beta, \kappa)$, obtained with a plane parallel to the β axis at the chosen value of κ, whose trace is illustrated in Figure 11c for two values of κ; these traces map in the curves of Figure 11d on the invariants' plane. Thus, the nonlinear mapping between the two planes allows to represent the evolution of the arrangement of the propagation bands.

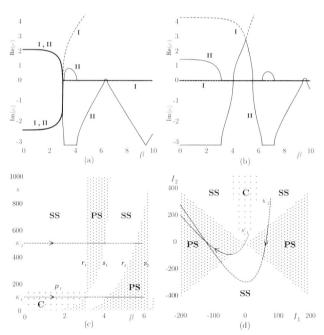

Figure 11. Real and imaginary parts of the propagation constants, (a) $\kappa = 100$, (b) $\kappa = 500$, and corresponding paths $\kappa =$const for increasing β, (c) on the physical plane, (d) on the invariants' plane.

Finally, several mechanical considerations can be associated either to the curves r and s on the invariants' plane or, equivalently, to their representation r_i, s_i on the physical parameters' plane. Mead (1975) presented the

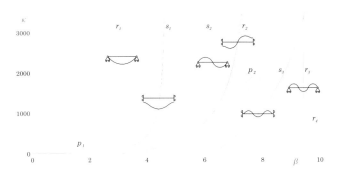

Figure 12. Bounding curves on the β-κ plane and natural frequencies of single elements.

main results pertaining multi-coupled periodic systems; besides highlighting the existence of the complex bands, he identified the natural frequencies of a single symmetric element of a bi-coupled system with the bounding frequencies of the propagation bands on the β-axis. Since this result holds for any κ, it follows that along the line r_i (s_i) are located both the natural frequencies corresponding to the symmetric modes of the single element when its translational (rotational) coordinates are locked and the natural frequencies corresponding to the anti-symmetric modes of the single element when its rotational (translational) coordinates are locked (see Figure 12). Figure 12 explains the role of the parameter κ in the sequence of the natural frequencies of the hinged-hinged and sliding-sliding beams. As expected, the parameter κ does not affect the natural frequencies of the former.

Beams with a mid-span lumped mass are next considered, for which the propagation zones of Figure 13 are found for different mass ratios δ. Comparison with Figure 9, representing the case $\delta = 0$ highlights he effect of the parameter δ. It can be noticed that, due to the mid-span location of the lumped mass, the value of δ does not affect the domains boundaries corresponding to the natural frequencies of the anti-symmetric modes when either rotational or translational coordinates are locked. Furthermore a contraction of the complex zones as δ increases can be noticed. For $\kappa = 0$ and $\delta \neq 0$ the propagation zones are of the SS type for all βs. Differently, for

$\kappa = 0$ and $\delta = 0$ such SS zones disappear; thereby we are left with a system whose period is arbitrary or, in other words, with a system behaving as being uniformly continuous where only PS zones can exist.

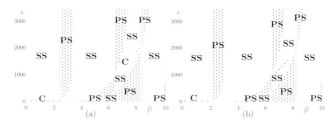

Figure 13. Propagation zones on the β-κ plane for the three-parameter model. (a): $\delta = 1$; (b): $\delta = 2$.

It might be of interest to investigate the representation of the propagation domains provided by a different choice of the control parameters. Figure 14 shows the propagation domains on the β-δ plane for two values of the springs stiffness κ. As it can be readily noticed the propagation properties undergo qualitative variations depending on the considered stiffness value; for instance, the complex region found for $\kappa = 125$ (Figure 14a) disappears for $\kappa = 500$ (Figure 14b). Therefore it can be concluded that the latter choice of physical parameters prevent from a comprehensive qualitative description of the dynamic behavior of such class of bi-coupled periodic elements.

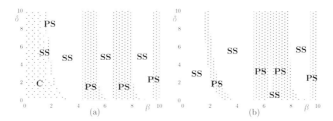

Figure 14. Propagation zones on the β-δ plane for the three-parameter model. (a): $\kappa = 125$; (b): $\kappa = 500$.

1.4   Three-coupled mechanical models

The three-coupled nature of the mechanical models that will be considered in the following sections results from the *sum* of mono-coupled and bi-coupled dynamics. The relevant equations of motion are indeed characterized by the coupling between either a longitudinal or a torsional (mono-coupled) dynamics and a transversal (bi-coupled) one. In order to derive the single element transfer matrix, it is essential to detect the end forces given the end displacements. This implies the solution of a boundary value problem of the form

$$
\begin{aligned}
\mathcal{M}\ddot{\mathbf{u}} + \mathcal{L}\mathbf{u} &= \mathbf{0} &&\text{in}\quad \mathcal{B}\\
\mathcal{K}_H \mathbf{u} &= \mathbf{d}_H &&\text{on}\quad \partial\mathcal{B}_H \quad H = L, R
\end{aligned}
\tag{19}
$$

where equations (19_1) are the field equations defined over the domain $\mathcal{B}$, representing the single element, and equations (19_2) are the kinematic boundary conditions at the left and right element interfaces. The terms $\mathcal{M}$ and $\mathcal{L}$ represent inertial and elastic linear differential operators, respectively, acting over the displacement field $\mathbf{u} = \mathbf{u}(x, t)$, while $\mathcal{K}_H$ is the linear kinematic boundary conditions operator acting on the left $\partial\mathcal{B}_L$ and right $\partial\mathcal{B}_R$ element interfaces. The equation (19_1) can be put in the form

$$
\begin{bmatrix} \mathcal{M}_{uu} & \mathcal{M}_{uv} \\ \mathcal{M}_{vu} & \mathcal{M}_{vv} \end{bmatrix} \begin{pmatrix} \ddot{u} \\ \ddot{v} \end{pmatrix} + \begin{bmatrix} \mathcal{L}_{uu} & \mathcal{L}_{uv} \\ \mathcal{L}_{vu} & \mathcal{L}_{vv} \end{bmatrix} \begin{pmatrix} u \\ v \end{pmatrix} = \begin{pmatrix} 0 \\ 0 \end{pmatrix} \quad \text{in } \mathcal{B} \tag{20}
$$

where the terms u and v represent the mono-coupled and bi-coupled kinematic variables, respectively. The solution, given by

$$
\begin{pmatrix} u \\ v \end{pmatrix} = \begin{pmatrix} a \\ b \end{pmatrix} e^{\beta x} e^{i\Omega t} \tag{21}
$$

gives rise to the eigenvalue problem $\mathbf{L}(\beta, \Omega)\,\mathbf{a} = \mathbf{0}$ in $\mathcal{B}$, where

$$
\mathbf{L} = \begin{bmatrix} L_{uu}(\beta) - \Omega^2 M_{uu} & L_{uv}(\beta) - \Omega^2 M_{uv} \\ L_{vu}(\beta) - \Omega^2 M_{vu} & L_{vv}(\beta) - \Omega^2 M_{vv} \end{bmatrix}, \quad \mathbf{a} = \begin{pmatrix} a \\ b \end{pmatrix} \tag{22}
$$

and the algebraic operators M and L have been introduced. The wave numbers β_i $i = 1, \ldots, 6$ can be determined by solving the characteristic equation:

$$
\beta^6 + J_1 \beta^4 + J_2 \beta^2 + J_3 = 0 \tag{23}
$$

where $J_i = J_i(\Omega)$ $(i = 1, 2, 3)$. Since equation (23) admits six roots, the general solution can be written as

$$
\begin{pmatrix} u \\ v \end{pmatrix} = \sum_{k=1}^{6} c_k \begin{pmatrix} a_k \\ b_k \end{pmatrix} e^{\beta_k x} e^{i\Omega t} \tag{24}
$$

where the coefficients a_k, b_k are grouped as follows:

$$\begin{cases} a_k = L_{uv}(\beta_k) - \Omega^2 M_{uv} \\ b_k = \Omega^2 M_{uu} - L_{uu}(\beta_k) \end{cases} \quad k = 1,2;$$

$$\begin{cases} a_k = \Omega^2 M_{vv} - L_{vv}(\beta_k) \\ b_k = L_{vu}(\beta_k) - \Omega^2 M_{vu} \end{cases} \quad k = 3,4,5,6 \tag{25}$$

so that the eigenvalue problem for uncoupled cases can still be solved. By substituting equations (24) into (19$_2$) and solving for c_k, the linear homogeneous functions $\mathbf{u}$ over $\mathbf{d}_H$ are obtained. By imposing the equilibrium at the interfaces through the dynamic boundary conditions:

$$\mathcal{D}_H \mathbf{u} = \mathbf{f}_H \quad \text{on } \partial \mathcal{B}_H \quad H = L, R \tag{26}$$

the end forces can be determined as

$$\begin{bmatrix} \mathbf{Z}_{LL} & \mathbf{Z}_{LR} \\ \mathbf{Z}_{RL} & \mathbf{Z}_{RR} \end{bmatrix} \begin{pmatrix} \mathbf{d}_L \\ \mathbf{d}_R \end{pmatrix} = \begin{pmatrix} \mathbf{f}_L \\ \mathbf{f}_R \end{pmatrix} \tag{27}$$

In equation (27) the matrix $\mathbf{Z} = [\mathbf{Z}_{HK}]$, with $H, K = L, R$, represents the frequency dependent dynamic stiffness matrix, where L and R refers to left and right side of the modular element, respectively. The transfer matrix $\mathbf{T}$, which relates the right end displacements and forces to the left ones, is eventually obtained as

$$\begin{pmatrix} \mathbf{d}_R \\ \mathbf{f}_R \end{pmatrix} = \begin{bmatrix} \mathbf{T}_{dd} & \mathbf{T}_{df} \\ -\mathbf{T}_{fd} & -\mathbf{T}_{ff} \end{bmatrix} \begin{pmatrix} \mathbf{d}_L \\ \mathbf{f}_L \end{pmatrix} \tag{28}$$

by using the well-known relations Pestel and Leckie (1963):

$$\mathbf{T}_{dd} = -\mathbf{Z}_{LR}^{-1}\mathbf{Z}_{LL}, \ \mathbf{T}_{df} = \mathbf{Z}_{LR}^{-1},$$

$$\mathbf{T}_{fd} = -\mathbf{Z}_{RL} + \mathbf{Z}_{RR}\mathbf{Z}_{LR}^{-1}\mathbf{Z}_{LL}, \ \mathbf{T}_{ff} = -\mathbf{Z}_{RR}\mathbf{Z}_{LR}^{-1} \tag{29}$$

Once the transfer matrix is obtained, its invariants can be expressed in terms of physical parameters; then, the traces of the surfaces given by equations (12) and (13) can be mapped into a plane of two control parameters duly selected. Among them, the frequency is the most significant one and is always selected so that the curves r, s, and $p_{1,2}$ provide the bounding frequencies of the propagation regions.

As singled out by Mead (1975), if the interest lies in the mechanical interpretation of the bounding frequency of multi-coupled periodic structures, it is then crucial to distinguish between two types of coupling coordinates. When symmetric structural elements vibrate in a symmetric mode, some of the matching pairs of coupling coordinates at either end have the same sign and magnitude (*type (i) coordinates*) whereas the remaining pairs have opposite sign and equal magnitude (*type (ii) coordinates*). The opposite occurs for anti-symmetric modes. For periodic structures of symmetric multi-coupled elements, the bounding frequencies of the wave propagation zones are identical to the natural frequencies of a single element with the two types of coupling coordinate either locked or free. In the following sections type (i) and type (ii) coordinates of each three-coupled mechanical model discussed will be identified and the relationship between bounding and natural frequencies will be shown to obey to Mead's results.

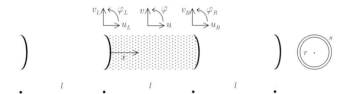

Figure 15. Periodic cylindrical shells.

Periodic axisymmetric cylindrical shells   The general derivation of the propagation regions for three-coupled structures is specialized in this section to deal with periodic pipes sketched in Figure 15. The pipes are modelled as thin cylindrical shells undergoing axisymmetric (*breathing*) modes of vibration and the periodicity is provided by evenly spaced stiffeners. The displacement field is defined by the longitudinal and transverse components $u(x)$ and $v(x)$, respectively, and by a rotation $\varphi(x) = v'(x)$, x being the abscissa. The parameters that are involved in the model are the following: Young modulus E, Poisson coefficient ν, shell thickness s, mass density ρ, radius r and element length l. Introducing the nondimensional quantities

$$\hat{x} = \frac{x}{r}, \quad \hat{u} = \frac{u}{r}, \quad \hat{v} = \frac{v}{r}, \quad \sigma = \frac{s}{r}, \quad \ell = \frac{l}{r}, \quad \tau = \omega t \qquad (30)$$

where $\omega^2 = E/\left[\rho r^2\left(1 - \nu^2\right)\right]$, the field equations can be written, omitting the hat, as

$$\ddot{u} - u'' - \nu v' = 0, \quad \ddot{v} + \frac{\sigma^2}{12}v'''' + v + \nu u' = 0 \qquad (31)$$

In view of the propagation properties analysis that follows on, it is worth stressing that, as shown by the equations (31), the coupling between longitudinal and transversal dynamics is governed by the Poisson ratio ν. Moreover, v represents the type (i) coordinate while u and φ are type (ii) coordinates. The kinematic boundary conditions are:

$$u(0) = u_L, \quad u(\ell) = u_R, \quad v(0) = v_L,$$
$$v(\ell) = v_R, \quad v'(0) = \varphi_L, \quad v'(\ell) = \varphi_R$$

(32)

By defining the equivalent axial and transversal stiffnesses provided by the circumferential stiffeners at the ends of the single element as $k_u/2$ and $k_v/2$, respectively, and setting

$$n_{L,R} = \frac{1-\nu^2}{Es} N_{L,R}, \quad q_{L,R} = \frac{12\left(1-\nu^2\right)r^2}{Es^3} Q_{L,R},$$
$$m_{L,R} = \frac{12\left(1-\nu^2\right)r}{Es^3} M_{L,R}$$
$$\kappa_u = \frac{1-\nu^2}{\sigma} \frac{k_u}{2E}, \quad \kappa_v = \frac{6(1-\nu^2)}{\sigma^3} \frac{k_v}{E}$$

(33)

the dynamic boundary conditions are given by:

$$[u' + \nu v \mp \kappa_u u]_{0,\ell} = \mp n_{L,R}, \quad [-v''' \mp \kappa_v v]_{0,\ell} = \mp q_{L,R},$$
$$v''_{0,\ell} = \mp m_{L,R}$$

(34)

Therefore, in order to derive the element transfer matrix according to the procedure set forth in section 2.5, the matrix $\mathbf{L}$ in equation (22) becomes

$$\mathbf{L} = \begin{bmatrix} -\beta^2 - \Omega^2 & -\nu\beta \\ \nu\beta & 1 + \frac{\sigma^2}{12}\beta^4 - \Omega^2 \end{bmatrix}$$

(35)

Starting from the eigenvalue problem governed by the matrix (35), the element transfer matrix is obtained and the invariants are expressed in terms of the physical parameters.

The propagation regions are mapped into planes of two control parameters where the curves r, s, $p_{1,2}$ in Figure 7 give rise to a number of branches. The selected parameter relevant to the results shown in Figures 16 and 17 are $\ell = 2.0$, $\sigma = 0.1$ and $\kappa_u = 50.0$.

In the limit case $\nu = 0$, the longitudinal and transversal dynamics are uncoupled so that the resulting propagation scenario can be interpreted as the superposition of the mono-coupled (bar) and the bi-coupled (beam on elastic foundations) scenarios. On the one hand, in the mono-coupled problem, the propagation regions can be either pass (P) or stop (S) and the curves bounding the pass regions, where the eigenvalues lie on the unit circle, can be readily determined by the condition $|tr(\mathbf{T})| = 2$. This condition applied to the transfer matrix associated to equation (31) with $\nu = 0$ and $\kappa_u = 0.0$, leads to $2\cos\Omega\ell = 2$, so that, for the bar, the longitudinal waves are always in a pass band, regardless of the value of Ω. On the other, the bi-coupled problem of the beam on elastic foundation resting on elastic translational supports behaves qualitatively like the Euler beam on elastic supports studied in Romeo and Luongo (2002). Thus, the overall propagation scenario is characterized by the superposition of the beam propagation regions, starting from the cut-on frequency ($\Omega > 1$) (Sorokin and Ershova, 2004), and the everywhere pass region of the bar. In Figure 16 the propagation zones

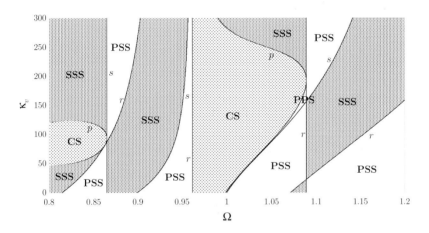

Figure 16. Propagation zones on the $\Omega - \kappa_v$ plane for $\nu = 0.5$.

are shown on the $\Omega - \kappa_v$ plane for $\nu = 0.5$ and $\kappa_u = 50.0$; the propagation regions representation obeys the three-coupled nature of the problem, as proven by the appearance of SSS and CS regions. Figure 17 shows the propagation regions in the $\Omega - \nu$ plane, for $\kappa_u = \kappa_v = 50.0$. Thus, the evolution of the propagation scenario as the coupling parameter ν varies in the range $0.0 \leq \nu \leq 0.5$, for given κ_u and κ_v, can be analyzed. As expected, when transversal and longitudinal dynamics are uncoupled ($\nu = 0$), the flexural natural frequencies of the hinged-hinged single beam, given by $\Omega_n = \left[1 + (n\pi/\ell)^4\sigma^2/12\right]^{1/2}$, correspond to bounding frequencies provided by either the branches r or s, namely $\Omega_1 = 1.0025$ (r), $\Omega_2 = 1.0398$ (s), $\Omega_3 = 1.1878$ (r). Moreover, the remaining branches of the same curves correspond to the sliding-sliding single beam natural frequencies and they tend asymptotically to the fixed-fixed case when $\kappa_v \to \infty$. The natural frequencies of either the free-free or fixed-fixed single bar are not displayed since they are located at higher values of Ω (i.e. $\Omega_n = n\pi/\ell$). As the coupling rises ($\nu \neq 0$) the three-coupled dynamics forces the bounding frequencies and the associated natural frequencies to bend. For example, the first three natural frequencies of the hinged-hinged single element decrease implying a "softening" effect of the coupling parameter. In particular, the first natural frequency decreases to the value $\Omega_1 = 0.8994$, the second to $\Omega_2 = 0.8143$ while the third to $\Omega_3 = 1.1116$; the crossing between the first and second natural frequency occurs at $\nu = 0.25$. The crossed curves in Figure 17 represent the first six natural frequencies of the two element periodic pipe when v (type (i) coordinate) is locked while u and φ (type (ii) coordinates) are free. Their paths on the $\Omega - \nu$ plane are always within either a PSS or a PPS region. As the periodic structure consists of two elements, two natural frequencies are found in each PSS region. Moreover, as expected, for each pair, one natural frequency overlaps with a bounding frequency provided by either r or s branches; for low values of ν ($\nu < 0.25$) such overlapping occurs according to the alternating sequence: r, s, r.

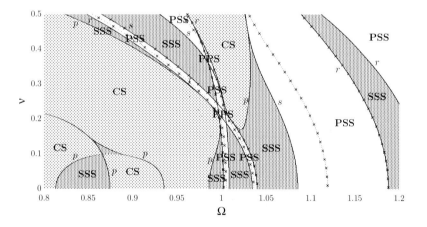

Figure 17. Propagation zones on the $\Omega - \nu$ plane for $\kappa_u = \kappa_v = 50.0$.

Periodic symmetric thin-walled beams   Thin-walled beams with symmetric cross-section undergoing flexural-torsional oscillations are analyzed in this section (see Figure 18a). The flexural-centre axis is taken as the x axis and the distance between the flexural centre F and the centroid C is

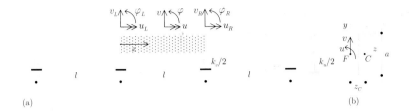

Figure 18. Periodic thin-walled beam.

equal to z_C, as sketched in Figure 18b. The displacement field is defined by the torsional rotation $u(x)$, transverse component $v(x)$, and by the flexural rotation $\varphi(x) = v'(x)$. The parameters entering the models are: mass density ρ, cross section area A, polar moment of inertia I_G ($I_F = I_G + A z_C^2$), Young modulus E, moment of inertia $I = I_z$, tangential elastic modulus G and torsional inertia J. Introducing the nondimensional quantities

$$\hat{x} = \frac{x}{a}, \quad \hat{v} = \frac{v}{a}, \quad \alpha = \frac{Aa^2}{I_F}, \quad \gamma = \frac{\alpha GJ}{EI}, \quad \ell = \frac{l}{a}, \quad \sigma = \frac{z_C}{a}, \quad \tau = \omega t \tag{36}$$

where $\omega^2 = EI/\left(\rho A a^4\right)$, the field equations read, omitting the hat, as

$$\ddot{u} - \alpha\,\sigma\,\ddot{v} - \gamma\,u'' = 0, \quad \ddot{v} - \sigma\,\ddot{u} + v'''' = 0 \tag{37}$$

where the warping torsional stiffness was neglected. In this case, it is worth pinpointing that, as shown by the equations (37), the coupling between torsional and transversal dynamics is governed by inertia terms through the parameter σ. For this model φ represents the type (i) coordinate while u and v are type (ii) coordinates.

The kinematic boundary conditions are still given by the expressions (32) and by setting

$$q_{L,R} = \frac{Q_{L,R}\,a^2}{EI}, \quad t_{L,R} = \frac{T_{L,R}\,a}{GJ},$$

$$m_{L,R} = \frac{M_{L,R}\,a}{EI}, \quad \kappa_u = \frac{k_u a}{GJ}, \quad \kappa_v = \frac{k_v a^3}{EI} \tag{38}$$

where k_u and k_v represent the rotational and transversal spring stiffnesses, the dynamic boundary conditions are:

$$[u' \mp \kappa_u u]_{0,\ell} = \mp t_{L,R}, \quad [-v''' \mp \kappa_v v]_{0,\ell} = \mp q_{L,R}, \quad v''_{0,\ell} = \mp m_{L,R} \tag{39}$$

Therefore, in order to derive the element transfer matrix according to the procedure described in section 2, the matrix $\mathbf{L}$ in equation (22) becomes

$$\mathbf{L} = \begin{bmatrix} -\gamma\,\beta^2 - \Omega^2 & \sigma\,\alpha\,\Omega^2 \\ \sigma\,\Omega^2 & \beta^4 - \Omega^2 \end{bmatrix} \tag{40}$$

Starting from the eigenvalue problem governed by the matrix (40), the element transfer matrix is obtained and the invariants are expressed in terms of physical parameters. The selected parameter relevant to the results shown in Figures 19 and 20 are $\ell = 5.0$, $\alpha = 1.0$ and $\gamma = 2.0$. In the $\sigma = 0.0$ borderline case the torsional and transversal dynamics are uncoupled. Here, we can make the same remarks as in the case of pipes mentioned above. In Figure 19 the propagation zones are shown on the $\Omega - \kappa_v$ plane for $\sigma = 2.0$ and $\kappa_u = 6.0$. Figure 20 shows the propagation regions in the $\Omega - \sigma$ plane, for

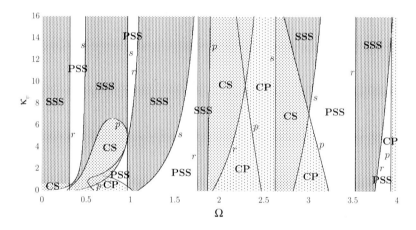

Figure 19. Propagation zones on the $\Omega - \kappa_v$ plane for $\sigma = 2.0$.

$\kappa_u = 6.0$ and $\kappa_v = 2.0$. Thus, the evolution of the propagation scenario as the coupling parameter σ varies in the range between $0.0 \leq \sigma \leq 2.0$, can be analyzed. For $\sigma = 0.0$ the flexural natural frequencies of the hinged-hinged single beam, $\Omega_n = (n\pi/\ell)^2$, correspond to the bounding frequencies given by the branches of curves r and s, such as $\Omega_1 = 0.395$ (r), $\Omega_2 = 1.579$ (s). The natural frequencies of the free-free (fixed-fixed) single bar ($\Omega_n = \sqrt{\gamma}\,n\pi/\ell$) also correspond to branches of curves r and s; the first two of them are located at $\Omega_1 = 0.888$ (r) and $\Omega_2 = 1.777$ (s). Moreover, the remaining branches of the same curves correspond to the sliding-sliding single beam natural frequencies. As noticed in the previous model, as the coupling parameter σ increases, the above mentioned uncoupled natural frequencies of the single element move along the r and s branches giving rise to involved paths. The lower natural frequencies of the periodic thin-walled beam composed by two elements when φ (type (i) coordinate) is locked while u and v (type (ii) coordinates) are free are represented in Figure 20 by the crossed curves. As expected such natural frequencies are forbidden within either SSS or CS regions. Since the periodic system is composed by two elements, two natural frequencies are found in each PSS region. Moreover, for each pair, one natural frequency overlaps with a bounding frequency provided by either r or s branches. In this case, crossing and veering phenomena in both bounding and natural frequencies can be observed around $\Omega = 1.2$ and $\Omega = 1.8$.

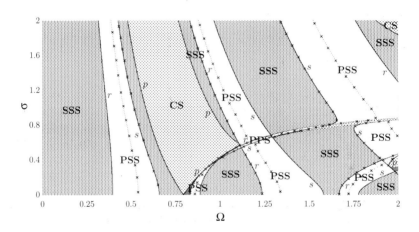

Figure 20. Propagation zones on the $\Omega - \sigma$ plane for $\kappa_u = 6.0$ and $\kappa_v = 2.0$.

1.5   Vibration reduction in piecewise periodic structures

Several authors have investigated the possibility of reducing transmitted vibrations in periodic structures. In Langley et al. (1997), the optimal design of beams on multiple supports to minimize vibration transmission and stress levels has been pursued. An optimization procedure has been employed to select suitable slight deviations from periodicity considering as design parameters the individual bay lengths and damping values. On the other hand, in Koo and Park (1998); Richards and Pines (2001), vibrations have been reduced by introducing intentional spatial periodicity in the otherwise not periodic structure and resorting to their propagation properties. In the first work, the authors propose to design periodically supported piping system. In the latter, aiming at reducing gear mesh vibration, a periodic shaft has been used to design stop bands in the frequency spectra that correspond to particular harmonics of the gear mesh contact dynamics; both numerical and experimental results have shown significant vibration reduction. Recently, in Baz (2001), pass and stop bands of mono-coupled periodic systems made up of masses connected by mechanical springs have been shown to be controllable by adding active piezoelectric springs; moreover, vibration localization have been obtained by randomly disordering the control gains of each active spring.

All of the above analyses have merely relied upon *ad hoc* numerical investigations on the propagation properties of the single periodic units in order to optimize the design. Moreover, the influence of the physical parameters on the amount of vibration reduction has not been fully investigated. As shown in section 1.4, bi-coupled periodic structures can be analysed on the basis of the transfer matrix characteristic equation and analytical maps of the single unit free-wave propagation domains (stop, pass and complex domains). From such maps, the role played by the physical (control) parameters in the wave propagation properties can be readily gathered and thereby exploited. Based on these results, aiming at reducing the transmitted vibrations, a design of optimal piecewise periodic structures is proposed in this work. In particular, the periodic cells properties and arrangement are tailored to localize the response around the excitation source within any assigned frequency range. The amount of vibration suppression along the periodic structure is also controlled as it can be described through iso-attenuation curves representing the contour plot of the real part of the propagation constants.

Analytical model and design strategy   Bi-coupled periodic structures whose repetitive elements, as sketched in Figure 1, are given by Euler beams of length l, flexural stiffness EI, resting on elastic supports with translational stiffness $k_t/2$, with distributed mass m, are considered. As described in section 1.3, for such periodic elements the state vector at the coupling point k is given by $\mathbf{z}_k = (v_k, \varphi_k, V_k, M_k)^T$, where v, φ and V, M represent the generalized displacement and forces components respectively. The selected control parameters governing the propagation properties of a cell are the nondimensional frequency β and spring stiffness κ defined in section 1.3. The dependence of the free wave propagation characteristics on the spring translational stiffness and frequency is described in Figure 22a where the propagation regions map is shown. In the pass regions waves propagate harmonically without attenuation, whereas in the stop regions waves decay; harmonic propagation with attenuation occurs in the complex regions. The branches r_i, s_i and p_i bounding the different type of regions have been

Figure 21. Uniform Euler beam on evenly spaced springs.

analytically derived in Section 2.4.

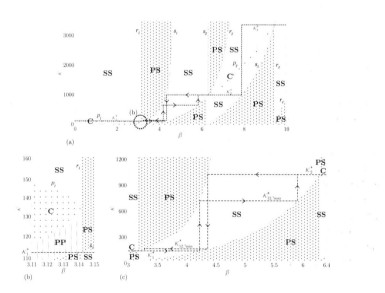

Figure 22. Propagation zones on the β-κ plane; (a) - - - optimal sequences of κ for increasing and decreasing frequencies, (b) closer view around the first PP, (c) closer view of the bridging values κ_{12}^*.

Aiming at reducing the transmitted vibrations induced by an excitation with known frequency bandwidth, the previous findings can be used to delineate a design strategy. This consists in realizing a piecewise periodic structure made up of the minimum number of dissimilar sections such that the union of the stop and complex propagation bands of each element covers the whole frequency range of interest, namely the excitation bandwidth. Such sequence of periodic sections will be referred to as *optimal* piecewise periodic structure. Starting from $\beta = 0$, a sequence of optimal values κ^* is determined as illustrated by the step-wise dashed line in Figure 2a. This is obtained by selecting the values κ_i^* relevant to the intersections between curves r_i and s_i (Figure 2b) as they assure the widest frequency ranges with the narrowest pass bands (or widest stop bands), as Figure 2a shows. To completely match the design frequency range with either SS or C bands, further optimal κ^* must be considered, such as any value $\kappa^*_{12,min} \leq \kappa^*_{12} \leq \kappa^*_{12,max}$, needed to bridge κ_1^* and κ_2^*. Once the optimal sequence κ^* has been determined in the frequency range of interest, the number of elements must be selected. This number should be sufficiently large to achieve the desired vibration reduction, according to the decaying properties of the component elements characterized by the modulus $\rho := |\lambda_i|$ of the transfer matrix eigenvalues. Therefore, the amount of vibration suppression along each element can be quantified through iso-attenuation curves representing the contour plot of the real part of the propagation constants, as shown in Figures 3 and 4. Two families of curves cross the SS domains (two real eigenvalues) while only one family crosses the PS zones (one real eigenvalue).

Iso-attenuation curves on the invariants' plane   As previously mentioned, the propagating or attenuating nature of the characteristic waves traveling through periodic systems is characterized by the modulus $\rho := |\lambda_i|$ of the transfer matrix eigenvalues. Except for the PP zone, where both the waves propagate, attenuation of one or both the characteristic waves occurs on the whole invariant plane. In this section, the attention is focused on describing the loci of the eigenvalues having the same modulus on the invariants' plane (*iso-attenuation curves*). From a geometrical point of view, the ensemble of these curves can be interpreted as the contour plot of the surface $\rho = \rho(I_1, I_2)$, shown in Figure 23a.

By setting $\lambda = \rho e^{i\vartheta}$ ($0 < \rho \leq 1$, $0 \leq \vartheta \leq \pi$) and separating the real and imaginary parts in equation (2), the following two equations are obtained

$$(\rho^2 \cos 3\vartheta + \cos \vartheta) I_1 + \rho \cos 2\vartheta \, I_2 = -\frac{1}{\rho} - \rho^3 \cos 4\vartheta \qquad (41)$$

$$(\rho^2 \sin 3\vartheta + \sin \vartheta) I_1 + \rho \sin 2\vartheta \, I_2 = -\rho^3 \sin 4\vartheta \qquad (42)$$

By fixing ρ and letting ϑ vary, equations (41,42) yield the parametric equations $I_1 = I_1(\vartheta)$, $I_2 = I_2(\vartheta)$ of the iso-ρ curves sought; they are depicted in Figure 23b. Inside the complex region C each point of the curves $\rho =$const is associated with a distinct value of $\vartheta \neq (0,\pi)$. When $\vartheta = (0,\pi)$ the equations (41,42) admit infinite solutions represented by the two families of straight lines with opposite slopes belonging to the PS and SS zones shown in Figure 4b. In particular, both the sets of straight lines cross the SS domains (two real eigenvalues) while only one set of lines (one real eigenvalue) crosses the PS zones. It can also be shown that the iso-attenuation curves encircle the PP domain as illustrated by the thick lines in Figure 23b. The set of equations (41,42) admits a unique solution for $0 < \vartheta < \pi$:

$$I_1(\rho, \vartheta) = -2\,(\rho + 1/\rho)\cos\vartheta \tag{43}$$
$$I_2(\rho, \vartheta) = 2\cos 2\vartheta + (\rho + 1/\rho)^2 \tag{44}$$

The curves (43,44) lie inside the complex region C (see Figure 23) and a distinct value of ϑ is associated with each point of such curves $\rho =$const. Differently, for $\vartheta = (0,\pi)$, the set of equations (41,42) provides infinite solutions consisting of the pairs (I_1, I_2) such that

$$\pm\rho(\rho^2 + 1)\,I_1 + \rho^2\,I_2 + \rho^4 + 1 = 0 \tag{45}$$

The curves (45) describe two families of straight lines, tangent to the parabola p, with negative and positive slope, respectively. The lines (45) belong to the PS and SS zones; in particular, both the families exist within the SS regions (two real eigenvalues) while only one family can exist within the PS region (one real eigenvalue). The same value $(0,\pi)$ of ϑ is associated with each point of such lines.

In order to show that the iso-ρ curves encircle the PP zone, it suffices to verify that, when $\vartheta \to (0,\pi)$ the parametric curves (43,44) meet the straight lines (45) for the same ρ value. Indeed, from (43,44),

$$\lim_{\vartheta\to(0,\pi)} I_1 = \mp 2(1/\rho + \rho) \qquad \lim_{\vartheta\to(0,\pi)} I_2 = 4 + 1/\rho^2 + \rho^2 \tag{46}$$

which satisfy equation (45). Since these values of I_1 and I_2 also satisfy the equation (8), the intersection points belong to the parabola p. Furthermore, the intersections of the iso-ρ belonging to the C region and those of the PS and SS regions occur along the curve p. For $\rho = 1$, the iso-attenuation curve is constituted by the curves r, s and the part of curve p between the points S and R indicated in Figure 4. In the vicinity of the PP zone (where $\mathcal{O}(I_1) = \mathcal{O}(I_2) = 1$) the strong gradient of ρ implies fast attenuation of the characteristic waves. Differently, for higher values of I_1, I_2, a small gradient of ρ is noticed.

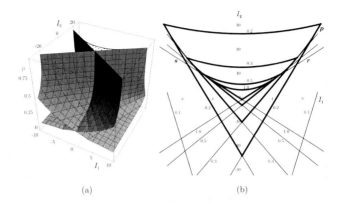

(a)   (b)

Figure 23. Modulus ρ of the eigenvalues of **T**: (a) surface $\rho = \rho(I_1, I_2)$, (b) iso-attenuation curves on the invariants' plane I_1-I_2.

The iso-attenuation curves crossing the PS and SS zones shown in Figure 23b can be transformed into the physical space (see Figure 24) and are given by

$$\kappa = \frac{2\beta^3}{\pm \dfrac{\rho \sin \beta}{1+\rho^2 \mp \cos \beta} \mp \dfrac{\rho \sinh \beta}{1+\rho^2 \mp \cosh \beta}} \tag{47}$$

In Figure 25 the iso-attenuation curves on the β-κ plane are shown around

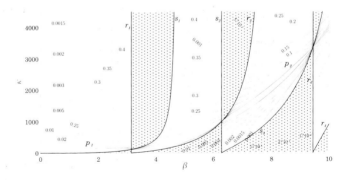

Figure 24. Iso-attenuation curves on the β-κ plane.

the second PP zone (see Figure 23b). They have been obtained by using equations (43,44) in the solution of equations (41,42) and solving for β and

κ. The original symmetry of Figure 23b is destroyed, whereas the strong gradient of the attenuation around the PP zone persists.   The outlined

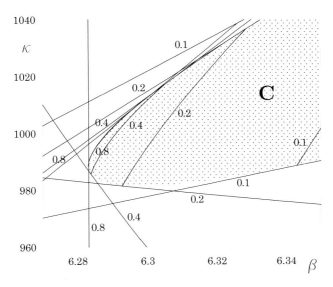

Figure 25. Iso-attenuation curves on the β-κ plane.

strategy prevents from vibration transmission at all frequencies within the design frequency range, except for the natural frequencies, where resonance takes place. The natural frequencies of the overall piecewise periodic structure lay within the union of either pass-pass and/or pass-stop bands of each section. Therefore piecewise periodic structures are characterised by lower modal density than uniformely periodic ones with the same number of elements, this entailing a further beneficial effect on the response. In the ideal undamped structure the reduction of vibration propagation along the structure implies narrowing of the resonances while the amplification remains ideally infinite. On the other hand, in real structures, a however small damping limits the resonance amplifications; therefore, not only the response at generic frequencies but also the resonance peaks are lowered.

Numerical results   Uniform undamped Euler beams supported by evenly spaced springs (Figure 1) have been considered in the numerical investigations. The ends are constrained by sliding supports and a harmonic excitation is applied at the left-end of the periodic beam. According to the design

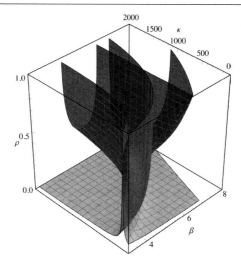

Figure 26. Modulus ρ of the eigenvalues of **T**: surface $\rho = \rho(\beta, \kappa)$.

strategy outlined in Section 2, piecewise periodicity is introduced to reduce transmitted vibrations in a given frequency range. Willing to reduce the transmitted vibrations induced by excitations with frequency starting from $\beta = 0$, the first optimal value is κ_1^* (Figure 2). Adopting this value of the spring translational stiffness a pass-pass region ($3.092 \leq \beta \leq \pi$) is found between a complex ($0 \leq \beta < 3.092$) and a stop-stop ($\pi \leq \beta \leq 4.220$) region; then, a pass-stop band ending at $\beta = 2\pi$ exists. Therefore by composing the whole structure with this type of element, attenuating vibrations up to $\beta = 3.092$ are assured. In order to extend the design frequency range beyond $\beta = 3.092$, the optimal value $\kappa_{12,max}^*$ must be taken into account, thus assuring vibration reduction up to $\beta = 5.908$. Further extension of the design frequency interval for vibration reduction is achieved by using three types of elements of stiffness, from left to right, κ_1^*, $\kappa_{12,max}^*$, κ_2^*, thus covering the frequency range $0 \leq \beta \leq 7.886$. If the design goal is to reduce transmitted vibrations starting from $\beta = 7.886$ to $\beta = 0$, then the optimal sequence from left to right would be κ_2^*, $\kappa_{12,min}^*$, κ_1^*. In principle any value $\kappa_{12,min}^* \leq \kappa_{12}^* \leq \kappa_{12,max}^*$ guarantees the overlay of the frequency range with either stop or complex bands. Thus both the problems of the choice of κ_{12}^* and of the spatial arrangement of the sections arise. Concerning this, it is worth noticing that the frequency response belonging to the stop band of the section closest to the excitation source exhibit the fastest spatial decay rate. Therefore, aiming at minimizing the spatial extension of the response

Table 1. Optimal sequence of κ ($0 \leq \beta \leq 11.308$) and associated controlled frequency range bounds;(a) increasing frequencies, (b) decreasing frequencies.

	κ^*	β
κ_1^*	113.750	3.092
$\kappa_{12,max}^*$	696.500	5.908
κ_2^*	995.914	7.886
κ_3^*	3348.137	11.308

(a)

	κ^*	β
κ_3^*	3348.137	7.272
κ_2^*	995.914	4.355
$\kappa_{12,min}^*$	136.823	3.265
κ_1^*	113.750	0

(b)

in the low frequency range, the sections must be ordered with increasing stiffnesses, selecting $\kappa_{12}^* = \kappa_{12,max}^*$. Otherwise, to minimize the extension of the high frequency response, the sections must be ordered with decreasing stiffnesses, selecting $\kappa_{12}^* = \kappa_{12,min}^*$. Indeed, such choices of κ_{12}^* extend at most upwards and downwards, respectively, the frequency range inhibited by the intermediate section.

The numerical values of the optimal sequence of $\kappa^\star$ for $0 \leq \beta \leq 11.308$ are reported in Table 2. The frequency value in each row represents the bound of the range controlled by using elements with $\kappa^\star$ up to the actual one.

For a six-span beam made up by equal elements κ_i^*, the displacement magnification factors D_A and D_B at the ends of the beam are shown in Figure 6a and b, respectively. As expected, by considering the range $0 \leq \beta \leq 6$, the vibration transmission from node A to node B takes place within the pass-pass and pass-stop bands where two groups of six natural frequencies are also found. The amount of attenuation varies as the frequency grows. Specifically, by focusing on the intervals between the spikes of the natural frequencies in the PS band in Figure 6b, the magnification factor D_B shows a decaying trend. Such behavior can also be inferred by the arrangement of the iso-attenuation curves illustrated in Figure 4. Figures 6c,d show the magnification factors D_A and D_B obtained for a six-span beam made up by two sections of three elements of stiffness κ_1^* (left) and $\kappa_{12,max}^*$ (right), respectively. The remarkable reduction of the transmitted vibrations obtained with this configuration is shown in Figure 6d. The noteworthy result is that, except for the natural frequencies of the ideal undamped system, the vibration reduction concerns all the frequencies of interest. Such result is evidenced by the narrow spikes around the natural frequencies.

Next, in Figure 7a,b,c, the displacement dynamic magnification fac-

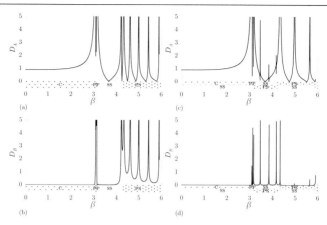

Figure 27. Displacement dynamic magnification factors. (a)-(b) $\kappa_i = \kappa_1^\star$, (c)-(d) $\kappa_{1,2,3} = \kappa_1^\star$, $\kappa_{4,5,6} = \kappa_{12,max}^*$; (a)-(c) node A, (b)-(d) node B.

tor D along the six-bays made up by three sections of two elements of stiffness κ_1^*, $\kappa_{12,max}^*$ and κ_2^* is shown for three adjacent frequency ranges ($3.00 \leq \beta \leq 4.35$, $4.35 \leq \beta \leq 6.30$, $6.30 \leq \beta \leq 8.0$, respectively). For such undamped structure, althought the peaks of the resonance amplificatons are ideally infinite, the dynamic amplification factor near resonances lowers and therefore resonance peaks narrow as one proceeds along the structure. In order to highlight the influence of the total number of elements, the dynamic amplification factor for an 18-bays beam is considered (Figure 7d,e,f). Since the dimensionless parameter κ must be kept constant, the dimensional length l and stiffness k_t must be decreased and increased, respectively, if compared with the 6-bays beam. In one hand these figures show a faster vibration attenuation than the 6-bays case; the attenuation occurring for all frequencies but the the natural ones. On the other hand, by increasing the number of elements the number of peaks increases as well. The plots refer to an excitation frequency sampling $\Delta\beta = 0.01$, and the decaying trends of the resonance sharp peaks show that the resonances become narrower than

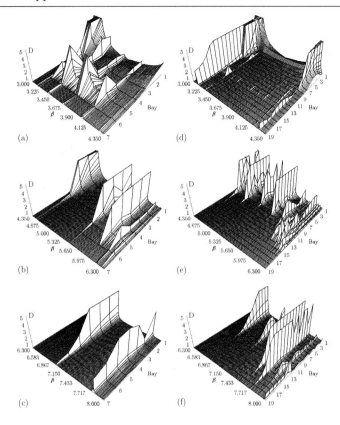

Figure 28. Displacement dynamic magnification factors of optimal 6-span (a-b-c) and 18-span (d-e-f) piecewise periodic beams in the range $0 < \beta < 7.886$. (a),(d) $3.0 < \beta < 4.35$; (b),(e) $4.35 < \beta < 6.30$; (c),(f) $6.30 < \beta < 8.0$.

the adopted sampling rate.

Further plots showing the effectiveness of the proposed design strategy are also reported in Figures 8 and 9. The propagation bands of each optimal element considered are shown in Figure 8a,c,e in terms of the propagation constant $\mu = \log \lambda$. In Figure 8b,d,f, sections of the dynamic magnification

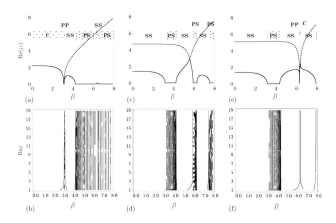

Figure 29. (a,c,e) Propagation bands of the first three optimal elements. (a) κ_1^*, (c) $\kappa_{12,max}^*$, (e) κ_2^*. (b,d,f) Corresponding sections of the dynamic amplification factor at $D = 1.5$ for 18 bays periodic beams.

factors at the value $D = 1.5$ for 18-bays uniformly periodic beams composed by equal elements κ_1^*, $\kappa_{12,max}^*$ and κ_2^*, respectively, are shown. It is rather evident that vibration transmission proceeds unaltered along the structure within the pass bands. On the other hand, as shown in Figure 9a, by arranging in series the elements (with propagation characteristics depicted in Figure 9b) only disturbances coinciding with the natural frequencies can propagate. From the modal analysis standpoint the vibration transmission inhibition can be explained through the modal shape localization. However, only the first few modes have been found to be localized (see e.g. Figure 10), therefore the high frequency inhibition could solely be explained through the cumulative effect of different closely spaced modes; that is to say, modal analysis does not highlight the phenomenon. The above mentioned localization pertaining only the frst few modes is in agreement with Luongo (1995) since periodic structure with a finite number of elements are considered and piecewise periodicity acts as spatially-dependent large imperfections in continuos structures. Damped vibrations are then considered by introducing an

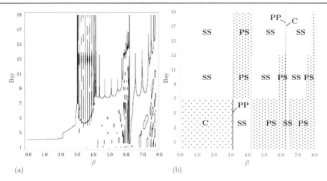

Figure 30. Propagation along the optimal piecewise periodic beam in the range $0 < \beta < 7.886$. (a) section of the displacement dynamic magnification at $D = 0.1$, (b) propagation regions.

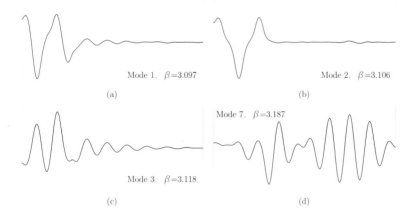

Figure 31. (a)-(b)-(c) Localized modal shapes; (d) Non-localized modal shape.

hysteretic damping factor η in the Euler beam equation of motion. Figure 11 shows the dynamic amplification factor for the damped 18 bays optimal piecewise periodic beam with $\eta = 0.01$. Besides the expected reduction of the resonance amplification, vibration transmission is now inhibited at all frequencies including the natural ones.

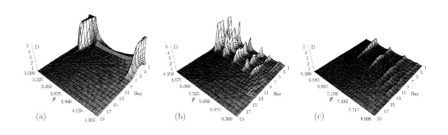

Figure 32. Displacement dynamic magnification factors of the optimal 18-span piecewise periodic damped beam in the range $0 < \beta < 7.886$; damping factor $\eta = 0.01$. (a) $3.0 < \beta < 4.35$; (b) $4.35 < \beta < 6.30$; (c) $6.30 < \beta < 8.0$.

The influence of the number of elements in each optimal section along the structure is eventually investigated. A set of numerical tests has been carried out on a beam assuming different combinations of the number of elements in each section while keeping fixed the overall number of 18 bays. Under damped forced vibrations, the integral of the response at the final node, evaluated on the design frequency range $0 < \beta < 7.886$, is reported in Table 2. In one hand it can be inferred that, once the optimal sequence is adopted, the number of elements in each section does not significantly affect the overall achievable vibration reduction since the differences in Table 2 concern small numbers. On the other hand it can be noticed that the optimization of the number of elements in each section provides with vibration reductions larger or comparable with those obtained by a 1% damping increase in non-optimized sequences. If piecewise periodicity is removed by composing the periodic beam with only one optimal section, the amount of vibration reduction decreases by at least two order of magnitudes.

Table 2. Integral of the response at the final node of 18-bays periodic structures $(0 \leq \beta \leq 7.886)$; optimal sequence κ_1^*-$\kappa_{12,max}^*$-κ_2^*.

Sections	Combinations	$\eta = 1\%$	$\eta = 2\%$
3	6-6-6	9.33E-3	5.27E-3
3	4-7-7	1.68E-2	1.03E-2
3	2-8-8	4.96E-2	3.31E-2
3	7-4-7	6.59E-3	3.87E-3
3	8-2-8	8.38E-3	4.53E-3
3	7-7-4	1.17E-2	6.95E-3
3	8-8-2	2.98E-2	1.63E-2
1	18-0-0	1.66	1.04
1	0-18-0	1.59	6.94E-1
1	0-0-18	4.96E-1	2.95E-1

1.6   Vibration analysis using wave vectors

The well known wave vector approach has been proposed von Flotow (1986); Yong and Lin (1989) for the analysis of the response of long un-damped periodic structures to overcome numerical difficulties arising in the transfer matrix formulation (Gupta, 1970; Faulkner and Hong , 1985) when the number of periodic units increases; in particular, the frequency de-pendent global transfer matrix, connecting the state variables at the ends, becomes ill-conditioned since the ratio between its maximum and minimum eigenvalue increases as well. This problem can be physically interpreted by considering that the transfer matrix implies rightwards transmission. Therefore, backward decaying waves appear numerically as growing waves in the opposite direction, thereby amplifying unavoidable numerical errors. By transforming traditional transfer matrices for state vectors to transfer matrices for wave vectors (wave coordinates) the computations always pro-ceed in the direction of wave motion.

The effectiveness of the wave vector approach has been confirmed by a number of applications pertaining trusses (Signorelli and von Flotow, 1988; Yong and Lin, 1992), generic structural networks (Cai and Lin, 1991), wave localization phenomena (Bouzit and Pierre, 2000) and piecewise periodic structures (Romeo and Luongo, 2003). Nevertheless, a number of authors have encountered numerical difficulties to determine the natural frequen-cies when solving the complex determinant resulting from the wave vector approach. For finite truss beams, Pierre and Chen Chen and Pierre (1991)

ascribed such difficulties to the high modal density. They observed that the
zeros of the real and imaginary parts of the characteristic equation were
not even close enough to derive natural frequencies yielding accurate mode
shapes; to circumvent the problem they transformed the latter determinant
into a real one. Moreover, Signorelli and von Flotow (1988) considered the
absolute value of the complex characteristic determinant stating that the
natural frequencies of the truss need to be identified whenever it "tends" to
vanish.

The causes of such drawbacks were investigated in Luongo and Romeo
(2005). It was shown that numerical difficulties are due to ill-positioning of
the mathematical problem when it is formulated in terms of complex waves
incoming in the structure. In fact, according to this method, natural frequencies are found as real solutions of complex characteristic equations, i.e.
as solution of a system of two equations in just one unknown. Aiming to
overcome such problems while retaining the advantages of the wave transfer
matrix framework, a modified version of the traditional wave vector computational scheme was proposed. The method deals with real quantities only,
thus avoiding the described pathology. In particular, while the traditional
method implies the transformation of frequency-dependent real transfer matrices for state vectors to complex transfer matrices for wave vectors, the
approach here proposed provides for transformation still to real matrices. In
essence the computational scheme adopts a basis of eigenvectors coinciding
with the real and imaginary parts of the transfer matrix complex eigenvectors thereby leading to a real block-diagonal matrix. The blocks associated
with eigenvalues with modulus greater than one are inverted, while those
associated with unit modulus eigenvalues are left unaltered because they do
not feed ill-conditioning. As a consequence a twofold objective is pursued.
In one hand, the obtained real boundary condition's matrix is more suited
for free vibration analysis; on the other hand, the stability of computations,
assured by proceeding in the direction of wave motion, is preserved. The
proposed transformation can also be interpreted as a change of coordinates
that brings the linear map $\mathbf{x}_{k+1} = \mathbf{T}\mathbf{x}_k$ into a normal form, $\mathbf{x}_k$ being the
state vector at the coupling point k and $\mathbf{T}$ the transfer matrix.

In the vibration analysis of periodic and piecewise periodic structures, summarized in this section according to the wave vector approach, structural
members are treated as waveguides which transmit a disturbance from one
location to another as wave motion (see Figure 33). According to this approach, the equations governing the problem are derived by meeting transmission, continuity and boundary conditions. The starting point of the
computational scheme is the transformation of state vectors to wave vectors. Next, continuity conditions at the interface of dissimilar cells as well

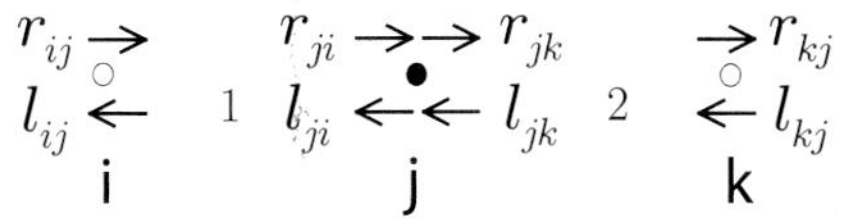

$$r_{ij} \underset{l_{ij}}{\overset{\to}{\leftarrow}} \underset{i}{\circ} \quad 1 \quad r_{ji} \underset{l_{ji}}{\overset{\to}{\leftarrow}} \underset{j}{\overset{\to}{\bullet}} \overset{\to}{\leftarrow} r_{jk} \quad 2 \quad \underset{k}{\overset{\to}{\circ}} r_{kj} \\ \leftarrow l_{kj}$$

Figure 33. Two-way junction of dissimilar elements and corresponding wave coordinates. ∘ external constraint, • internal constraint.

as boundary conditions are introduced in terms of wave coordinates.
For piecewise periodic structures the generic section α is composed by N_α elements connecting nodes i and j; according to the transfer matrix approach the state vector at the coupling point j is related to the state vector at the coupling point i by

$$\begin{pmatrix} \mathbf{x}_j \\ \mathbf{f}_j \end{pmatrix} = \mathbf{T}_\alpha^{N_\alpha} \begin{pmatrix} \mathbf{x}_i \\ \mathbf{f}_i \end{pmatrix} \tag{48}$$

where, if the elements are coupled through n degrees of freedom, $\mathbf{T}_\alpha$ is the $2n \times 2n$ frequency dependent transfer matrix of the single element of section α. A peculiar property of the transfer matrix is that its eigenvalues are reciprocal pairs $(\lambda, 1/\lambda)$. The $2n$ state vector can be transformed to wave vector through the matrix $\mathbf{U}_\alpha$ whose columns are eigenvectors of $\mathbf{T}_\alpha$ as follows

$$\begin{pmatrix} \mathbf{x}_i \\ \mathbf{f}_i \end{pmatrix} = \mathbf{U}_\alpha \begin{pmatrix} \mathbf{r}_{ij} \\ \mathbf{l}_{ij} \end{pmatrix} \tag{49}$$

As depicted in Figure 5, the terms $\mathbf{r}$ and $\mathbf{l}$ hints at right-going and left-going waves, respectively. They represent the amplitudes of the eigenvectors associated to reciprocal pairs with $|\lambda| < 1$ and $|\lambda| > 1$, respectively; if $|\lambda| = 1$, the two reciprocal eigenvalues are complex conjugate and the associated amplitudes split in $\mathbf{r}$ and $\mathbf{l}$. Substituting equation (49) in (48) gives

$$\begin{pmatrix} \mathbf{r}_{ji} \\ \mathbf{l}_{ji} \end{pmatrix} = \mathbf{U}_\alpha^{-1} \mathbf{T}_\alpha^{N_\alpha} \mathbf{U}_\alpha \begin{pmatrix} \mathbf{r}_{ij} \\ \mathbf{l}_{ij} \end{pmatrix} = \begin{bmatrix} \mathbf{\Lambda}_\alpha^{N_\alpha} & 0 \\ 0 & \mathbf{\Lambda}_\alpha^{-N_\alpha} \end{bmatrix} \begin{pmatrix} \mathbf{r}_{ij} \\ \mathbf{l}_{ij} \end{pmatrix} \tag{50}$$

where $\mathbf{\Lambda}_\alpha$ is a diagonal matrix collecting the eigenvalues with $|\lambda| \leq 1$. Rearranging equation (50) to follow the waves propagation direction,

$$\begin{pmatrix} \mathbf{r}_{ji} \\ \mathbf{l}_{ij} \end{pmatrix} = \begin{bmatrix} \mathbf{\Lambda}_\alpha^{N_\alpha} & 0 \\ 0 & \mathbf{\Lambda}_\alpha^{N_\alpha} \end{bmatrix} \begin{pmatrix} \mathbf{r}_{ij} \\ \mathbf{l}_{ji} \end{pmatrix} \tag{51}$$

The state vector at the interface of dissimilar cells, such as the coupling point j between nodes i and k (see Figure 5), must be continuous in the sense that displacements must be equal and forces must be in equilibrium, as long as no excitation is applied at the interface. Such continuity requirement is expressed relating the outgoing waves at the interfaces of cells to the incoming ones as follows

$$\begin{pmatrix} \mathbf{l}_{ji} \\ \mathbf{r}_{jk} \end{pmatrix} = \begin{bmatrix} \mathbf{S}_{lr} & \mathbf{S}_{ll} \\ \mathbf{S}_{rr} & \mathbf{S}_{rl} \end{bmatrix} \begin{pmatrix} \mathbf{r}_{ji} \\ \mathbf{l}_{jk} \end{pmatrix} \qquad (52)$$

where the wave scattering matrix $\mathbf{S}$ at the interface j has been partitioned in reflection $(\mathbf{S}_{lr}, \mathbf{S}_{rl})$ and transmission $(\mathbf{S}_{ll}, \mathbf{S}_{rr})$ submatrices.

Let's refer to the two-way junction shown in Figure 5 excited by a harmonic forces $\mathbf{f}_i$ applied at node i; let $N_1 = N$ and $N_2 = M$ be the number of elements of sections 1 and 2, respectively. The boundary conditions at the ends of the periodic chains allow to express the outgoing waves in terms of the incoming ones by means of reflection matrices $\mathbf{R}$ as follows

$$\mathbf{l}_{ij} = \mathbf{R}_i \mathbf{r}_{ij} + \mathbf{D} \mathbf{f}_i \quad , \quad \mathbf{r}_{kj} = \mathbf{R}_k \mathbf{l}_{kj} \qquad (53)$$

Without loss of generality, the scattering matrix $\mathbf{S}$ is derived with reference to the interface j of the scheme shown in Figure 1a. The transformation from state to wave vectors at node j located between nodes i and k gives

$$\begin{pmatrix} \mathbf{x}_{ji} \\ \mathbf{f}_{ji} \end{pmatrix} = \mathbf{U}_1 \begin{pmatrix} \mathbf{r}_{ji} \\ \mathbf{l}_{ji} \end{pmatrix} \quad ; \quad \begin{pmatrix} \mathbf{x}_{jk} \\ -\mathbf{f}_{jk} \end{pmatrix} = \mathbf{U}_2 \begin{pmatrix} \mathbf{r}_{jk} \\ \mathbf{l}_{jk} \end{pmatrix} \qquad (54)$$

Continuity and equilibrium imply that $(\mathbf{x}_{ji}, \mathbf{f}_{ji}) = (\mathbf{x}_{jk}, -\mathbf{f}_{jk})$ so that

$$\mathbf{U}_1 \begin{pmatrix} \mathbf{r}_{ji} \\ \mathbf{l}_{ji} \end{pmatrix} = \mathbf{U}_2 \begin{pmatrix} \mathbf{r}_{jk} \\ \mathbf{l}_{jk} \end{pmatrix} \qquad (55)$$

In order to express the outgoing waves in terms of the incoming ones, the matrices $\mathbf{U}_i$, $i = 1, 2$ are partitioned

$$\mathbf{U}_i = \begin{bmatrix} \mathbf{U}_{i,11} & \mathbf{U}_{i,12} \\ \mathbf{U}_{i,21} & \mathbf{U}_{i,22} \end{bmatrix} \quad i = 1, 2 \qquad (56)$$

and the following matrices are assembled

$$\mathbf{S}_A = \begin{bmatrix} \mathbf{U}_{1,12} & -\mathbf{U}_{2,11} \\ \mathbf{U}_{1,22} & -\mathbf{U}_{2,21} \end{bmatrix} \quad ; \quad \mathbf{S}_B = \begin{bmatrix} -\mathbf{U}_{1,11} & \mathbf{U}_{2,12} \\ -\mathbf{U}_{1,21} & \mathbf{U}_{2,22} \end{bmatrix} \qquad (57)$$

leading to the sought $(2n \times 2n)$ scattering matrix

$$\mathbf{S}_A^{-1} \mathbf{S}_B = \mathbf{S} \qquad (58)$$

The reflection matrices $\mathbf{R}$ are also derived referring to the boundary condition of nodes i and k in Figure 1. Such matrices are obtained by expressing the boundary conditions in terms of wave coordinates

$$\begin{aligned} \mathbf{B}_{r,i}\mathbf{r}_{ij} &+ \mathbf{B}_{l,i}\mathbf{l}_{ij} = \mathbf{f}_i \\ \mathbf{B}_{r,k}\mathbf{r}_{kj} &+ \mathbf{B}_{l,k}\mathbf{l}_{kj} = \mathbf{0} \end{aligned} \tag{59}$$

where the entries in the matrices $\mathbf{B}$ are given by elements of the matrices $\mathbf{U}_i$ varying according to the type of constraint. Equation (53) is obtained by rearranging equation (59); the reflections matrices $\mathbf{R}$ and the matrix $\mathbf{D}$ are given by

$$\mathbf{R}_i = -\mathbf{B}_{l,i}^{-1}\mathbf{B}_{r,i} \quad ; \quad \mathbf{R}_k = -\mathbf{B}_{r,k}^{-1}\mathbf{B}_{l,k} \quad ; \quad \mathbf{D} = \mathbf{B}_{l,i}^{-1} \tag{60}$$

By combining equations (51,52,53) any arrangement of elements at a junction can be solved in terms of the unknown wave vectors. The core of the computational scheme consists of solving the above-mentioned sets of equations by condensing the whole unknown wave vector in only the components entering the domain. Using equations (51) and (52) to eliminate the outgoing waves at the ends and all the waves at node j, the resolving equations (53) are obtained in terms of waves entering the domain

$$\begin{bmatrix} -\mathbf{R}_i + \boldsymbol{\Lambda}_1^N \mathbf{S}_{lr}\boldsymbol{\Lambda}_1^N & \boldsymbol{\Lambda}_1^N \mathbf{S}_{ll}\boldsymbol{\Lambda}_2^M \\ \boldsymbol{\Lambda}_2^M \mathbf{S}_{rr}\boldsymbol{\Lambda}_1^N & -\mathbf{R}_k + \boldsymbol{\Lambda}_2^M \mathbf{S}_{rl}\boldsymbol{\Lambda}_2^M \end{bmatrix} \begin{pmatrix} \mathbf{r}_{ij} \\ \mathbf{l}_{kj} \end{pmatrix} = \begin{pmatrix} \mathbf{D}\mathbf{f}_i \\ \mathbf{0} \end{pmatrix} \tag{61}$$

The wave coordinates at the intermediate nodes can be derived and transformed back to obtain the response expressed by state variables; next, the response of the elements is readily evaluated as a function of the ends' displacements and rotations. The homogeneous form of the problem (61) furnishes the natural frequencies and the "mode" $(\mathbf{r}_{ij}, \mathbf{l}_{kj})^T$.

When the groups of elements between nodes i and k are identical, the scattering matrix becomes simply

$$\mathbf{S} = \begin{bmatrix} \mathbf{0} & \mathbf{I} \\ \mathbf{I} & \mathbf{0} \end{bmatrix} \tag{62}$$

meaning that the continuity at the junction must be met not only by the state vector but also by the wave vector. Consequently, the system (61) becomes

$$\begin{bmatrix} -\mathbf{R}_i & \boldsymbol{\Lambda}^{N+M} \\ \boldsymbol{\Lambda}^{M+N} & -\mathbf{R}_k \end{bmatrix} \begin{pmatrix} \mathbf{r}_{ij} \\ \mathbf{l}_{kj} \end{pmatrix} = \begin{pmatrix} \mathbf{D}\mathbf{f}_i \\ \mathbf{0} \end{pmatrix} \tag{63}$$

Table 1: Standing waves in a free-free rod.

Solution	Boundary conditions	Characteristic eq.
	Real Form	
$u = a\cos\beta x + b\sin\beta x$ $a, b \in \mathbb{R}$ (a_1)	$\begin{bmatrix} 0 & 1 \\ -\sin\beta l & \cos\beta l \end{bmatrix} \begin{pmatrix} a \\ b \end{pmatrix} = 0$ (b_1)	$\sin\beta l = 0$ (c_1)
	Complex Form	
$u = r_A\,e^{i\beta x} + l_A e^{-i\beta x}$ $r_A, l_A \in \mathbb{C}$ (a_2)	$\begin{bmatrix} 1 & -1 \\ e^{i\beta l} & -e^{-i\beta l} \end{bmatrix} \begin{pmatrix} r_A \\ l_A \end{pmatrix} = 0$ (b_2)	$e^{i\beta l} - e^{-i\beta l} = 0$ (c_2)
	Modified Complex Form	
$u = r_A\,e^{i\beta x} + l_B e^{-i\beta(x-l)}$ $r_A, l_B \in \mathbb{C}$ (a_3)	$\begin{bmatrix} 1 & -e^{i\beta l} \\ e^{i\beta l} & -1 \end{bmatrix} \begin{pmatrix} r_A \\ l_B \end{pmatrix} = 0$ (b_3)	$e^{2i\beta l} - 1 = 0$ (c_3)

Real and complex formulations With the aim of introducing the problem arising in the wave vector computational scheme outlined in the previous section, let us consider the free axial vibrations of a free-free rod of length l. They are governed by the following boundary value problem:

$$u''(x) + \beta^2 u(x) = 0 \quad x \in [0, l]$$
$$u'(0) = 0, \ u'(l) = 0 \tag{64}$$

in which β is the eigenvalue, proportional to the time-frequency ω. Solutions to equation (64) can be put in different forms, here named Real, Complex and Modified Complex Forms, resumed in Table 1. If the field $u(x)$ is expressed in the Real Form (equation (a_1)), the boundary conditions (64a) lead to the real eigenvalue problem (b_1), whose real characteristic equation (c_1) admits the roots:

$$\vartheta = \frac{j\pi}{N} \quad \text{with} \ j = 0, 1, \ldots \tag{65}$$

Alternatively, the solution can be expressed in the Complex Form (a_2). In it, $e^{i\beta x}$ and $e^{-i\beta x}$ are steady waves (conventionally) propagating rightward and leftward, respectively. Their complex amplitudes, r_A and l_A, represent the contribution of each wave to the displacement at the left end A of the road, where $e^{i\beta x}$ and $e^{-i\beta x}$ assume unit values. By enforcing boundary conditions, the equations (b_2) are drawn. In them, the columns of the matrix are linear combinations of the columns of the matrix in equation (b_1), hence the same real characteristic equation (c_2) follows. Therefore, the Real and Complex Forms are equivalent. Finally, a Modified Complex Form of the solution is considered (equation (a_3)), derived from equation (a_2) by letting $l_A := l_B e^{i\beta l}$. This change in the complex amplitudes is often performed in order to avoid numerical ill-conditioning in the boundary conditions when l is large and β is allowed to assume complex values (e.g. for long beams resting on Winkler soil, or for long tubes with radially simmetric loads). In fact, the modification renders of the same order of magnitude complex amplitudes otherwise very different from each other. By again enforcing boundary conditions, equations (b_3) are drawn. They could be obtained directly from equations (b_2) through the product of the second column of the matrix by the factor $e^{i\beta l}$, which transforms l_A in l_B. Since the factor is complex, the associated characteristic equation (c_3), differently from the previous cases, is also complex. It leads to two real equations in the unique real unknown beta, namely:

$$\cos 2\beta l = 1 \quad , \quad \sin 2\beta l = 0 \qquad (66)$$

The problem (66) appears as an overdetermined nonlinear system, since it requires satisfying two simultaneous equations with just one unknown. It will be referred to as *ill-posed*, since its true nature of determined system is disguised. In principle, no solutions exist for such problem; however, it is easy to check that equations (66) do admit solutions, and these all coincide with the solutions (65). This circumstance is a consequence of the fact that, since a column of the matrix in equation (b_2) has been altered by $e^{i\beta l}$, also its characteristic equation (c_2) has been multiplied by the same factor, so that equation (c_3) admits the factorized form:

$$e^{i\beta l}(e^{i\beta l} - e^{-i\beta l}) = 0 \qquad (67)$$

Since the factor never vanishes, equation (67) admits the same roots of equation (c_2) and no other ones. However, in more complex cases than that at the hand, it is not easy to factorize the characteristic equation, and one has to solve a problem of the following form:

$$f(\omega) + i\, g(\omega) = 0 \qquad (68)$$

where ω is the unknown. There are two ways to solve this equation: (a) to search for complex roots $\omega = \gamma + i\delta$ of the simultaneous equation

$\mathrm{Re}(f)-\mathrm{Im}(g) = 0$, $\mathrm{Im}(f)+\mathrm{Re}(g) = 0$, by retaining only the real solutions $\gamma \neq 0$, $\delta = 0$; (b) to search for real roots ω of each equation $f(\omega) = 0$ and $g(\omega) = 0$, by keeping only the common roots. Numerical problems arise in both the approaches, since it is difficult to decide wheter roots with small γ's (in the approach (a)) or roots very closed each other (in the approach (b)) must be retained or rejected. Such difficulties was encountered in the works cited in the Introduction, as a consequence of the ill-position of the problem.

In conclusion, the complexification of the characteristic equation does not depend on the complexification of the amplitudes, but rather by their rearrangement. The expedient of the change of the variables is a peculiar aspect of the complex waves coordinate method; it permits to avoid numerical ill-conditioning, but entails ill-positioning. In the next section a variant of the method is illustrated, able to avoid both ill-conditioning and ill-positioning of the problem.

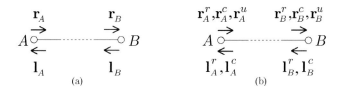

Figure 34. Wave vector components: (a) complex; (b) real.

Real wave vector approach   For periodic structures, the state vectors $\mathbf{x}_k = (\mathbf{d}_k, \mathbf{f}_k)$, listing generalized displacements $\mathbf{d}$ and forces $\mathbf{f}$ at the coupling point k, is related to the state vector $\mathbf{x}_{k+1}$, at the following coupling point $k+1$, through the transfer matrix $\mathbf{T}$, i.e. $\mathbf{x}_{k+1} = \mathbf{T}\mathbf{x}_k$. For elements coupled through n degrees of freedom, $\mathbf{T}$ is a real $2n \times 2n$ frequency dependent matrix and, if the structure is composed by N elements connecting nodes A and B, then

$$\begin{pmatrix} \mathbf{d}_B \\ \mathbf{f}_B \end{pmatrix} = \mathbf{T}^N \begin{pmatrix} \mathbf{d}_A \\ \mathbf{f}_A \end{pmatrix} \tag{69}$$

A peculiar property of the transfer matrix is that its eigenvalues are reciprocal pairs $(\lambda, 1/\lambda)$. The case of distinct eigenvalues will be considered throughout the analysis.

According to the well known complex wave vectors approach (see Appendix), the state vectors can be transformed to complex wave vectors through a matrix whose columns are the complex eigenvectors of $\mathbf{T}$. The obtained wave vector $(\mathbf{r}_k\,,\,\mathbf{l}_k)^T$, at the generic coupling point k, has components $\mathbf{r}$ and $\mathbf{l}$ hinting at right-going and left-going waves, respectively (see Figure 1a). They represent the amplitudes of the eigenvectors associated to reciprocal pairs with $|\lambda| < 1$ and $|\lambda| > 1$, respectively; if $|\lambda| = 1$, the two reciprocal eigenvalues are complex conjugate and the associated amplitudes split in $\mathbf{r}$ and $\mathbf{l}$.

An alternative transformation can be adopted relying on a modified basis of eigenvectors. Let's consider the generic case of $\mathbf{T}$ with $2n_r$ real eigenvalues $\lambda = \rho,\, 1/\rho$, $2n_c$ complex eigenvalues $\lambda = \rho\, e^{\pm i\theta}$, $1/\rho\, e^{\mp i\theta}$ with $\rho < 1$ and $2n_u$ complex eigenvalues with unit modulus $\lambda = e^{\pm i\theta}$. Let also $\mathbf{u}_h$ and $\mathbf{u}_h^*$, $h = 1,\ldots,n_r$, be the real eigenvectors of $\mathbf{T}$ associated with the real eigenvalues ρ_h and $1/\rho_h$, respectively; $\mathbf{z}_j$ and $\mathbf{z}_j^*$, $j = 1,\ldots,n_c$, be the complex eigenvectors associated with the eigenvalues $\rho_j\, e^{\pm i\theta_j}$, $1/\rho_j\, e^{\mp i\theta_j}$, respectively; $\mathbf{y}_l$, $l = 1,\ldots,2n_u$, be the complex eigenvectors associated with the eigenvalues $e^{\pm i\theta_l}$. By setting up the invertible matrix $\mathbf{U}$ in the form

$$\mathbf{U} = \left[\mathbf{u}_h\,,\,\mathrm{Re}(\mathbf{z}_j)\,,\,\mathrm{Im}(\mathbf{z}_j)\,,\,\mathrm{Re}(\mathbf{y}_l)\,,\,\mathrm{Im}(\mathbf{y}_l)\,,\,\mathbf{u}_h^*\,,\,\mathrm{Re}(\mathbf{z}_j^*)\,,\,\mathrm{Im}(\mathbf{z}_j^*)\right] \quad (70)$$

it is well known that is possible to obtain a matrix

$$\mathbf{U}^{-1}\mathbf{T}\mathbf{U} = \mathrm{diag}\left(\lambda_h\,,\,\mathbf{C}_j\,,\,\mathbf{P}_l\,,\,\lambda_h^{-1}\,,\,\mathbf{C}_j^{-1}\right) \quad (71)$$

representing a real transfer matrix for wave vectors. In equation (71) $\mathbf{C}_j$, $1 \le j \le n_c$, are 2×2 blocks:

$$\mathbf{C}_j = \rho_j \begin{bmatrix} \cos\theta_j & \sin\theta_j \\ -\sin\theta_j & \cos\theta_j \end{bmatrix} \quad (72)$$

and $\mathbf{P}_l$, $1 \le l \le 2n_u$, are submatrices equal to $\mathbf{C}_j$ with $\rho = 1$. The state vector transformation can now be expressed through a new change of coordinates to real wave vector using the real matrix $\mathbf{U}$ as

$$\begin{pmatrix} \mathbf{d}_k \\ \mathbf{f}_k \end{pmatrix} = \mathbf{U}\,(\mathbf{r}_k^r\,,\,\mathbf{r}_k^c\,,\,\mathbf{r}_k^u\,,\,\mathbf{l}_k^r\,,\,\mathbf{l}_k^c)^T \quad (73)$$

where the real wave vector at node k has been partitioned into five subvectors $\mathbf{r}_k^r$, $\mathbf{r}_k^c$, $\mathbf{r}_k^u$, $\mathbf{l}_k^r$ and $\mathbf{l}_k^c$, representing the amplitudes of the eigenvectors $\mathbf{u}_h$, $\mathbf{z}_j$, $\mathbf{y}_l$, $\mathbf{u}_h^*$ and $\mathbf{z}_j^*$, respectively (see Figure 1b). Substituting equation

(73) in (69) gives

$$
\begin{pmatrix} \mathbf{r}_B^r \\ \mathbf{r}_B^c \\ \mathbf{r}_B^u \\ \mathbf{l}_B^r \\ \mathbf{l}_B^c \end{pmatrix} = \begin{bmatrix} \mathbf{S}^N & 0 & 0 & 0 & 0 \\ 0 & \mathbf{C}^N & 0 & 0 & 0 \\ 0 & 0 & \mathbf{P}^N & 0 & 0 \\ 0 & 0 & 0 & \mathbf{S}^{-N} & 0 \\ 0 & 0 & 0 & 0 & \mathbf{C}^{-N} \end{bmatrix} \begin{pmatrix} \mathbf{r}_A^r \\ \mathbf{r}_A^c \\ \mathbf{r}_A^u \\ \mathbf{l}_A^r \\ \mathbf{l}_A^c \end{pmatrix}
\tag{74}
$$

where the square submatrices in the global wave transfer matrix $\mathbf{U}^{-1}\mathbf{T}^N\mathbf{U}$ are defined as $\mathbf{S}^N = \mathrm{diag}(\rho_1^N, \dots, \rho_h^N)$, $\mathbf{C}^N$ is the $2n_c$ submatrix with diagonal 2×2 blocks, $\mathbf{C}_j^N$, $1 \leq j \leq n_c$, given by

$$
\mathbf{C}_j^N = \rho_j^N \begin{bmatrix} \cos N\theta_j & \sin N\theta_j \\ -\sin N\theta_j & \cos N\theta_j \end{bmatrix}
\tag{75}
$$

and $\mathbf{P}^N$ is a matrix equal to $\mathbf{C}^N$ with $\rho = 1$. In equation (74) symbols $\mathbf{S}$, $\mathbf{C}$ and $\mathbf{P}$ refer to submatrices listing eigenvalues belonging to stop, complex and pass bands, respectively. The real wave vectors entering equations (73) and (74), whose complex counterparts enter equations (54) and (55) in Appendix, represent the main computational difference between the proposed wave vector approach and the traditional one.

Equation (74) can be rearranged to follow the real wave propagation direction,

$$
\begin{pmatrix} \mathbf{r}_B \\ \mathbf{l}_A \end{pmatrix} = \begin{bmatrix} \mathbf{\Lambda}_r^N & 0 \\ 0 & \mathbf{\Lambda}_l^N \end{bmatrix} \begin{pmatrix} \mathbf{r}_A \\ \mathbf{l}_B \end{pmatrix}
\tag{76}
$$

where $\mathbf{\Lambda}_r^N = \mathrm{diag}(\mathbf{S}^N, \mathbf{C}^N, \mathbf{P}^N)$, $\mathbf{\Lambda}_l^N = \mathrm{diag}(\mathbf{S}^N, \mathbf{C}^N)$, $\mathbf{r}_H = (\mathbf{r}_H^r, \mathbf{r}_H^c, \mathbf{r}_H^u)$ and $\mathbf{l}_H = (\mathbf{l}_H^r, \mathbf{l}_H^c)$, $H = A, B$. It should be noticed that submatrices $\mathbf{P}^N$ are not rearranged, since they are not affected by ill-conditioning, thanks to the unit modulus of their eigenvalues. Therefore, rightward propagation of the associated $\mathbf{r}^u$ waves is considered, differently from the complex wave method.

The boundary conditions at the ends of the periodic structure expressed in terms of wave coordinates read

$$
\mathbf{R}_H \mathbf{r}_H + \mathbf{L}_H \mathbf{l}_H = \mathbf{B}_H \mathbf{f}_H \qquad H = A, B
\tag{77}
$$

where $\mathbf{B}_H$ are boolean matrices and $\mathbf{R}_H$ and $\mathbf{L}_H$ are $n \times (n_r + n_c + 2n_u)$ and $n \times x(n_r + n_c)$ rectangular submatrices of $\mathbf{U}$, respectively, varying according to the type of constraint and $\mathbf{f}_H$ are assigned ends' forces and/or displacements. Substituting eventually equations (76) into (77) leads to the non-homogeneous problem

$$\left[\begin{array}{cc} \mathbf{R}_A & \mathbf{L}_A\boldsymbol{\Lambda}_l^N \\ \mathbf{R}_B\boldsymbol{\Lambda}_r^N & \mathbf{L}_B \end{array} \right] \left(\begin{array}{c} \mathbf{r}_A \\ \mathbf{l}_B \end{array} \right) = \left(\begin{array}{c} \mathbf{B}_A\mathbf{f}_A \\ \mathbf{B}_B\mathbf{f}_B \end{array} \right) \tag{78}$$

The wave coordinates at the intermediate nodes can be derived and transformed back to obtain the response expressed by state variables; next, the response of the elements is readily evaluated as a function of the ends' displacements. However, while in the complex wave vectors approach the real state vectors are obtained by summing up products of complex conjugate eigenvectors and amplitudes, in the real wave vectors approach, the real state vectors are obtained by taking straight summation over products of real eigenvectors and amplitudes. Since the real eigenvectors represent the real and imaginary parts of the complex ones, the associated real amplitudes are half the real and imaginary parts of the complex amplitudes.

Alike the complex wave formulation, the real one can be readily extended to piecewise periodic structures by introducing the scattering matrix arising at the interface of dissimilar cells as given by equation (58).

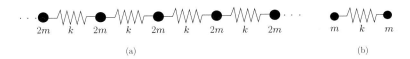

(a)   (b)

Figure 35. Mono-coupled system: (a) spring-mass chain; (b) symmetric single unit.

A simple illustrative example In this section the proposed procedure is illustrated for a spring-mass chain (mono-coupled system) already studied in Faulkner and Hong (1985) and shown in Figure ??a. It represents the discrete counterpart of the problem discussed in Section 2. By considering the spring-mass chain composed by N symmetric elements shown in Figure ??b, the transfer matrix reads

$$\mathbf{T} = \left[\begin{array}{cc} \cos\vartheta & -1 \\ \sin^2\vartheta & \cos\vartheta \end{array} \right] \tag{79}$$

where $\cos\vartheta = 1 - \beta^2$ and $\sin\vartheta = \beta\sqrt{2 - \beta^2}$, $\beta = (m/k)^{1/2}\omega$ being the frequency parameter. The transfer matrix eigenvalues are $\lambda_{1,2} = e^{\pm iN\vartheta}$ and the pass band frequency range is $0 \leq \vartheta \leq \pi$. In the following subsections, the natural frequencies, modal shapes and forced response are determined by using several approaches, namely: transfer matrix, complex wave vectors and real wave vectors.

Natural frequencies

(a) *Transfer Matrix Method*
The transfer matrix of the whole chain is obtained by recursive product:

$$\mathbf{T}^N = \left[\begin{array}{cc} \cos N\vartheta & -\frac{\sin N\vartheta}{\sin \vartheta} \\ \sin N\vartheta \sin \vartheta & \cos N\vartheta \end{array} \right] \tag{80}$$

Then, by imposing the boundary conditions, the natural frequencies are obtained from the following equation

$$\sin N\vartheta \sin \vartheta = 0, \quad \text{from which} \quad \vartheta = \frac{j\pi}{N}, \quad \text{with} \ \ j = 0, 1, \ldots, N-1 \tag{81}$$

Equation (81) is plotted in Figure (**??**).

(b) *Complex Wave Vector Method*
By applying equation (54), the complex wave vector method provides with the following state vector transformation

$$\left(\begin{array}{c} d_k \\ f_k \end{array} \right) = \left[\begin{array}{cc} \frac{i}{\sin \vartheta} & -\frac{i}{\sin \vartheta} \\ 1 & 1 \end{array} \right] \left(\begin{array}{c} r_k \\ l_k \end{array} \right) \tag{82}$$

Next, by imposing the boundary conditions (equations (57,58)), a set of two equations is obtained

$$\left[\begin{array}{cc} 1 & e^{iN\vartheta} \\ e^{iN\vartheta} & 1 \end{array} \right] \left(\begin{array}{c} r_A \\ l_B \end{array} \right) = \left(\begin{array}{c} 0 \\ 0 \end{array} \right) \tag{83}$$

The characteristic equation of the 2×2 boundary conditions' matrix is

$$(\cos \vartheta + i \sin \vartheta)^{2N} = \cos 2N\vartheta + i \sin 2N\vartheta = 1 \tag{84}$$

The natural frequencies are given by the zeros of *both* its real and imaginary parts, represented in Figure (**??**):

$$\begin{array}{ll} \cos 2N\vartheta = 1, & \text{from which} \ \ \vartheta = \frac{j\pi}{N} \\ \sin 2N\vartheta = 0, & \text{from which} \ \ \vartheta = \frac{j\pi}{2N} \end{array} \quad \text{with} \ \ j = 0, 1, \ldots, N-1 \tag{85}$$

and therefore coincide with the roots of equation (81). It must be noted that, real and imaginary parts of the characteristic equation cannot be individually zeroed since, while the former shows erroneous double roots, the latter has zeros that are not roots of the set.

(c) *Real Wave Vector Method*
According to equation (73), the state vector transformation is now given by

$$\begin{pmatrix} d_k \\ f_k \end{pmatrix} = \begin{bmatrix} 0 & \frac{1}{\sin \vartheta} \\ 1 & 0 \end{bmatrix} \begin{pmatrix} r_{1,k} \\ r_{2,k} \end{pmatrix} \tag{86}$$

By imposing the boundary conditions (equations (77,78) in Appendix), a set of two equations is obtained

$$\begin{bmatrix} 1 & 0 \\ \cos N\vartheta & \sin N\vartheta \end{bmatrix} \begin{pmatrix} r_{1,A} \\ r_{2,A} \end{pmatrix} = \begin{pmatrix} 0 \\ 0 \end{pmatrix} \tag{87}$$

The characteristic equation of the 2x2 boundary conditions' matrix, plotted in Figure (??), and the natural frequencies are:

$$\sin N\vartheta = 0, \quad \text{from which} \quad \vartheta = \frac{j\pi}{N} \quad \text{with} \quad j = 0, 1, \ldots, N-1 \tag{88}$$

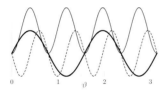

Figure 36. Natural frequencies of the four-bay spring-mass system: (thin line) Transfer Matrix; (thick and dashed lines) Real and Imaginary parts of the Complex Wave Vector method; (heavy line) Real Wave Vector.

By summarizing, the example has shown the same issue discussed in Section 2. In particular, the complex wave vector method leads to a complex characteristic equation (of the type of equation (68)) in the real unknown ω. When numerical solution is required, computational problems arise; in contrast, the real wave vector method is not affected by these drawbacks.

Modal Shapes

(a) *Transfer Matrix Method*
Modal shapes are derived by first setting $\mathbf{x}_A = (1,0)^T$ and then obtaining $\mathbf{x}_k = \mathbf{T}^k \mathbf{x}_A$, from which the modal vector at the generic node k reads

$$\mathbf{x}_k = (\cos k\vartheta \,, \, \sin k\vartheta \, \sin \vartheta)^T \tag{89}$$

(b) *Complex Wave Vector Method*

From equation (83a), setting $r_A = -1/2\,i\sin\vartheta$, $l_B = 1/2\,i\sin\vartheta\,e^{-iN\vartheta}$ follows, and using rightward and leftward transmission, the complex wave coordinates at node k are obtained

$$r_k = r_A e^{ik\vartheta} \quad , \quad l_k = l_B e^{i(N-k)\vartheta} \tag{90}$$

Then, by substituting into equation (82), the state vector reads

$$\begin{pmatrix} d_k \\ f_k \end{pmatrix} = \begin{bmatrix} \frac{i}{\sin\vartheta} & -\frac{i}{\sin\vartheta} \\ 1 & 1 \end{bmatrix} \begin{pmatrix} a_k + i\,b_k \\ a_k - i\,b_k \end{pmatrix} \tag{91}$$

where

$$a_k = \frac{1}{2}\sin\vartheta\sin k\vartheta \quad , \quad b_k = -\frac{1}{2}\sin\vartheta\cos k\vartheta. \tag{92}$$

and the result (89) is recovered. It must be noted that since the real state is expressed by a linear combination of two complex conjugate eigenvectors, the associated amplitudes must also be complex conjugate, i.e. $l_k = \bar{r}_k$. Therefore the only independent quantities are the real (a_k) and imaginary (b_k) parts of r_k.

(c) *Real Wave Vector Method*

From equation (87a), $r_{1,A} = 0$, $r_{2,A} = \sin\vartheta$ and using rightward trasmission, the real wave coordinates at node k are obtained

$$\begin{pmatrix} r_1 \\ r_2 \end{pmatrix}_k = \begin{bmatrix} \cos k\vartheta & \sin k\vartheta \\ -\sin k\vartheta & \cos k\vartheta \end{bmatrix} \begin{pmatrix} r_1 \\ r_2 \end{pmatrix}_A \tag{93}$$

Then, by substituting into equation (86), the state vector expression reads

$$\begin{pmatrix} d_k \\ f_k \end{pmatrix} = \begin{bmatrix} 0 & \frac{1}{\sin\vartheta} \\ 1 & 0 \end{bmatrix} \begin{pmatrix} 2a_k \\ -2b_k \end{pmatrix} \tag{94}$$

where a_k and b_k are given by the equation (92), and the resulting state vector $\mathbf{x}_k$ coincides with equation (89). In equation (94) the real state is expressed by a linear combination of two real eigenvectors, so that the associated amplitudes are also real and independent each other. By comparing (91) and (94) the relation between complex and real amplitudes, according to the two described approaches, can be readily deduced.

Forced Vibrations

The forced response to a horizontal force applied at the left hand boundary A of the chain is determined.

(a) *Transfer Matrix Method*
From (69), with $\mathbf{T}^N$ given by (80), by imposing the boundary conditions $f_A = F$ and $f_B = 0$, the state vector at the generic node k reads

$$\mathbf{x}_k = \left(\frac{F\cos(N-k)\vartheta}{\sin\vartheta\sin N\vartheta} \,,\, \frac{F\sin(N-k)\vartheta}{\sin N\vartheta} \right)^T \qquad (95)$$

(b) *Complex Wave Vector Method*
Taking into account the force F at the boundary A, equations (83) become

$$\begin{bmatrix} 1 & e^{iN\vartheta} \\ e^{iN\vartheta} & 1 \end{bmatrix} \begin{pmatrix} r_A \\ l_B \end{pmatrix} = \begin{pmatrix} F \\ 0 \end{pmatrix} \qquad (96)$$

Then, by following the same steps described in the above section for deriving the modal shapes, an expression analogous to (91) is obtained, where

$$a_k = F\,\frac{\sin(N-k)\,\vartheta}{2\sin N\vartheta} \quad,\quad b_k = -F\,\frac{\cos(N-k)\,\vartheta}{2\sin N\vartheta} \qquad (97)$$

(c) *Real Wave Vector Method*
The equations (87) modify:

$$\begin{bmatrix} 1 & 0 \\ \cos N\vartheta & \sin N\vartheta \end{bmatrix} \begin{pmatrix} r_{1,A} \\ r_{2,A} \end{pmatrix} = \begin{pmatrix} F \\ 0 \end{pmatrix} \qquad (98)$$

By following also in this case the previously described steps, an expression analogous to (94) is obtained, in which a_k and b_k are given by the equations (97). Comments similar to the free response case hold. Further numerical results can be found in Luongo and Romeo (2005).

2   Nonlinear periodic structures

A question that may arise in the study of periodic systems is the effect of nonlinearities on wave propagation. In order to start answering to the latter complex question, this part of the chapter is devoted to the dynamic analysis of nonlinear periodic systems where nonlinear maps are used to describe the dynamic behavior of both, discrete and continuous mechanical models. The dynamics of nonlinear periodic mechanical systems has been studied by Vakakis and King (1995) by addressing monocoupled periodic systems of infinite extent with material nonlinearities; two different asymptotic approaches have been devised for studying standing (stop-band) and traveling (pass-bands) waves and amplitude dependent frequencies bounding nonlinear propagation and attenuation zones have been found. In (Davies and Moon, 1996) an array of elastic oscillators coupled through buckling sensitive elastica has been addressed both numerically and experimentally. The existence of transition from soliton-like motions to spatially and temporally disordered motions due to a sudden excitation has been shown relying on a modified Toda lattice model. A first extension of the transfer matrix approach to post-buckling problems of periodic systems has been proposed in Luongo (1995). Then, in (Luongo and Romeo, 2006), a perturbation method was applied to the transfer matrix of a chain of continuous nonlinear beams; the main content of this work will be reported in what follows. Regardless of the specific physical context, one-dimensional chains of nonlinear oscillators are the paradigmatic models used to represent more complex periodic structures; these models will be addressed in the last part of the chapter. The dynamics of various types of one-dimensional chains has been extensively addressed in the literature. Depending on the application, the coupled oscillators constituting the chains can be characterized by different sources of nonlinearity affecting either the oscillators (on-site nonlinearity) or their coupling (nonlinear nearest neighbor interaction). A further distinction among the chain models can be based on the number of degrees of freedom considered for the single oscillator. Moreover, the analytical treatment can be carried out either by introducing a continuum approximation or by keeping the discrete nature of the original second order nonlinear difference equations governing the dynamics of the chains. It must be pointed out that the continuum approximation assumption, which holds well for wavelengths longer than the length of any chain element, is well suited for lattice models, mostly arising in condensed-matter physics. In contrast, if one is mainly interested in mechanical applications where large elements make up the periodic structures, once the elements are reduced to oscillators, discrete formulation ought to be considered. Being interested in dynamical phenomena with the same length scale of the interoscillator distance, discrete nonlinear systems will be considered, with the main goal of investigating analytically the modification of the boundary of the linear propagation/attenuation zones due to the nonlinearities. Chains of oscillators with cubic nonlinearities were studied in Umberger et al. (1989); Manevitch (2001), through numerical and asymptotic approaches, respectively. Nonlinear chains were also addressed by Chakraborty and Mallik (2001) where a perturbation approach was devised for oscillators coupled by cubic springs. The use of nonlinear maps was proposed by Wan and Soukoulis (1990) and Hennig and Tsironis (1999): the governing difference equations were regarded as symplectic nonlinear transformations relating the amplitudes in adjacent chain sites (n, $n + 1$) thereby considering a dynamical system where the location index n plays the role of the discrete time. By doing so, wave propagation becomes synonymous of stability: finding regions of propagating wave solutions (pass bands) is equivalent to finding regions of linearly stable map solutions. According to this approach, chains of nonlinear oscillators will be treated following the main results presented in Romeo and Rega (2006) and Romeo and Rega (2008).

2.1   Continuous beam on sliding supports

Following Luongo and Romeo (2006), in this section a chain of N shear indeformable beams with distributed mass m, resting on movable supports, as schematically depicted in Figure (37), is considered. After deriving the nonlinear continuous beam mode, the nonlinear element transfer function is obtained through a perturbation method. By iterating the nonlinear element map, a consistent system transfer function is obtained and free and forced oscillations are studied.

The single beam, of length l and flexural stiffness EI, is also axially indeformable. The displacement field is defined by the longitudinal and transverse components $u(s)$ and $v(s)$, respectively, and by a rotation $\varphi(s)$ of the beam cross section, s being the abscissa. The conditions of zero transverse shear and elongation, imply, respectively,

$$\sin \varphi = v' \quad ; \quad \epsilon := \sqrt{(1 + u')^2 + v'^2} - 1 = 0 \tag{99}$$

from which the curvature κ follows:

$$\kappa := \varphi' = \frac{v''}{\sqrt{1 - v'^2}} \tag{100}$$

The equation of motion are derived from the stationarity of the constrained

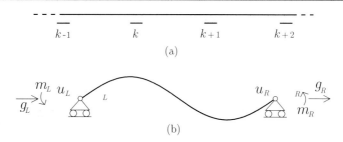

(a)

(b)

Figure 37. (a) Chain of nonlinear beams; (b) inextensible and shear indeformable single beam.

Hamiltonian

$$\bar{H} = \int_{t_1}^{t_2}\int_0^l \left[\tfrac{1}{2}m(\dot{u}^2 + \dot{v}^2) - \tfrac{1}{2}EI\kappa^2 - \lambda\epsilon\right] ds\, dt +$$
$$g_L u_L + m_L \varphi_L + g_R u_R + m_R \varphi_R \tag{101}$$

where $u_H, \varphi_H, g_H, m_H, (H = L, R)$ are the end displacements and forces
and the lagrangian parameter $\lambda(s)$ has been introduced. By considering the
series expansions, $\kappa \simeq v'' \left(1 + \tfrac{v'^2}{2}\right)$, $\epsilon \simeq u' + \tfrac{1}{2}v'^2$, the stationarity condition
leads to the set of partial differential equations

$$v'''' + \beta^4 \ddot{v} + \left[v'\left(v'v''\right)'\right]' - (\lambda v')' = 0$$
$$\beta^4 \ddot{u} - \lambda' = 0 \tag{102}$$
$$u' + \tfrac{1}{2}v'^2 = 0$$

where the following nondimensional quantities have been used:

$$\bar{v} = \tfrac{v}{l} \ , \quad \bar{u} = \tfrac{u}{l} \ , \quad \bar{s} = \tfrac{s}{l} \ , \quad \tau = \omega t \ , \quad \beta^4 = \tfrac{m\omega^2}{EI}l^4 \ , \quad \bar{\lambda} = \tfrac{\lambda l^2}{EI} \ ,$$
$$\bar{m} = \tfrac{ml}{EI} \ , \quad (\,\cdot\,) = \tfrac{d}{d\tau} \tag{103}$$

where ω represents a frequency to be specified later and the overbars have
been omitted. By prescribing the force g_R and all the displacements except
u_R, the boundary conditions follow

$$v(0, \tau) = 0 \quad , \quad v'(0, \tau) + \tfrac{1}{6}v'^3(0, \tau) = \varphi_L$$
$$v(1, \tau) = 0 \quad , \quad v'(1, \tau) + \tfrac{1}{6}v'^3(1, \tau) = \varphi_R \tag{104}$$
$$u(0, \tau) = u_L \quad , \quad \lambda(1, \tau) = g_R$$

where the series expansion $\varphi \simeq v' + \frac{1}{6}v'^3$ has been used. Having solved the boundary value problem (143,145), by using the moment-curvature relation, the unknown variables at both ends are obtained

$$
\begin{aligned}
m_L &= -\left[v'' + \tfrac{1}{2}v''v'^2\right]_{0,\tau} \\
m_R &= \left[v'' + \tfrac{1}{2}v''v'^2\right]_{1,\tau} \\
u_R &= u(1,\tau) \\
g_L &= -\lambda(0,\tau)
\end{aligned}
\tag{105}
$$

In view of the transfer matrix formulation it is convenient to parameterize the solution in terms of the left-hand variables $(\varphi_L, m_L, g_L, u_L)$; therefore equations $(145_4,145_6)$ are replaced by equations $(146_1,146_4)$.

Nonlinear element transfer function   To solve the nonlinear problem (143), with the proper boundary conditions (145) and (146), use is made of the harmonic balance method. By retaining only the first harmonic in $v(s,\tau)$, equations (143_3) and (143_2) ordinately provide with the time dependence of $u(s,\tau)$ and $\lambda(s,\tau)$, namely:

$$
v(s,\tau) = \hat{v}(s)\cos\tau \ , \quad u(s,\tau) = \hat{u}(s)\cos^2\tau \ , \quad \lambda(s,\tau) = \hat{\lambda}(s)\cos 2\tau \tag{106}
$$

Consequently, ω in equation (144) is the frequency of the (prevailing) transversal motion. The space-dependent functions $\hat{v}(s)$, $\hat{u}(s)$, and $\hat{\lambda}(s)$ are then expanded in series of an artificially introduced small bookkeeping parameter ε, having the meaning of transversal motion amplitude, i.e. $\varepsilon = O(\|\hat{v}(s)\|)$. Aiming at considering only significant (non-vanishing) quantities, incomplete series are adopted, according to the following considerations. If equations (143) admit the solution (v,u,λ), they also admit the solution $(-v,u,\lambda)$, which describes a state of the beam symmetrical with respect to its undeformed axis. Therefore. by changing the sign of the amplitude ε, v must change sign (i.e. it is an odd-function of ε), whereas u and λ must keep their sign (i.e. they are even-functions of ε). By also taking into account equations (145) and (146), the following unknowns series expansions and ordering of the boundary terms are introduced

$$
\begin{aligned}
\hat{v} &= \varepsilon v_1 + \varepsilon^3 v_3 + \dots \ , \quad \hat{u} = \varepsilon^2 u_2 + \dots \ , \quad \hat{\lambda} = \varepsilon^2 \lambda_2 + \dots \\
\varphi_H &= \varepsilon\tilde{\varphi}_H \ , \quad u_H = \varepsilon^2 \tilde{u}_H \ , \quad g_H = \varepsilon^2 \tilde{g}_H \quad (H = L, R)
\end{aligned}
\tag{107}
$$

From equations (107) it turns out that the longitudinal displacement u and the axial force λ are second order variables with respect to the transversal displacement v. They allow the kinematical inextensibility condition (143_3) and the equilibrium in the longitudinal direction (143_2) to be satisfied in the nonlinear range.

By substituting equations (107) in equations (143), equations $(145_{1,2,3,5})$ and $(146_{1,4})$, and by omitting hat and tilde, the perturbation equations follow

$$\epsilon: \quad v_1'''' - \beta^4 v_1 = 0$$

$$v_1(0) = 0 \ , \ v_1'(0) = \varphi_L$$
$$v_1(1) = 0 \ , \ v_1''(0) = -\frac{m_L}{EI}$$

$$\epsilon^2: \quad \begin{aligned} u_2' + \tfrac{1}{2}{v_1'}^2 &= 0 \\ \lambda_2' - 2\beta^4 u_2 &= 0 \end{aligned}$$

$$u_2(0) = u_L$$
$$\lambda_2(0) = -g_L \qquad (108)$$

$$\epsilon^3: \quad \begin{aligned} v_3'''' - \beta^4 v_3 = \\ -\tfrac{1}{2}\left(\lambda_2 v_1'\right)' - \tfrac{3}{4}\left[v_1'\left(v_1'v_1''\right)'\right]' \end{aligned}$$

$$v_3(0) = 0 \ , \ v_3(1) = 0$$
$$v_3'(0) + \tfrac{1}{6}{v_1'}^3(0) = 0$$
$$v_3''(0) + \tfrac{1}{2}{v_1'}^2(0)v_1''(0) = 0$$

By solving the equations (108) according to the increasing powers of ϵ, the corresponding solutions are obtained in the form:

$$\epsilon: \quad v_1(s) = \varphi_L F_\varphi(s) + m_L F_m(s)$$

$$\epsilon^2: \quad \begin{aligned} u_2(s) &= u_2(0) - \tfrac{1}{2}\int_0^s {v_1'}^2 \, ds = \\ & u_L + \varphi_L^2 F_{\varphi\varphi}(s) + m_L^2 F_{mm}(s) + \varphi_L m_L F_{\varphi m}(s) \\ \lambda_2(s) &= \lambda_2(0) + 2\beta^4 \int_0^s u_2 \, ds = -g_L + 2\beta^4 \left[u_L s \right. \\ & \left. + \varphi_L^2 G_{\varphi\varphi}(s) + m_L^2 G_{mm}(s) + \varphi_L m_L G_{\varphi m}(s)\right] \end{aligned} \qquad (109)$$

$$\epsilon^3: \quad \begin{aligned} v_3(s) &= \varphi_L^3 F_{\varphi\varphi\varphi}(s) + \varphi_L^2 m_L F_{\varphi\varphi m}(s) + \varphi_L m_L^2 F_{\varphi mm}(s) \\ & + m_L^3 F_{mmm}(s) + g_L \left[\varphi_L H_\varphi(s) + m_L H_m(s)\right] \\ & + u_L \left[\varphi_L K_\varphi(s) + m_L K_m(s)\right] \end{aligned}$$

where F_α, $F_{\alpha\alpha}$, $F_{\alpha\alpha\alpha}$, $G_{\alpha\alpha}$, H_α and K_α, with $\alpha = \varphi$, m, are combinations of circular, hyperbolic and polynomial functions. As expected, $v_1(s)$ and $v_3(s)$ are linear and cubic, respectively, with respect to φ_L, m_L. Moreover $v_3(s)$ is bilinear in the products of φ_L and m_L with the second order variables u_L, g_L.

Given the perturbative expansions, the remaining equations ($145_{4,6}$,$146_{2,3}$) are rewritten as

$$\varepsilon \varphi_R = \varepsilon v_1'(1) + \varepsilon^3 \left[v_3'(1) + \tfrac{1}{6} v_1'^{\,3}(1) \right]$$

$$\varepsilon m_R = -\varepsilon v_1''(1) - \varepsilon^3 \left[v_3''(1) + \tfrac{1}{2} v_1''(1) v_1'^{\,2}(1) \right]$$

$$\varepsilon^2 u_R = \varepsilon^2 u_2(1)$$

$$\varepsilon^2 g_R = -\varepsilon^2 \lambda_2(1)$$

(110)

They represent the nonlinear relationship between right-hand and left-hand variables. Continuity and equilibrium at the generic node k imply $d_{R,k} = d_{L,k+1} \doteq d_k$ ($d = \varphi, u$) and $f_{R,k} = -f_{L,k+1} \doteq f_k$ ($f = m, g$), respectively, where the second index labels the element. By substituting in equation (110) $v_1(s)$, $v_3(s)$, $u_2(s)$ and $\lambda_2(s)$, as given by equations (109), the element transfer function, linking the state variables at adjacent interfaces, is obtained

$$\varepsilon \begin{pmatrix} \varphi \\ m \end{pmatrix}_{k+1} = \varepsilon \begin{bmatrix} t_{dd} & t_{df} \\ t_{fd} & t_{ff} \end{bmatrix} \begin{pmatrix} \varphi \\ m \end{pmatrix}_k + \varepsilon^3 \begin{pmatrix} \mathcal{T}_\varphi(\varphi_k, m_k, u_k, g_k) \\ \mathcal{T}_m(\varphi_k, m_k, u_k, g_k) \end{pmatrix} \quad (111)$$

$$\varepsilon^2 \begin{pmatrix} u \\ g \end{pmatrix}_{k+1} = \varepsilon^2 \begin{bmatrix} 1 & 0 \\ -2\beta^4 & 1 \end{bmatrix} \begin{pmatrix} u \\ g \end{pmatrix}_k + \varepsilon^2 \begin{pmatrix} \mathcal{T}_u(\varphi_k, m_k) \\ \mathcal{T}_g(\varphi_k, m_k) \end{pmatrix} \quad (112)$$

where t_{pq}, $(p,q) = (d,f)$, are the usual entries of the mono-coupled linear transfer matrix Yong and Lin (1989) and $\mathcal{T}_\alpha$, $(\alpha = \varphi, m, u, g)$, are the components of the nonlinear part of the element transfer function.

Nonlinear system transfer function In order to link the state variables at the end of the chain, the map (111,112) must be iterated consistently with the approximation order. Towards this goal, the discrete counterpart of the perturbation scheme adopted for the continuous problem is followed. By setting $\mathbf{v} = (\varphi, m)$ and $\mathbf{u} = (u, g)$, after reabsorbing ε, the equations (111,112) are rewritten as

$$\mathbf{v}(k+1) - \mathbf{T}_v \mathbf{v}(k) = \mathcal{T}_v \left[\mathbf{v}(k), \mathbf{u}(k) \right]$$

$$\mathbf{u}(k+1) - \mathbf{T}_u \mathbf{u}(k) = \mathcal{T}_u \left[\mathbf{v}(k) \right]$$

(113)

By defining $\mathbf{T}_{u_0} = \mathbf{T}_u(\beta_0)$ and $\mathbf{T}_{v_0} = \mathbf{T}_v(\beta_0)$, and performing the following series expansions in analogy to equation (107)

$$\mathbf{v} = \epsilon\mathbf{v}_1 + \epsilon^3\mathbf{v}_3 \qquad \mathbf{u} = \epsilon^2\mathbf{u}_2 \qquad \beta = \beta_0 + \epsilon^2\beta_2$$

$$\mathbf{T}_\alpha = \mathbf{T}_{\alpha 0} + \epsilon^2\beta_2\mathbf{T}'_{\alpha 0} \quad \text{with} \quad \alpha = u, v \tag{114}$$

the following perturbation equations are obtained

$$\begin{aligned}
\epsilon &: \quad \mathbf{v}_1(k+1) - \mathbf{T}_{v_0}\mathbf{v}_1(k) = 0 \\
\epsilon^2 &: \quad \mathbf{u}_2(k+1) - \mathbf{T}_{u_0}\mathbf{u}_2(k) = \mathcal{T}_u\left[\mathbf{v}_1(k)\right] \\
\epsilon^3 &: \quad \mathbf{v}_3(k+1) - \mathbf{T}_{v_0}\mathbf{v}_3(k) = \beta_2\mathbf{T}'_{v_0}\mathbf{v}_1(k) + \mathcal{T}_v\left[\mathbf{v}_1(k), \mathbf{u}_2(k)\right]
\end{aligned} \tag{115}$$

The solutions of the difference equations (115) are

$$\begin{aligned}
\mathbf{v}_1(k) &= \mathbf{T}_{v_0}^k\mathbf{v}_1(0) \\
\mathbf{u}_2(k) &= \mathbf{T}_{u_0}^k\mathbf{u}_2(0) + \sum_{j=0}^{k-1}\mathbf{T}_{u_0}^j\mathcal{T}_u\left[\mathbf{T}_{v_0}^{k-1-j}\mathbf{v}_1(0)\right] \\
\mathbf{v}_3(k) &= \mathbf{T}_{v_0}^k\mathbf{v}_3(0) + \sum_{j=0}^{k-1}\beta_2\mathbf{T}_{v_0}^j\mathbf{T}'_{v_0}\mathbf{T}_{v_0}^{k-1-j}\mathbf{v}_1(0) \\
&\quad + \sum_{j=0}^{k-1}\mathbf{T}_{v_0}^j\mathcal{T}_v\left[\mathbf{T}_{v_0}^{k-1-j}\mathbf{v}_1(0), \mathbf{u}_2(k-1-j)\right]
\end{aligned} \tag{116}$$

By introducing the global transfer matrix $\mathbf{S}_v \doteq \mathbf{T}_{v_0}^N$, the state at the end of the periodic beams is then obtained from equation (116) for $k = N$:

$$\begin{aligned}
\mathbf{v}_1(N) &= \mathbf{S}_v\mathbf{v}_1(0) \\
\mathbf{u}_2(N) &= \mathbf{S}_u\mathbf{u}_2(0) + \mathcal{S}_u\left(\mathbf{v}_1(0), \mathbf{0}\right) \\
\mathbf{v}_3(N) &= \mathbf{S}_v\mathbf{v}_3(0) + \beta_2\mathbf{S}'_v\mathbf{v}_1(0) + \mathcal{S}_v\left(\mathbf{v}_1(0), \mathbf{U}_2\right)
\end{aligned} \tag{117}$$

where

$$\begin{aligned}
\mathcal{S}_\alpha(\mathbf{x}, \mathbf{Y}) &\doteq \sum_{j=0}^{N-1}\mathbf{T}_{\alpha 0}^j\mathcal{T}_\alpha\left[\mathbf{T}_{v_0}^{N-1-j}\mathbf{x}, \mathbf{y}(N-1-j)\right] \qquad \alpha = u, v \\
\mathbf{S}'_v &\doteq \sum_{j=0}^{N-1}\mathbf{T}_{v_0}^j\mathbf{T}'_{v_0}\mathbf{T}_{v_0}^{N-1-j}
\end{aligned} \tag{118}$$

with $\mathbf{Y} = [\mathbf{y}(0), \dots, \mathbf{y}(N-1)]$.

Frequency-amplitude relationship   An illustrative example consisting of a hinged-roll supported chain of beams loaded by a couple m_0 acting on the left-most support, is considered. The boundary conditions read

$$m(0) = \pm m_0, \quad m(N) = 0, \quad u(0) = 0, \quad g(N) = 0, \qquad (119)$$

where the $\pm$ in equation (119_1) is due to the force being either in-phase or counter-phase with the response. By using equations $(114_1, 114_2)$ the boundary conditions for the perturbation equation (117) follow

$$
\begin{array}{llll}
\varepsilon : & m_1(0) = m_0 & , & m_1(N) = 0 \\
\varepsilon^2 : & u_2(0) = 0 & , & g_2(N) = 0 \\
\varepsilon^3 : & m_3(0) = 0 & , & m_3(N) = 0
\end{array}
\qquad (120)
$$

The frequency-amplitude relation for free vibrations ($m_0 = 0$) is sought. To define the amplitude of the modal shape, the normalization condition $\varphi(0) = \varepsilon\pi$ is chosen, with ε being a scaling factor representing the mid-span displacement to length ratio in the linear approximation. Consequently,

$$\varphi_1(0) = \pi \text{ and } \varphi_3(0) = 0 \qquad (121)$$

are the normalization conditions for the perturbation equations $(117_1, 117_3)$ The order ϵ equations (117_1) read

$$
\epsilon : \quad
\begin{pmatrix} \varphi_1(N) \\ m_1(N) \end{pmatrix}
=
\begin{bmatrix} S_{\varphi\varphi} & S_{\varphi m} \\ S_{m\varphi} & S_{mm} \end{bmatrix}
\begin{pmatrix} \varphi_1(0) \\ m_1(0) \end{pmatrix}
\qquad (122)
$$

From the equation (122_2), by accounting for the boundary condition $(120_{1,2})$

$$S_{m\varphi}(\beta_0) = 0 \qquad (123)$$

provides with the roots β_{0_k} corresponding to the natural frequencies of the small amplitude oscillations. By iterating (116_1), with $\mathbf{v}_1(0) = (\pi, 0)$, the linear modal shape is obtained.
The equations (117_2) read

$$
\epsilon^2 : \quad
\begin{pmatrix} u_2(N) \\ g_2(N) \end{pmatrix}
=
\begin{bmatrix} S_{uu} & S_{ug} \\ S_{gu} & S_{gg} \end{bmatrix}
\begin{pmatrix} u_2(0) \\ g_2(0) \end{pmatrix}
+
\begin{pmatrix} \mathcal{S}_u(\mathbf{v}_1(0), \mathbf{0}) \\ \mathcal{S}_g(\mathbf{v}_1(0), \mathbf{0}) \end{pmatrix}
\qquad (124)
$$

By taking into account the boundary condition $(120_{3,4})$, equation (124_2) yields the axial force at the left end of the beams

$$g_2(0) = -\frac{\mathcal{S}_g(\mathbf{v}_1(0), \mathbf{0})}{S_{gg}} \qquad (125)$$

consequently $\mathbf{u}_2(0) = (0, g_2(0))$. By iterating equation (116_2), starting from the known values of $\mathbf{u}_2(0)$ and $\mathbf{v}_1(0)$, $\mathbf{u}_2(k)$ is evaluated at any node of the chain. Equations (117_3) read

$$
\epsilon^3 : \quad
\begin{pmatrix} \varphi_3(N) \\ m_3(N) \end{pmatrix}
=
\begin{bmatrix} S_{\varphi\varphi} & S_{\varphi m} \\ S_{m\varphi} & S_{mm} \end{bmatrix}
\begin{pmatrix} \varphi_3(0) \\ m_3(0) \end{pmatrix}
$$
$$
+ \beta_2
\begin{bmatrix} S'_{\varphi\varphi} & S'_{\varphi m} \\ S'_{m\varphi} & S'_{mm} \end{bmatrix}
\begin{pmatrix} \varphi_1(0) \\ m_1(0) \end{pmatrix}
\tag{126}
$$
$$
+
\begin{pmatrix} \mathcal{S}_{\varphi}(\mathbf{v}_1(0), \mathbf{U}_2) \\ \mathcal{S}_m(\mathbf{v}_1(0), \mathbf{U}_2) \end{pmatrix}
$$

From the equation (126_2), by using the boundary conditions $(120_{1,5,6})$ and the normalization conditions (121), the natural frequency correction is derived as

$$
\beta_2 = -\frac{\mathcal{S}_m(\mathbf{v}_1(0), \mathbf{U}_2)}{\pi \, S'_{m\varphi}}
\tag{127}
$$

Finally, by iterating (116_3) with $\mathbf{v}_3(0) = (0,0)$, the nonlinear correction $\mathbf{v}_3(k)$ to the linear mode is obtained. In forced resonant motions the frequency $\beta = \beta_0 + \epsilon^2 \beta_2$ is prescribed, with β_2 representing a detuning parameter. At first, the boundary conditions $(120_{1,2})$ are introduced in equation (122_2) to obtain the first order approximation of the left end rotation $\varphi_1(0)$. From equation (116_1) the first approximation of the chain response is obtained, then, by using the boundary conditions $(120_{3,4})$, $g_2(0)$ is determined through equation (124_2). Once $\mathbf{u}_2(0)$ and $\mathbf{v}_1(0)$ are known, $\mathbf{u}_2(k)$ at any node of the chain is obtained by iterating equation (116_2). From equation (126_2), the boundary conditions $(120_{1,5,6})$ and choosing $\varphi_3(0) = 0$ as normalization condition, the frequency-amplitude relation follows

$$
\beta_2 = -\frac{\mathcal{S}_m(\mathbf{v}_1(0), \mathbf{U}_2)}{\varphi_1(0) S'_{m\varphi} \pm m_0 S'_{mm}}
\tag{128}
$$

Numerical results and final remarks A chain of four simply supported beams, each of length l, with the left end hinged, is considered. The frequency-amplitude relation (114_3) is evaluated by using either equation (127) or (128), for the free and forced case, respectively. The perturbation parameter ϵ entering equation (114_3) is expressed either $\epsilon = a/\pi$ or $\epsilon = a/\varphi_1(0)$ in the two cases, where $a \doteq \varphi(0)$ is the amplitude. The backbone curves of the free motion of the chain are first drawn. Figures

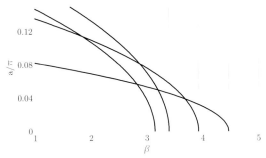

Figure 38. First four natural frequencies-amplitude relationship for a 4-bay beam; thick lines: nonlinear; thin lines: linear.

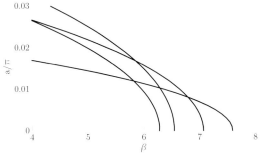

Figure 39. Second four natural frequencies-amplitude relationship for a 4-bay beam; thick lines: nonlinear; thin lines: linear.

38 and 39 show these curves for the natural frequencies of the first four modes and for modes from the fifth to the eighth, respectively. The two sets of four curves have been shown separately as they belong to the first ($\pi < \beta < 4.730$) and second ($2\pi < \beta < 7.853$) pass-band, respectively, of the corresponding linear model. As the amplitude of oscillations increases, all the natural frequencies trespass the lower frequency boundary of the corresponding linear pass band. The curves relevant to the third, fourth, seventh and eighth modes are characterized by slopes higher than that of the neighbouring ones, entailing crossings among several backbone curves. A softening type of behaviour can be observed. This is due to the prevailing effect of the inertia nonlinearity with respect to the hardening effect of the elastic one Luongo et al. (1986). Moreover, such softening effect increases with the number of beams in the chain, as shown in Figure 40 where the first (dimensional) natural frequency corrections versus the amplitude of the free oscillation are drawn for increasing number of beams entering

the chain. Such a behaviour is explained by the fact that each beam of
the chain, in addition to the longitudinal displacements caused by the own
transversal displacements, undergoes a longitudinal translation equal to the
sum of the shortenings of all the preceding beams. Therefore, the longer
the chain, the more important the longitudinal kinetic energy. The curve
corresponding to the single beam coincides with the one obtained in Luongo
et al. (1986), where the nonlinear dynamics of beams on movable supports
have been studied. Figures 41 and 42 show the first four modal shapes;

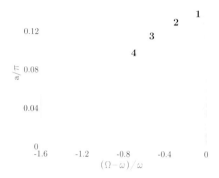

Figure 40. First natural frequencies-amplitude relationship for a chain
with number of beams increasing from 1 to 4 (Ω, ω: nonlinear and linear
frequency).

the comparison between the linear and nonlinear modes are depicted for increasing amplitude of oscillations. The nonlinear shortening effect destroys
the symmetry of the linear modes; it can be best noticed at the right end
of the chain.

As far as the forced oscillations is concerned, Figure 43 shows the frequency-
response curves for the first four modes separately, for different intensities
of the couple m_0. As shown in Figure 44, where the frequency-response
curves are drawn next to each other, such intensity must be limited in order
for the present unimodal analysis to be meaningful.

   The free and forced harmonic response of a chain of nonlinear beams
resting on sliding supports has been determined. The periodicity of the
system has been exploited by using an asymptotic approximation of the
nonlinear transfer function. In analogy to the linear case, the approach leads
to an algebraic set of equations whose dimensions are equal to the number of
coupling degrees of freedom. Starting from the single inextensible and shear
undeformable beam transfer function, the global transfer function has been

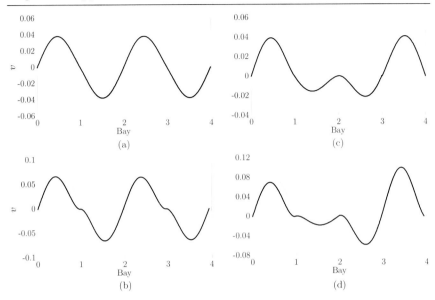

Figure 41. First and second mode of a 4-bay beam ; (a) First mode ($\beta = \pi$) $a = 0.04\,\pi$, (b) First mode ($\beta = \pi$) $a = 0.08\,\pi$, (c) Second mode ($\beta = 3.393$) $a = 0.04\,\pi$, (d) Second mode ($\beta = 3.393$) $a = 0.08\,\pi$.

derived in order to obtain the frequency-amplitude relationship. Due to the movable supports, the system is not symmetric with respect to longitudinal displacement and forces that enter the problem as second order variables. For the beam model considered it has been found that the softening effect due to the longitudinal inertia increases with the number of elements of the chain. Future investigations are in order to extend the procedure to multi-modal analysis to assess the influence of closely spaced modes.

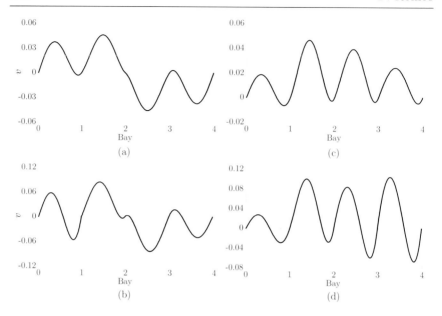

Figure 42. Third and fourth mode of a 4-bay beam ; (a) Third mode ($\beta = 3.926$) $a = 0.04\,\pi$, (b) Third mode ($\beta = 3.926$) $a = 0.06\,\pi$, (c) Fourth mode ($\beta = 4.464$) $a = 0.02\,\pi$, (d) Fourth mode ($\beta = 4.464$) $a = 0.035\,\pi$.

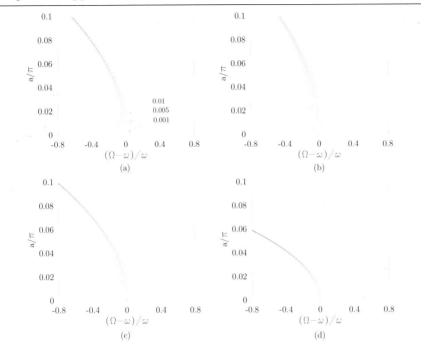

Figure 43. Frequency-response relationship for a 4-bay beam with $m_0 = 0.0, 0.001, 0.005, 0.01$; (a) first mode, (b) second mode, (c) third mode, (d) fourth mode (Ω, ω: nonlinear and linear frequency).

Figure 44. Frequency-response relationship in the first linear pass band for a 4-bay beam; $m_0 = 0.01$.

2.2   Chains of mono-coupled nonlinear oscillators

Monocoupled chains of oscillators with cubic nonlinearity are studied
at first on the basis of the results presented in Romeo and Rega (2006).
Pass and stop band regions are analytically determined for period-q orbits
as they are governed by the eigenvalues of the linearized 2D map arising
from linear stability analysis of periodic orbits. The interest in this model
lies in its mathematical representation as it is equivalent to a fictitious bi-
coupled chain of nonlinear oscillators. The nonlinearity causes amplitude
dependent pass and stop bands, and the occurrence of vibration modes in
the linear attenuation zones observed in literature Manevitch et al. (2005)
can thereby be explained. Numerical investigations carried out by nonlinear map iteration are presented. They are mainly aimed at characterizing
the bounded solutions occurring within the passing band, where, besides
periodic orbits, quasiperiodic and chaotic orbits do exist. Good agreement
between the approximate analytical prediction of the propagation regions
and the numerical evidence is found.

A mechanical model for a chain of linearly-coupled nonlinear oscillators,
schematically depicted in Figure 1, has been chosen in the form:

$$m\ddot{q}_n + k_1 q_n + k_3 q_n^3 + k(2q_n - q_{n-1} - q_{n+1}) = 0 \qquad (129)$$

The equation of motion (129) is characterized by on-site cubic nonlinearity,
as e.g. in Manevitch (2001) and Umberger et al. (1989), describing a chain of
Hamiltonian oscillators. Moreover it is worth noticing that equation (129)
is also formally equivalent to the discrete nonlinear Schrödinger equation
(DNLS) in real domain considered in Wan and Soukoulis (1990).

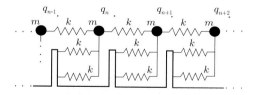

Figure 45. Monocoupled nonlinear spring-mass chain.

Periodic solutions of equation (129) are sought for by assuming the time
harmonic solution $q_n = a_n \cos(\omega t)$ (harmonic balance with only the first

harmonic), and setting

$$\alpha = \frac{m\omega^2 - k_1}{k} - 2 \quad \text{and} \quad \beta = -\frac{k_3}{k} \tag{130}$$

the following second-order difference equation for the stationary amplitude is obtained

$$(\alpha + \beta a_n^2)a_n + a_{n+1} + a_{n-1} = 0 \tag{131}$$

Equation (131), relating the amplitudes a in adjacent chain sites $n - 1$, n and $n + 1$, can be rewritten in matrix form as

$$\begin{pmatrix} a_{n+1} \\ a_n \end{pmatrix} = \begin{bmatrix} -\alpha - \beta a_n^2 & -1 \\ 1 & 0 \end{bmatrix} \begin{pmatrix} a_n \\ a_{n-1} \end{pmatrix} \tag{132}$$

or $\mathbf{a}_{n+1} = \mathbf{T}(a_n)\mathbf{a}_n$, where $\mathbf{T}(a_n)$ is the nonlinear transfer matrix. In the linear case, $\mathbf{T}$ represents a symplectic linear transformation and its reciprocal eigenvalues λ satisfy $\lambda_1\lambda_2 = 1$. As well known from Signorelli and von Flotow (1988), such eigenvalues govern the stationary wave transmission properties: if the eigenvalues lie on the unit circle, then free waves propagate harmonically without attenuation (pass band, $\mathbf{P}$); if the eigenvalues are real, then free waves decay without oscillations (stop band, $\mathbf{S}$). In the more general nonlinear case, $\mathbf{T}(a_n)$ belongs to the class of area preserving maps Wan and Soukoulis (1990) such that $\det(\mathbf{D}T(a_n)) = 1$, where $\mathbf{D}T$ is the Jacobian or tangent map with reciprocal eigenvalues. Therefore, in order to study the stationary wave transmission properties of the one-dimensional nonlinear chain (129) of lenght N, it is convenient to rely on the eigenvalues of the linearized map equations in the neighborhood of an orbit ranging from (a_0, a_1) to (a_{N-1}, a_N). Indeed, according to Hennig and Tsironis (1999), the transformation (132) can be considered as a dynamical system where the chain index n plays the role of discrete time n, so that the analysis of the transmission properties is equivalent to the stability analysis of the orbits. The linear stability of a given orbit is investigated by introducing a small complex-valued perturbation u_n; linearizing the map equations, a second-order difference equation for the perturbation is obtained

$$u_{n+1} + u_{n-1} + (\alpha + \beta a_n^2)u_n = 0 \tag{133}$$

Equation (133) can be put in matrix form leading to the two-dimensional Jacobian

$$\mathbf{D}T = \begin{bmatrix} -\alpha - 3\beta a_n^2 & -1 \\ 1 & 0 \end{bmatrix} \tag{134}$$

The interest lies in the linear stability of spatially periodic orbits $a_{n+q} = a_n$ with cycle length q. The eigenvalues of $\mathbf{D}T$ are determined by its trace,

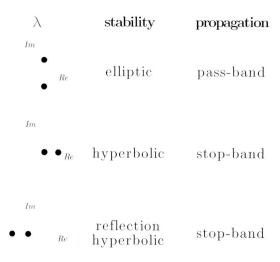

Figure 46. Stability-propagation equivalence based on the eigenvalues of the Jacobian $\mathbf{D}T$.

so the stability of period-q orbits is described by $Tr(\mathbf{D}T^q)$, where $\mathbf{D}T^q$ is given by the matrix product

$$\mathbf{D}T^q = \prod_{n=0}^{q-1} \mathbf{D}T(a_n) \tag{135}$$

If the eigenvalues lie on the unit circle, then stable elliptic periodic cycles or oscillating solutions (pass band, $\mathbf{P}$) occur; if the eigenvalues are real, then unstable hyperbolic periodic cycles or exponentially increasing solutions (stop band, $\mathbf{S}$) occur (Hennig and Tsironis, 1999), see Figure 46. By setting $a_{n+1} = x_{n+1}$ and $a_n = y_{n+1}$, we can express the map (132) as

$$\mathbf{T}: \begin{aligned} x_{n+1} &= -y_n - x_n(\beta x_n^2 + \alpha) \\ y_{n+1} &= x_n \end{aligned} \tag{136}$$

For period-1 orbits the curves bounding the propagation regions, where the eigenvalues lie on the unit circle, can be determined by the condition $|Tr(\mathbf{D}T)| = 2$. Having introduced the change of variables $(x, y) \rightarrow$

$(x/\sqrt{\beta}, y/\sqrt{\beta})$, such boundaries are given by

$$r := \quad \{(x, \alpha) \mid 3x^2 + \alpha + 2 = 0\}$$
$$s := \quad \{(x, \alpha) \mid 3x^2 + \alpha - 2 = 0\}$$

(137)

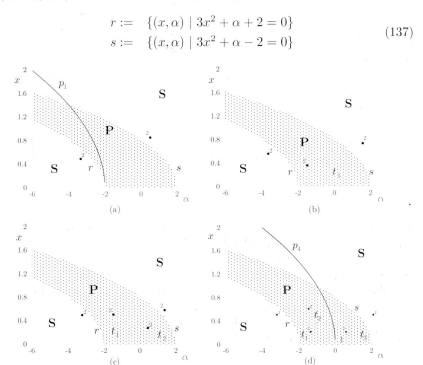

Figure 47. Propagation region of period-q orbits: a) $q = 1$; b) $q = 2$; c) $q = 3$; d) $q = 4$.

and are shown in Figure 47a in the x-positive half-plane. The curves r and s represent hyperbolic ($\lambda_1 = \lambda_2 = 1$) and reflection hyperbolic ($\lambda_1 = \lambda_2 = -1$) boundaries, respectively. In Figures 47b,c,d further curves t_i, lying inside the pass band region, are depicted; they are determined by satisfying the condition $|Tr(\mathbf{D}T^q)| = 2$ for $q = 2, 3, 4$ and their number depends on the periodicity of the orbit, i.e. t_i with $i = 1, \ldots, q - 1$. While the curves r are always hyperbolic boundaries, the curves s are either hyperbolic with reflection boundaries, for q odd, or hyperbolic boundaries, for q even. Whenever a period-q orbit crosses a curve t_i it temporarily loses its stability or, equivalently, does not propagate, through either a saddle-node or a period-doubling bifurcation for i even or i odd, respectively. Such alternating pattern of the internal curves t_i implies that, while the elliptic even-period orbits become hyperbolic at the upper pass-band boundary, the odd-period ones, at the same boundary, become hyperbolic with reflection. It must be emphasized that the periodicity q governs the new born internal curves without affecting the overall propagation region. As expected, the nonlinearity ($\beta \neq 0$) implies a propagation region depending on the amplitude of oscillations x (see Figure 48), in contrast to the linear case, where the propagation region is given by $|\alpha| \leq 2$.

In order to find periodic orbits of the map $\mathbf{T}$ defined by equation (132), it is convenient to exploit its reversibility, therefore it can be cast into the product of two involutions $\mathbf{T} = \mathbf{T}_1 \mathbf{T}_2$, given by

$$
\begin{aligned}
\mathbf{T}_1 \quad &: \quad \begin{aligned} x_{n+1} &= y_n \\ y_{n+1} &= x_n \end{aligned} \\[2mm]
\mathbf{T}_2 \quad &: \quad \begin{aligned} x_{n+1} &= x_n \\ y_{n+1} &= (-\beta x_n^2 - \alpha)x_n - y_n \end{aligned}
\end{aligned}
\tag{138}
$$

The involutions (138) satisfy $\mathbf{T}_1^2 = \mathbf{T}_2^2 = \mathbf{I}$ so that the inverse of $\mathbf{T}$ can be expressed as $\mathbf{T}^{-1} = \mathbf{T}_2 \mathbf{T}_1$. The sets of fixed points of $\mathbf{T}_1$ and $\mathbf{T}_2$ form two curves in the phase plane, $x = y$ and $y = -1/2\,x(\beta x^2 + \alpha)$ respectively, called dominant symmetry lines. The latter curves are very useful to determine periodic orbits since a periodic orbit initially on the symmetry line stays on the line as the parameters are changed Wan and Soukoulis (1990). For instance, period-1 and period-4 orbits, representing the fixed points of $\mathbf{T}$ and $\mathbf{T}^4$, respectively, born on the symmetry line of $\mathbf{T}_1$, are determined as

$$
\begin{aligned}
\text{Period-1} \quad &: \quad x = y = \pm\sqrt{-2 - \alpha} \\
\text{Period-4} \quad &: \quad x = y = \pm\sqrt{-\alpha}
\end{aligned}
\tag{139}
$$

and are superimposed in Figures 47a and 47d, respectively. The propagation properties of the periodic orbits (139) will be numerically investigated in the next section.

It can be noticed that, regardless of the periodicity q, the bounded orbits region coincides with that of the period-1 case whose boundaries are given by the curves of equations (137). However, the loss of stability through the upper bound involves different bifurcations if even-period or odd-period orbits are considered. Namely, while odd-period ones lose stability via period-doubling bifurcation, such as point A at $\alpha = -4$ in the following Figure 49 for the period-1 case, the even-period orbits lose stability via saddle-node bifurcation, such as point B at $\alpha = -1$ for the period-4 case.

Numerical results vs. analytical predictions   Periodic orbits of map $\mathbf{T}$ in equation (139) are numerically investigated in this section. In the

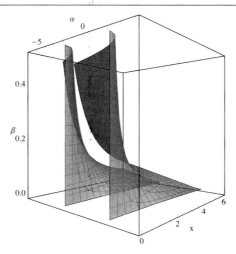

Figure 48. Amplitude dependent propagation band vs increasing nonlinearity.

numerical procedure only the orbits with amplitude smaller than a selected upper limit are considered bounded; it has been verified that the stability or boundedness zone boundary is not affected by the upper limit value as long the number of iterations exceeds the order of 10^2. In Figure 49 bifurcation diagrams for period-1 and period-4 orbits obtained by iterating the map **T** starting with initial conditions along the curves of equations (139), are shown. According to the analytical predictions, bounded periodic orbits are found within the **P** zone as indicated by the continuous lines, whereas after crossing the upper boundary unbounded orbits take place, as shown by the occurrence of empty intervals along the curves. The eigenvalues obtained by the numerical iteration of the maps confirm for odd- and even-period orbits the different type of loss of stability analytically predicted.

Yet, it is worth noticing that, within the propagation zone, not only regular solutions but also a rich variety of non-regular bounded orbits can occur (Wan and Soukoulis, 1990). In Figure 50 several orbits occurring numerically in the neighbourhood of the intersections of period-1 and period-4 orbits with internal critical thresholds, when varying initial conditions, are shown on the phase plane of **T**. Temporary loss of stability of period-1 orbits according to a tripling bifurcation at $\alpha = -3.5$ and a quadrupling bifurcation at $\alpha = -3$ are shown in Figures 50a and 50b, respectively. These bifurcated solutions occur where the period-1 orbits cross the curves t_2 of $\mathbf{T}^3$ (point A in Figure 50a) and $\mathbf{T}^4$ (point B in Figure 50b), respectively, where saddle-node bifurcations are analytically predicted. In particular, Figure 50a shows the fixed point of the map $\mathbf{T}$ (period-1 orbits) on the symmetry line at $x = y = 1.224$ splitting into a period-3 orbit, as well as the ensuing quasiperiodic orbits diverging along three paths when varying the initial conditions. In Figure 50b the fixed point at $x = y = 1$ splits into a period-4 orbit and then quasiperiodic orbits and chaotic layers occur before the onset of unbounded solutions. In turn, Figure 50c, besides a number of neighboring quasiperiodic orbits, shows two pairs of doubled numerical solutions at $\alpha = -\sqrt{2}/2$, consistent with the analytical crossing of the reference period-4 orbit with the period doubling bifurcation curve t_3 of $\mathbf{T}^4$ (point C in Figure 50c). It must be noticed that since period-2 orbits do not exist, period doubling bifurcations of period-1 orbits at the intersection with t_1 of either $\mathbf{T}^2$ or $\mathbf{T}^3$ cannot occur.

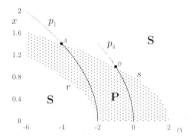

Figure 49. Numerical bifurcation diagram for period-1 and period-4 orbits.

2.3   Chains of bi-coupled nonlinear oscillators

By extending the results of the previous section concerning mono-coupled oscillators (Romeo and Rega, 2006), where essentially bidimensional maps were involved, four-dimensional nonlinear maps are introduced in this section to analyse free wave propagation properties of bi-coupled oscillators with cubic nonlinearities. As far as multi-coupled nonlinear chains are concerned, in Sayadi and Pouget (1991); Pouget (2005), the nonlinear dynamics of a lattice model involving internal degrees of freedom were studied using a continuum approximation. In order to model molecular crystals, longitudinal and transverse displacements of the mass centre of the molecular group and two rigid-body rotational motions of the electric dipoles were considered.

The bi-coupled model here considered consists of a chain of linearly coupled mechanical oscillators characterized by on-site cubic nonlinearities in

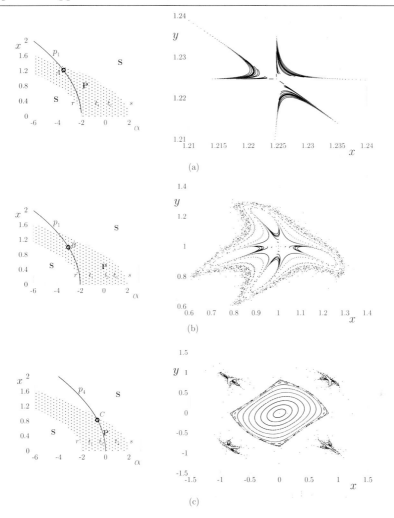

Figure 50. Bifurcations of analytical periodic orbits at internal critical thresholds (left) and corresponding numerical phase portraits with varying initial conditions (right) around:(a) tripling bifurcation of period-1 orbits (point A); (b) quadrupling bifurcation of period-1 orbits (point B); (c) period-doubling bifurcation of period-4 orbits (point C).

both the longitudinal and rotational degrees of freedom. Pass, stop and complex regions are analytically determined for period-q orbits as they are governed by the eigenvalues of the linearized map arising from linear stability analysis of periodic orbits. By varying the parameters governing both the coupling between the two d.o.f. and the nonlinearity, a variable scenario of propagation regions can be obtained and the transition between mono- and bi-coupled behaviors can be described. In both cases the nonlinearity causes amplitude dependent propagation regions. The analytical findings concerning the propagation properties are then complemented and compared with numerical results obtained through nonlinear map iteration. The latter are mainly focused on the bounded solutions occurring within the pass-pass band, where, besides periodic orbits, quasiperiodic and chaotic orbits may occur in each d.o.f. with a possibly rich variety of overall response. The model of the chain and the nonlinear map approach are described at first. The approach is then used to derive the analytical boundaries of the propagation regions and to carry out a parametric analysis pertaining the influence of both coupling and nonlinearities. The numerical validation of the analytical predictions highlighting interesting features of bounded solutions are discussed by relying on periodic orbits paths, bifurcation diagrams and basins of attraction.

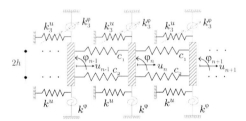

Figure 51. Bi-coupled chain of nonlinear oscillators.

Modeling and nonlinear map approach   The chain of oscillators schematically depicted in Figure 51 represents the bi-coupled periodic system under investigation. The bi-coupled nature arises from the presence of longitudinal (u) and rotational (φ) degrees of freedom. The oscillators' lumped translational (m_u) and rotational (m_φ) masses are coupled by two linear longitudinal springs (c_1, c_2) at a distance $2h$, whereas both linear (k^u, k^φ) and cubic (k_3^u, k_3^φ) on-site springs act on both degrees of freedom. The relevant Eqs. of motion for the generic n-th oscillator read:

$$m_u \ddot{u}_n + k^u u_n + k_3^u u_n^3 \quad + (c_1 + c_2)(2u_n - u_{n-1} - u_{n+1})$$
$$+ \; h(c_2 - c_1)(2\varphi_n - \varphi_{n-1} - \varphi_{n+1}) = 0 \quad (140)$$

$$m_\varphi \ddot{\varphi}_n + k^\varphi \varphi_n + k_3^\varphi \varphi_n^3 \quad + h(c_2 - c_1)(2u_n - u_{n-1} - u_{n+1})$$
$$+ \; h^2(c_1 + c_2)(2\varphi_n - \varphi_{n-1} - \varphi_{n+1}) = 0 \quad (141)$$

Introducing the nondimensional variable $\tilde{u} = u/h$ and assuming the time harmonic solutions $\tilde{u}_n = a_n \cos(\omega t)$, $\varphi_n = b_n \cos(\omega t)$, Eqs. (140,141) can be rewritten as second order nondimensional nonlinear difference equations in the longitudinal and rotational stationary amplitudes a_n and b_n, in the form

$$(\alpha_u + \beta_u a_n^2)a_n + a_{n+1} + a_{n-1} + \gamma(2b_n - b_{n-1} - b_{n+1}) = 0$$
$$(\alpha_\varphi + \beta_\varphi b_n^2)b_n + b_{n+1} + b_{n-1} + \gamma(2a_n - a_{n-1} - a_{n+1}) = 0 \quad (142)$$

where the parameters α_h, β_h $(h = u, \varphi)$ and γ, defined as

$$\alpha_u = \frac{m_u \omega^2 - k^u}{c_1 + c_2} - 2 \quad \beta_u = -\frac{3}{4}\frac{k_3^u h^2}{c_1 + c_2} \quad \gamma = -\frac{c_2 - c_1}{c_1 + c_2}$$
$$\alpha_\varphi = \frac{m_\varphi \omega^2 - k^\varphi}{h^2(c_1 + c_2)} - 2 \quad \beta_\varphi = -\frac{3}{4}\frac{k_3^\varphi}{h^2(c_1 + c_2)} \quad (143)$$

group the longitudinal and rotational linear and nonlinear parameters, respectively, as well as the coupling one. In particular, α_u and α_φ play the role of equivalent linear frequencies in the map representation. After straightforward algebraic manipulations Eqs. (142) can be put in the compact form $\mathbf{a}_{n+1} = \mathbf{T}(a_n, b_n)\,\mathbf{a}_n$, where $\mathbf{T}(a_n, b_n)$ is the 4×4 nonlinear transfer matrix and the amplitude vector is defined as $\mathbf{a}_{n+1} = (a_{n+1}, a_n, b_{n+1}, b_n)^T$. In the linear case, the transfer matrix $\mathbf{T}$ represents a symplectic linear transformation and its four reciprocal eigenvalues are, in pairs, λ_i and λ_i^{-1} with $i = 1, 2$. Therefore forward and backward waves always exist in pairs and have both the same propagation properties. It follows that two eigenvalues λ_1 and λ_2 such that $|\lambda_i| \leq 1$ $(i = 1, 2)$, completely define the propagation properties of a bi-coupled periodic structure; they will be here referred to as *principal* eigenvalues (Romeo and Luongo, 2002). Such eigenvalues govern the stationary wave transmission properties: if the generic pair of eigenvalues lie on the unit circle, then free waves propagate harmonically without attenuation (pass band, $\mathbf{P}$); if the pair is real, then free waves decay without oscillations (stop band, $\mathbf{S}$). Since the coupling coordinates are more than one, there exist further frequency bands characterized by harmonic propagation with attenuation (complex bands, $\mathbf{C}$) where pairs of complex conjugate eigenvalues, with modulus different from 1, exist.

In the more general nonlinear case, $\mathbf{T}(a_n, b_n)$ belongs to the class of area preserving maps (Wan and Soukoulis, 1990) such that $\det(\mathbf{DT}(a_n, b_n)) = 1$, where $\mathbf{DT}$ is the Jacobian or tangent map with reciprocal eigenvalues. In this case, the stationary wave transmission properties of the one-dimensional nonlinear chain (140,141) of length N are identified by relying on the eigenvalues of the linearized map equations in the neighborhood of an orbit ranging from $\mathbf{a}_1$ to $\mathbf{a}_N$. In other words, the transformation (142) is considered as a dynamical system where the chain index n plays the role of discrete time n (Hennig and Tsironis, 1999), so that the analysis of the transmission properties is equivalent to the stability analysis of the orbits. The linear stability of a given orbit is investigated by introducing a small perturbation; then, linearizing the map equations, two second-order difference equations for the perturbation are obtained. From the latter equations, the following four dimensional tangent map is deduced:

$$\mathbf{DT}(a_n, b_n) = \begin{bmatrix} \frac{-\alpha_u - 3\beta_u a_n^2 - 2\gamma^2}{1 - \gamma^2} & -1 & -\frac{(\alpha_\varphi + 3\beta_\varphi b_n^2 - 2)\gamma}{1 - \gamma^2} & 0 \\ 1 & 0 & 0 & 0 \\ -\frac{(\alpha_u + 3\beta_u a_n^2 - 2)\gamma}{1 - \gamma^2} & 0 & \frac{-\alpha_\varphi - 3\beta_\varphi b_n^2 - 2\gamma^2}{1 - \gamma^2} & -1 \\ 0 & 0 & 1 & 0 \end{bmatrix} \quad (144)$$

Propagation Region Parametric Analysis   On the basis of the tangent map (144), the linear stability or, equivalently, the propagation properties of spatially periodic orbits $\mathbf{a}_{n+q} = \mathbf{a}_n$, with cycle length q, are studied through the eigenvalues of $\mathbf{DT}^q$, defined by the matrix product

$$\mathbf{DT}^q = \prod_{n=0}^{q-1} \mathbf{DT}(a_n, b_n) \quad (145)$$

In essence while in the linear case stationary wave transmission is governed by the transfer matrix principal eigenvalues, in the nonlinear case periodic orbits transmission is governed by the tangent map principal eigenvalues. Therefore, according to the propagation properties description of the previous section, if both the eigenvalues lie on the unit circle, then stable elliptic periodic cycles or oscillating solutions (pass-pass band, **PP**) occur; if at least one of the eigenvalues is real, then unstable hyperbolic periodic cycles or exponentially increasing solutions (pass-stop band, **PS** or stop-stop band, **SS**) occur. Lastly, harmonically increasing solutions take place if complex conjugate eigenvalues, with modulus different from 1, exist (complex band **C**). The type of eigenvalues depend upon the invariants I_1, I_2 of the Jacobian reflexive characteristic equation, $\det[\mathbf{DT} - \lambda\,\mathbf{I}] = 0$. More specifically, in the invariants' plane, three curves r, s and p exist which define the domains where the eigenvalues are of the same type (Romeo and Luongo, 2002):

$$r := \{(I_1, I_2) \mid 2 - 2I_1 + I_2 = 0\}, \quad s := \{(I_1, I_2) \mid 2 + 2I_1 + I_2 = 0\}$$

$$p := \{(I_1, I_2) \mid 8 + I_1^2 - 4I_2 = 0\}$$

(146)

In order to transform the invariant space into the physical space, the invariants I_1 and I_2 are expressed as functions of the stationary amplitudes and of the linear, nonlinear and coupling parameters defined by Eqs. (143), as follows

$$I_1 = \frac{\alpha_u + \alpha_\varphi + 3\beta_u a_n^2 + 3\beta_\varphi b_n^2 + 4\gamma^2}{1 - \gamma^2}$$

(147)

$$I_2 = \frac{2 + (\alpha_\varphi + 3\beta_\varphi b_n^2)(\alpha_u + 3\beta_u a_n^2) - 6\gamma^2}{1 - \gamma^2}$$

(148)

Given the variety of parameters entering the model, different aspects of the

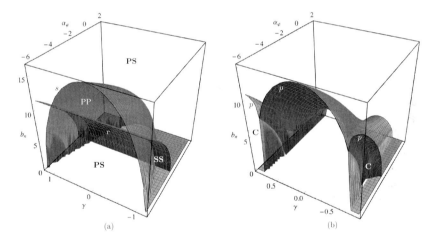

(a)

(b)

Figure 52. Influence of coupling on the propagation regions for $\alpha_u = -2.01, \beta_u = \beta_\varphi = 0.01, a_n = 1.0$. a) Pass/Stop regions; b) Complex regions.

propagation properties of the chain of oscillators can be analysed. Among them, the transitions between different behaviours such as the mono- and bi-coupled ones or the linear and nonlinear ones are of main concern. To this end the following sections are devoted to a parametric analysis in which the influence on the propagation regions of the coupling between the two degrees of freedom as well as of the nonlinearity is discussed.

Influence of coupling By substituting Eqs. (147) and (148) into Eqs. (146), the surfaces bounding each type of propagation region are obtained, as shown in Fig. 52, where all of the theoretically possible propagation regions of period-1 orbits are shown in the γ-b_n-α_φ space. It must be noticed that, due to the equivalent structure of the two equations (142), the same propagation scenario shown in Figure 52 would hold by replacing b_n and α_φ with a_n and α_u. The regions shown have been obtained assuming equal nonlinearities β_u and β_φ. In particular, to ease the graphics readability, pass/stop regions are solely shown in Figure 52a whereas the complex ones indicated in Figure 52b. As expected, the regions are symmetric with respect to the $\gamma = 0$ plane because the propagation is indifferent to clockwise or counter-clockwise rotations of the generic oscillator. It is worth emphasizing that, due to the nonlinearity, the bounding surfaces depend on the amplitude b_n and that the complex regions exist only for $|\gamma| > 1$.

The coupling parameter γ defined by equations (143) establishes the relationship between the longitudinal and rotational degrees of freedom through the stiffness of the two springs connecting the oscillators, namely c_1 and c_2. Several cases have to be discussed, depending on the value of the sole parameter γ. The cases $\gamma = \pm 1$ imply $c_2 = 0$ or $c_1 = 0$, respectively; they correspond to singular constitutive laws such that the matching invariants go to infinity. The case $|\gamma| > 1$ implies $|-c_2 + c_1| > c_1 + c_2$ and it has no physical meaning because only positive stiffnesses c_1 and c_2 are considered. Therefore it turns out that, in spite of their mathematical admissibility, complex regions are physically meaningless. As opposite, the values $|\gamma| < 1$ are plausible and the corresponding wave propagation is governed by combinations of pass and stop regions as already shown, by way of example, in Figure 52a. The limit case $\gamma = 0$ occurs when the two longitudinal springs are equal; such case yields two independent mono-coupled chains of oscillators formally equivalent to those studied in Romeo and Rega (2006). Willing to analyze the influence of the coupling parameters on the propagation properties of period-1 orbits, the latter must be sought. To this end it is convenient to set $a_{n+1} = x_{n+1}, a_n = y_{n+1}, b_{n+1} = w_{n+1}, b_n =$

z_{n+1} and to express the map $\mathbf{a}_{n+1} = \mathbf{T}(\mathbf{a}_n)\,\mathbf{a}_n$ (equation 142) as

$$x_{n+1} = \frac{-w_n\gamma\left(2 + w_n^2\beta_\varphi + \alpha_\varphi\right) - x_n\left(\alpha_u + x_n^2\beta_u + 2\gamma^2\right) + y_n\left(\gamma^2 - 1\right)}{1 - \gamma^2}$$

$$y_{n+1} = x_n \tag{149}$$

$$w_{n+1} = \frac{-x_n\gamma\left(2 + x_n^2\beta_u + \alpha_u\right) - w_n\left(\alpha_\varphi + w_n^2\beta_\varphi + 2\gamma^2\right) + z_n\left(\gamma^2 - 1\right)}{1 - \gamma^2}$$

$$z_{n+1} = w_n$$

From equations (10) it follows that period-1 orbits (p_1), representing the fixed points of $\mathbf{T}$, are given by

$$x = y = \pm\sqrt{\frac{-2 - \alpha_u}{\beta_u}}, \quad w = z = \pm\sqrt{\frac{-2 - \alpha_\varphi}{\beta_\varphi}} \tag{150}$$

Figure 53 shows in the α_φ-a_n-b_n space a comparison between propagation regions corresponding to the uncoupled case $\gamma = 0$ (Figure 53a) and a coupled case $\gamma = 0.6$ (Figure 53b); the other fixed parameters are $\alpha_u = -2.21$ and $\beta_u = \beta_\varphi = 0.01$. Besides the two surfaces r and s dividing $\mathbf{PP}$ and $\mathbf{PS}$ regions, the further surface q_1 represents the locus of the period-1 orbits given by Eqs. (150). As expected, in the uncoupled case the pass-pass region $\mathbf{PP}$, within which the period-1 orbits are bounded, does not depend upon a_n; in other words the stationary longitudinal and rotational amplitudes do not interact and thus do not affect the shape of the propagation regions. As opposite, in the coupled case, the downward bending of the surface s around $a_n = 7.5$ shows that the extent of the $\mathbf{PP}$ region depends on both the longitudinal and rotational amplitudes.

In the limiting uncoupled case, the system reduces to two parallel mono-coupled chains of oscillators each characterized by the two-dimensional Jacobian given by (134).

Influence of nonlinearity The model under investigation, described by equations (142), is characterized by two nonlinear coefficients, namely β_u and β_φ, accounting for the ratio between the strengths of the on-site cubic spring in the longitudinal and rotational degrees of freedom and those of the connecting linear springs, respectively. In this section the parametric analysis is devoted to assess the influence of these nonlinear parameters on the propagation regions.

At first, the same values for the two coefficients are considered ($\beta_u = \beta_\varphi = \beta$) and we look at the intersection curves of the bounding surfaces r and s in Figure 52a with a generic b_n plane, e.g. $b_n = 0$, at different β values. In particular, Figure 54 shows the arrangement of the propagation

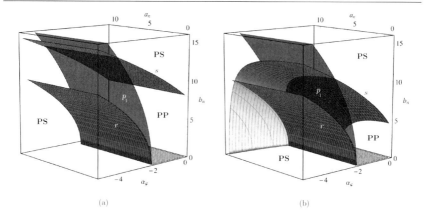

Figure 53. Period-1 orbits versus propagation region for $\alpha_u = -2.21$, $\beta_u = \beta_\varphi = 0.01$. a) uncoupled $\gamma = 0$; b) coupled $\gamma = 0.6$.

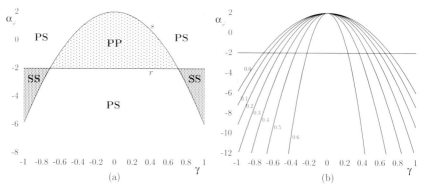

Figure 54. Influence of nonlinearity on the propagation regions; $\alpha_u = -0.01$, $b_n = 0$ and $a_n = 1.0$. a) $\beta = 0$; b) $0 \le \beta \le 0.6$.

regions on the α_φ-γ plane for zero rotational amplitude ($b_n = 0$), $a_n = 1$ and $\alpha_u = -0.01$. Figure 54a represents the linear reference case ($\beta = 0$) highlighting the location of **SS**, **PS** and **PP** regions. The evolution of such boundaries as the nonlinearity β increases is shown in Figure 54b. It can be noticed that the bounded region **PP** shrinks as the nonlinearity increases; this trend is only due to the change of curve s whilst the boundary r is not affected by the strength of the nonlinearity. For the chosen parameters, the bounded region vanishes for $\beta = 0.67$, where the s boundary changes the sign of its curvature without crossing anymore the r boundary.

A further parametric analysis is carried out in the space β-b_n-α_φ for different amplitudes a_n of the longitudinal stationary response, by considering a coupled case ($\gamma = 0.3$) since the propagation region (**PP**) is independent of a_n in the uncoupled one (see Figure 53a). Figures 55a and 55b show the results relevant to the cases $a_n = 0.6$ and $a_n = 1.2$, respectively, with $\alpha_u = -0.01$. As observed before on a different space, the increasing nonlinearity strength β reduces the **PP** region. Moreover, by comparing Figures 55a and 55b , it can be inferred that, due to the coupling, the larger the amplitude a_n the lower the value of β where the **PP** vanishes and turns into a **SS** region, such as around $\beta = 0.6$ in Figure 55b. The combined

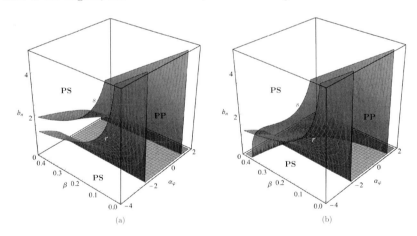

Figure 55. Influence of nonlinearity on the propagation regions; $\alpha_u = -0.01$, $\gamma = 0.3$. a) $a_n = 0.6$; b) $a_n = 1.2$.

effect of varying both the nonlinear coefficients (β_u, β_φ) and the sole coupling parameter (γ) is shown in the β_u-β_φ-γ space for fixed values of the response amplitude (56). In particular, Figs. 56a and 56b refer to the cases $a_n = b_n = 0.6$ and $a_n = b_n = 1.2$, respectively; by comparing them, the pass-pass region reduction as the oscillation amplitudes increase can be observed. Indeed Fig. 56b shows that, besides the marked curvature of the intersection of surfaces r and s, the β_φ scale is four times smaller than in Fig. 56a. As expected the propagation regions keep the symmetry with respect to the plane $\gamma = 0$.

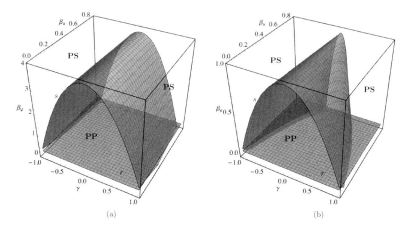

(a)   (b)

Figure 56. Combined influence of coupling and nonlinearity on the propagation regions; $\alpha_u = \alpha_\varphi = -2.21$. a) $a_n = b_n = 0.6$; b) $a_n = b_n = 1.2$.

Numerical Validation   This section is mainly devoted to validate the analytical predictions and properly complement them through numerical investigations. Results are presented in terms of basins of attraction at specific bifurcation points, where the richness of competing solutions is also highlighted, and periodic orbits paths of map **T** and its iterates. Bifurcation diagrams are also presented. In the numerical iteration procedure only the orbits with amplitude smaller than a selected upper limit are considered bounded; it has been verified that the stability or boundedness zone boundary is not affected by the upper limit value as long as the number of iterations exceeds the order of 10^2.

Results pertaining to the uncoupled case are first presented. Initially, a period-1 quadrupling bifurcation and a period-4 period-doubling bifurcation are addressed in Figures 57 and 58, respectively. The corresponding analytical bifurcation predictions of periodic orbits intersecting internal critical thresholds are depicted in Figures 57a (point **A**) and 58a (point **B**) while numerical basins of attraction showing the occurrence of a variety of regular and nonregular bounded solutions are shown in Figures 57b and 58b. In particular Figure 57b refers to a localized loss of stability (boundedness) of period-1 orbits according to a quadrupling bifurcation. Phase portraits relevant to orbits belonging to the different coloured regions of the basin of attraction and their spectral content are also shown side by side in Figure 57. The fixed point at $x = y = 1$ splits into a period-4 orbit (central black region) the spectrum of which shows the corresponding isolated peak. Then, one closed quasiperiodic orbit (white region) is shown followed by quasiperiodic orbits made up by four closed loops (cyan region and external black regions); a weakly chaotic layer (purple region) occurring before the onset of unbounded solutions (violet region) is eventually shown in the last row. Accordingly, the frequency spread around the period-4 peak increases. In Figure 58b the basin of attraction relevant to the period doubling bifurcation of period-4 orbits occurring at $\alpha = -\sqrt{2}/2$ is shown. It is consistent with the analytical crossing of the reference period-4 orbit (p_4) with the period doubling bifurcation internal curve s of $\mathbf{T}^4$ (point $\mathbf{B}$ in Figure 58a). Also in this case, samples of different orbits and the corresponding spectral content are shown in the same figure. As expected, the peak at half the frequency of period-4 orbits confirms the occurrence of the bifurcation. Figure 59 shows two further interesting basins of attraction for the uncoupled case corresponding to two further localized losses of stability of period-1 orbits, namely, a tripling bifurcation (Figure 59a) occuring at $T1$ point in the following Firgure 59c and a period doubling bifurcation (Figure 59b) occurring at $PD1$ point in the following Figure 60b.

In Figures 60 numerical periodic orbits paths and analytical saddle-node (r) and period-doubling (s) thresholds are overlapped in the uncoupled case. The numerical periodic orbits are plotted in three colours: the attractors in blue, the flip saddles in red and the regular saddles in green. Orbits up to period-4 are considered, thus the figure is composed by four graphics referring to different iterates of the map $\mathbf{T}$, namely a) $\mathbf{T}$, b) $\mathbf{T}^2$, c) $\mathbf{T}^3$ and d) $\mathbf{T}^4$.

As shown by Figure 60a, the locus of the bounded numerical period-1 orbits (blue) coincides with the analytical one, namely, they become unbounded (red) just after crossing the analytical upper boundary s where period-doubling bifurcation occurs ($PD1$). In Figure 60b bounded period-1 and period-2 orbits are both shown (blue); in particular, the latter arises just from the period-1 bifurcation occurring at the intersection with the $\mathbf{T}$ period-doubling s boundary ($PD1$) at $\alpha = -4.0$; then the bounded period-2 orbits become unbounded (red) at the intersection with its own (i.e., the $\mathbf{T}^2$) internal period-doubling threshold s ($PD2$). The results shown in Fig. 11c are concerned with $\mathbf{T}^3$ and shows period-3 orbits emanating from the tripling bifurcation ($T1$) of the period-1 ones occurring at the crossing of

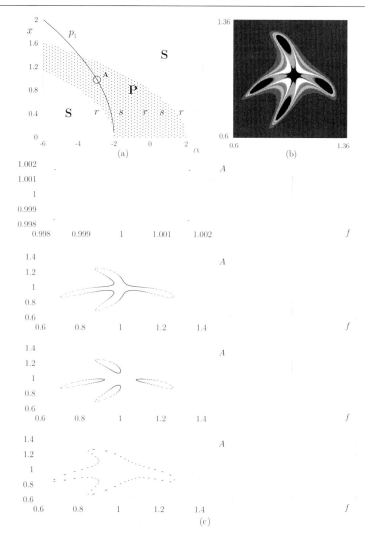

(a)

(b)

(c)

Figure 57. Quadrupling bifurcation of period-1 orbits (uncoupled case): a) period-1 orbits and internal thresholds; b) basin of attraction at $\alpha = -3.0$; c) phase portraits and frequency power spectra of different orbits.

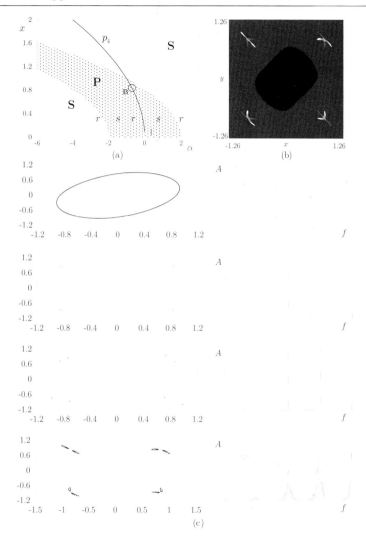

Figure 58. Period doubling bifurcation of period-4 orbits (uncoupled case):
a) period-4 orbits and internal thresholds; b) basin of attraction at $\alpha = -\sqrt{2}/2$; c) phase portraits and frequency power spectra of different orbits.

the internal saddle-node threshold r at $\alpha = -3.5$; only two out of the three period-3 branches (regular saddles in green) can be seen in the $\alpha - x$ plane due to the symmetry line $x = y$, as shown in the corresponding basin of attraction in Figure 59a. The scenario presented in Figure 60d concerns the map $\mathbf{T}^4$ where period-4 orbits bifurcation paths can also be investigated. Again, the locus of the bounded numerical period-4 orbits emanating from $\alpha = 0$ coincides with the analytical one and the orbits become unbounded (regular saddle) after crossing the analytical upper boundary r where new born orbits settle down. Moreover, as expected, two newly born branches of period-4 orbits can be noticed around $PD2$ of the $\mathbf{T}^2$ map. Also the quadrupling bifurcation of period-1 orbits ($Q1$, coinciding with point A in Figure 57a at $\alpha = -3$ can be observed. Four period-4 branches originate from $Q1$, however once again the above mentioned symmetry about the line $x = y$ hides one of them as can be inferred by looking at the relevant basin in Figure 57b. Further numerical investigations were performed in

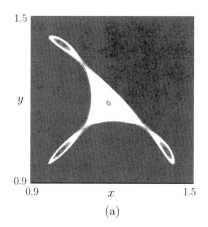

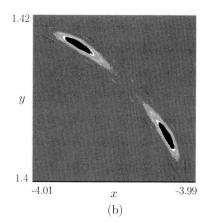

(a)   (b)

Figure 59. Basins of attraction at further localized losses of stability of period-1 orbits (uncoupled case): a) tripling bifurcation; b) period doubling bifurcation.

order to draw global indications about the richness of generic, non periodic, orbits. To this end, a number of bifurcation diagrams at different amplitudes x were carried out in order to distinguish between bounded and unbounded solutions. The thick horizontal segments in Figure 61 represent a qualitative estimate of the extent of the zone of generic bounded solutions against the analytical propagation region strictly holding for periodic solutions. The closer view in the insert refers to the bifurcation diagram obtained at $x = y = 0.5$; among the various solutions found for $-3 \le \alpha \le 1$, the period-1 and period-4 ones, occurring in agreement with the analytical predictions, are highlighted. Two more bifurcation diagrams are shown in Figure 62: they represent closer views around the period-1 quadrupling bifurcation $Q1$ and the period-4 saddle-node bifurcation $SN4$ both occurring at the relevant analytically determined intersections. The numerical inves-

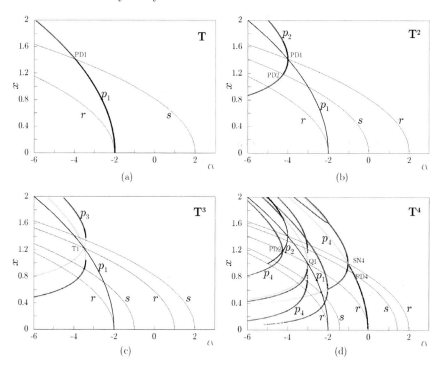

Figure 60. Periodic orbits numerical paths and bifurcations (uncoupled case): a) $\mathbf{T}$; b) $\mathbf{T}^2$; c) $\mathbf{T}^3$; d) $\mathbf{T}^4$.

tigations pertaining the uncoupled case so far discussed were also extended to the coupled case. Thus, by selecting a value of the coupling degree, bifurcation paths, basins of attraction and bifurcation diagrams were numerically derived as shown in Figures 63, 64 and 65, respectively.

The previous description of the numerical bifurcation paths can be extended to the corresponding paths for the coupled case ($\gamma = 0.6$) shown in Figure 63. As highlighted in Section 3.1 (Figure 52a) the width of the propagation region depend upon the amount of coupling γ; more specifically, the widest correspond to the uncoupled case ($\gamma = 0.0$). As the coupling increases,

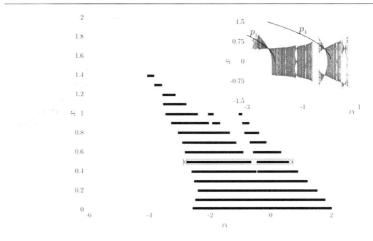

Figure 61. Global bounded solutions obtained from numerical bifurcation diagrams and closer view at $x = 0.5$ (uncoupled case).

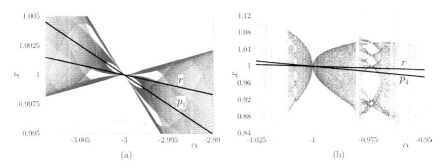

Figure 62. Numerical bifurcation diagrams against the corresponding crossing of analytical thresholds (uncoupled case): a) period-1 orbits quadrupling bifurcation ($Q1$); b) period-4 orbits saddle-node bifurcation ($SN4$).

the propagation zone shrinks because the upper boundary approaches the lower one which, in turn, does not change. This modification can be noticed by comparing Figures 60 and 63 where, for zero amplitude, such upper boundary decreases from $\alpha = 2$ to $\alpha_\varphi = 0.553$. Apart from quantitative differences, the numerical results in terms of paths and bifurcations are similar to the coupled case. A good agreement between the numerical findings and the analytical prediction is kept, although the higher dimensions of the map give rise to more involved patterns of the regular saddle numerical orbits.

The basins of attraction presented in Figure 64 refer to the period-doubling bifurcation ($PD4$) of period-4 orbits for the coordinates w,z (Figure 64b); the corresponding scenario for the x, y coordinates is also shown in Figure 64a. The basin portraits show that both phase planes are characterized by a symmetry line, namely $x = y$ and $w = z$. In Figure 65, two close views of significant bifurcation points pinpointed in Figures 63a and 63d are shown. Figure 65a refers to the period-doubling $PD1$ whereas Figure 65b to the quadrupling bifurcation $Q1$.

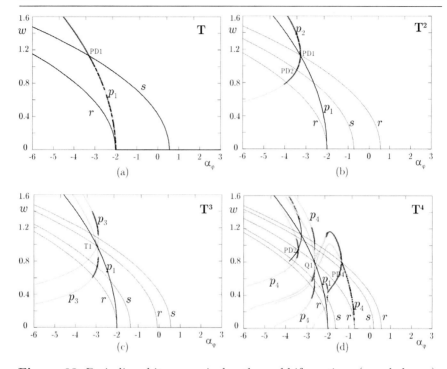

Figure 63. Periodic orbits numerical paths and bifurcations (coupled case): $\gamma = 0.6$, $\alpha_u = -2.01$, $\beta_u = \beta_\varphi = 0.01$. a) $\mathbf{T}$; b) $\mathbf{T}^2$; c) $\mathbf{T}^3$; d) $\mathbf{T}^4$.

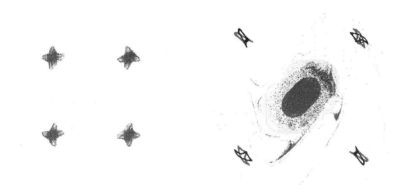

Figure 64. Period doubling bifurcation ($PD4$) of period-4 orbits (coupled case): $\gamma = 0.6$, $\alpha_u = -2.01$, $\alpha_\varphi = -1.19$, $\beta_u = \beta_\varphi = 0.01$. Cross-section of basin of attraction with a) x-y plane, b) w-z plane.

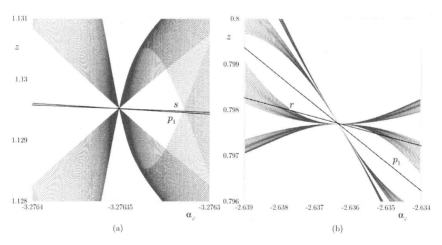

Figure 65. Numerical bifurcation diagrams against the corresponding crossing of analytical thresholds (coupled case): a) period doubling of period-1 orbits ($PD1$); b) period-1 quadrupling ($Q1$).

Bibliography

G. Sen Gupta. Natural flexural waves and the normal modes of periodically-supported beams and plates, *J. of Sound and Vibration*,13:89–101, 1970.

M.G. Faulkner and D.P. Hong. Free vibrations of mono-coupled periodic system, *J. of Sound and Vibration*, 99:29–42, 1985.

D.J. Mead. Wave propagation and natural modes in periodic systems: I mono-coupled systems, *J. of Sound and Vibration* , 40:1–18, 1975.

D.J. Mead. Wave propagation and natural modes in periodic systems: II bi-coupled systems, with and without damping, *J. of Sound and Vibration*, 40:19–39, 1975.

J. Signorelli and A.H. von Flotow. Wave propagation, power flow, and resonance in a truss beam, *J. of Sound and Vibration*, 126:127–144, 1988.

Y. Yong and Y.K. Lin. Propagation of decaying waves in periodic and piecewise periodic structures of finite length, *J. of Sound and Vibration*, 129:99–118, 1989.

D.J. Mead. Free wave propagation in periodically supported infinite beams, *J. of Sound and Vibration*, 11:181–197, 1970.

D. Bouzit and C. Pierre. Wave localization and conversion phenomena in multi-coupled multi-span beams, *Chaos, Solitons and Fractals*, 11:1575–1596, 2000.

D.J. Mead. A new method of analyzing wave propagation in periodic structures; applications to periodic Timoshenko beams and stiffened plates, *J. of Sound and Vibration*, 104: 9–27, 1986.

A.S. Bansal. Free waves in periodically disordered systems: natural and bounding frequencies of unsymmetric systems and normal mode localization, *J. of Sound and Vibration*, 207:365–382, 1997.

G.H. Koo and Y.S. Park, Vibration reduction by using periodic supports in a piping system, *J. of Sound and Vibration*, 210:53–68, 1998.

J. Guckenheimer and P.J. Holmes. *Nonlinear oscillations, dynamical systems and bifurcations of vector fields*, Springer-Verlag, New York, 1983.

W.X. Zhong and F.W. Williams. Wave problems for repetitive structures and symplectic mathematics, *Proc. Instn Mech. Engrs, Part C*, 206:371–379, 1992.

W.X. Zhong and F.W. Williams. Physical interpretation of the symplectic orthogonality of the eigensolutions of a Hamiltonian or symplectic matrix, *Computers & Structures*, 49:749–750, 1993.

E.C. Pestel and F.A. Leckie. *Matrix Methods in Elastomechanics*. New York: McGraw-Hill, 1963.

R.S. Langley, N.S. Bardell, and P.M. Loasby. The optimal design of near-periodic structures to minimize vibration transmission and stress levels. *J. of Sound and Vibration*, 207:627–646, 1997.

D. Richards and D.J. Pines. Passive reduction of gear mesh vibration using a periodic drive shaft. *42nd AIAA SDM Conference*, Seattle WA, 2001.

A. Baz. Active control of periodic structures. *J. of Vibration and Acoustics*, 123:472–479, 2001.

A.H. von Flotow. Disturbance propagation in structural networks. *J. of Sound and Vibration*, 106:433–450, 1986.

G.Q. Cai and Y.K. Lin. Wave propagation and scattering in structural networks. *J. of Engrg. Mechanics*, 117:1555–1574, 1991.

A. Luongo. Mode localization in dynamics and buckling of linear imperfect continuous structures. *Nonlinear Dynamics*, 25:133–156, 2001.

Y. Yong and Y.K. Lin. Dynamic response analysis of truss-type structural networks: a wave propagation approach. *J. of Sound and Vibration*, 156:27–45, 1992.

W.J. Chen and C. Pierre. Exact linear dynamics of periodic and disordered truss beams: localization of normal modes and harmonic waves. *Proc. of 32nd AIAA/ASME/ASCE/AHS/ASC Struct, Struct Dyn, and Mat Conf*, Baltimore MD, April 1991.

S.V. Sorokin and O.A. Ershova. Plane wave propagation and frequency band gaps in periodic plates and cylindrical shells with and without heavy fluid loading. *Journal of Sound and Vibration*, 278:501–526, 2004.

F. Romeo and A. Luongo. Invariant representation of propagation properties for bi-coupled periodic structures, *J. of Sound and Vibration*, 257:869–886, 2002.

F. Romeo and A. Luongo. Vibration reduction in piecewise bi-coupled periodic structures. *J. of Sound and Vibration*, 268:601–615, 2003.

A. Luongo and F. Romeo. Real wave vectors for dynamic analysis of periodic structures. *J. of Sound and Vibration*, 279:309–325, 2005.

F. Romeo and A. Luongo. Wave propagation in three-coupled periodic structures . *J. of Sound and Vibration*, 301:635–648, 2007.

A.F. Vakakis, M.E. King, Nonlinear wave transmission in a mono-coupled elastic periodic system, *J. of Acoust. Soc. Am.*, 98:1534–1546, 1995.

M.A. Davies and F.C. Moon. Transition from soliton to chaotic motion following sudden excitation of a nonlinear structure, *J. of Applied Mechanics*, 63:445–449, 1996.

A. Luongo. A transfer matrix perturbation approach to the buckling analysis of nonlinear periodic structures, 10th *ASCE Conference*, Boulder, Colorado, 505–508, 1995.

A. Luongo and F. Romeo. A transfer-matrix perturbation approach to the dynamics of chains of nonlinear sliding beams, *J. of Vibration and Acoustics*, vol. 128, pp. 190-196, 2006.

A. Luongo, G. Rega, F. Vestroni. On nonlinear dynamics of planar shear indeformable beams, *J. of Applied Mechanics* , 53:619–624, 1986.

Y. Wan and C.M. Soukoulis. One-dimensional nonlinear Schrödinger equation: A nonlinear dynamical approach. *Physical Review A*, 41:800–809, 1990.

D. Hennig and G.P. Tsironis. Wave transmission in nonlinear lattices, *Physics Reports*, 307:333–432, 1999.

L.I. Manevitch. The description of localized normal modes in a chain of nonlinear coupled oscillators using complex variables, *Nonlinear Dynamics*, 25:95–109, 2001.

L.I. Manevitch, O.V. Gendelman and A.V. Savin. Nonlinear normal modes and chaotic motions in oscillatory chains, *IUTAM Symposium on Chaotic Dynamics and Control of Systems and Processes in Mechanics* (Eds. G. Rega, F. Vestroni), Springer,59–68, 2005.

K.D. Umberger, C. Grebogi, E. Ott and B. Afeyan. Spatiotemporal dynamics in a dispersively coupled chain of nonlinear oscillators, *Physical Review A*, 39:4835–4842, 1989.

L. Brillouin. *Wave propagation in periodic structures*, New York: Dover, 1953.

G. Chakraborty, A.K. Mallik. Dynamics of a weakly non-linear periodic chain, *Int. J. of Non-linear Mechanics*, 36, 375–389, 2001.

J. Pouget. Non-linear lattice models: complex dynamics, pattern formation and aspects of chaos, *Philosophical Magazine*, 85:4067–4094, 2005.

F. Romeo, G. Rega. Wave propagation properties in oscillatory chains with cubic nonlinearities via nonlinear map approach, *Chaos Solitons & Fractals*, 27:606–617, 2006.

F. Romeo and G. Rega. Propagation properties of bi-coupled nonlinear oscillatory chains: analytical prediction and numerical validation, *Internat. J. Bifur. Chaos Appl. Sci. Engrg.*, 18:1983–1998, 2008.

M.K. Sayadi, J. Pouget. Soliton dynamics in microstructured lattice model, *J. Phys. A*, 24:2151-2172, 1991.

Methodologies for Nonlinear Periodic Media

Alexander F. Vakakis

W. Grafton and Lillian B. Wilkins Professor,
Department of Mechanical Science and Engineering University of Illinois,
Urbana – Champaign avakakis@illinois.edu

Abstract We discuss analytical methodologies for analyzing waves in weakly or strongly nonlinear periodic media. These include techniques for studying standing and traveling waves in weakly nonlinear periodic media; methodologies for analyzing primary wave transmission in layered elastic media with strong nonlinearities; and new concepts and methods for traveling waves, solitary waves and pulse attenuation in strongly nonlinear homogeneous and heterogeneous granular media with no pre-compression. The field of nonlinear periodic media is a promising area of research with potential of diverse applications in science and engineering.

1 From Linear to Nonlinear Periodic Media

Systems with spatial periodicity occur often in engineering practice from the macro- to nano-scale. Typical examples are turbine blade assemblies, space trusses, rib skin structures in aircrafts, ordered granular media, but also multi-walled carbon nanotubes. Under the assumption of linearity analytical and computational methods have been developed for analyzing the dynamics of periodic systems (Brillouin, 1946), with typical approaches based on local/global transfer-matrices (Knopoff, 1964; Schmidt and Jensen, 1985; Mal, 1988; Cetinkaya, 1995; El-Raheb, 1993), spectral methods (Rizzi and Doyle, 1992) and wave modes (Mead, 1975, 1986). An interesting feature of the dynamics of linear periodic systems is that they act as filters due to the existence of propagation and attenuation zones (PZs and AZs) in the frequency – wavenumber space. In PZs there exist traveling waves with capacity to transfer energy to the far field, whereas in AZs only near-field standing waves can be realized. In addition, scattering and mode conversions of waves at the interfaces between periodic sets are realized leading under certain conditions to waves localized at these interfaces. Finally, it has been shown that mode localization and motion confinement can occur in weakly disordered and weakly coupled periodic systems (Hodges, 1982;

Pierre and Dowell, 1987), giving rise to the well known Anderson localization phenomenon (Anderson, 1958).

On the other hand, the study of the dynamics of nonlinear periodic media dictates the development of special and highly 'individualistic' methods of analysis due to the complexity of their dynamics (Toda, 1989). An indication of the complex dynamical phenomena that may be encountered in nonlinear periodic media is provided by mentioning the existence of energy-depended nonlinear PZs and AZs (Asfar and Nayfeh, 1981; Vakakis and King, 1995; Boechler and Daraio, 2009), and of solitary waves, solitons and localized standing or traveling breathers (Lomdahl, 1985; Flytzanis et al., 1985; Nesterenko, 2001); of nonlinear wave scattering at defects and at interfaces between periodic sets leading to complex wave dynamics and new multi-frequency waves (Pilipchuck et al., 1996; Pilipchuck and Vakakis, 1998; Goodman et al., 2004; Nesterenko et al., 2005); the occurrence of nonlinear mode localization, bifurcations and wave confinement even in perfectly ordered periodic media, i.e., even in the absence of structural disorder (Vakakis et al., 1993, 1996; Aubrecht and Vakakis, 1996); the realization of multitudes of nonlinear resonances and resonance captures, such as fundamental, subharmonic, superharmonic and combination resonances (Nayfeh and Mook, 1984; Vakakis and King, 1998); the possibility of nonlinear wave interactions, of spatial and temporal chaos, and of energy transfers across different spatial and temporal scales (Potapov et al., 2001; Zhou et al., 2005); the excitation of relaxation oscillations in periodic media with essentially nonlinear disorders (Vakakis, 2010); the occurrence of targeted nonlinear energy transfers to essentially nonlinear substructures (i.e., substructures with non-linearizable stiffness nonlinearities) (Vakakis et al., 2008); and the capacity of nonlinear media for passive nonlinear wave guidance and redirection,energy trapping and adaptive stress wave tailoring (e.g., of essentially nonlinear granular media) (Daraio et al., 2006, 2010). Since it is not possible to discuss all of these interesting dynamical phenomena in this chapter, we will focus in selected analytical techniques for studying some of the previous nonlinear phenomena in (even strongly) nonlinear periodic media. For a more in-depth study of the nonlinear dynamics of periodic systems the reader is referred to the aforementioned and to references therein.

We start by providing a methodology for computing AZs and PZs of nonlinear periodic systems. In similarity to linear theory, nonlinear waves with frequencies in PZs can propagate unattenuated through the periodic medium, whereas in nonlinear AZs standing waves are realized representing near field solutions of the problem. However, we will show that depending on substructure coupling, distinctly nonlinear dynamical phenomena can arise, such as wave localization and solitary waves; such solutions cannot be realized in linear periodic systems. In a later section we provide further examples that highlight the generation of spatial chaos in ordered granular media, which are a special class of spatially periodic, highly discontinuous and strongly nonlinear media.

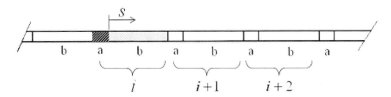

Figure 1. The nonlinear one-dimensional bi-layered system.

2   Propagation and Attenuation Zones (PZs and AZs) in a Strongly Coupled Nonlinear Periodic Medium

We demonstrate the methodology for computing nonlinear AZs and PZs by considering the one-dimensional bi-layered system of Figure 1 composed of linearly elastic layers (a) coupled to the weakly nonlinear elastic layers (b). The analysis follows (Vakakis and King, 1995). We assume that layers (b) possess the nonlinear elastic constitutive law, $\sigma = E_{b1} u_s + \varepsilon E_{b3} u_s^3$, $0 < \varepsilon \ll 1$, where σ is the axial stress and u_s is the partial derivative of the axial deformation with respect to the local spatial variable s. Moreover, we assume that the linear layers (a) possess approximately stiffness behavior, by requiring that $(l/c)_a \to 0, (AE/c)_a \to 0$, and $1/(l/c)_a(AE/c)_a \to k_a$, where l denotes layer length, c velocity of sound, and A, E cross section and elastic modulus, respectively. Then, the periodic medium degenerates to a bi-infinite set of nonlinear layers (b) coupled by means of linear springs k_a (cf. Figure 2), with governing normalized partial equations of motion expressed as (Vakakis and King, 1995),

$$u_{\tau\tau}^{(i)}(x,\tau) = u_{xx}^{(i)}(x,\tau) + \varepsilon e\, u_x^{(i)2}(x,\tau)\, u_{xx}^{(i)}(x,\tau) \tag{1}$$

with boundary conditions,

$$\begin{aligned}
u_x^{(i)}(0,\tau) + (\varepsilon e/3)\, u_x^{(i)3}(0,\tau) &= K\left[u^{(i)}(0,\tau) - u^{(i-1)}(1,\tau)\right] \\
u_x^{(i)}(1,\tau) + (\varepsilon e/3)\, u_x^{(i)3}(1,\tau) &= K\left[u^{(i+1)}(0,\tau) - u^{(i)}(1,\tau)\right]
\end{aligned} \tag{2}$$

where $i=0,\pm1,\pm2,...,$ $x=s/l_b[0,1]$ is a local nondimensional spatial variable for each layer (b), $\tau=(E_{b1}/\rho_b)^{1/2}/l_b^{-1}t$ is nondimensional time,

$$e = 3E_{b3}E_{b1}^{-1}l_b^{-2}$$

and $K = k_a l_b/A_b E_{b1}$; $u^{(i)}(x,\tau)$ denotes the axial deformation of the i-th nonlinear layer (b).

This is a set of weakly nonlinear partial differential equations with coupling provided through the boundary conditions (2). The weak stiffness nonlinearity [introduced through the constitutive law for layers (b)] is scaled by the small parameter ε, which is the perturbation parameter for the following asymptotic analysis.

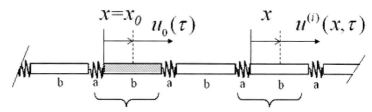

$$u^{(i)}(x,\tau) = U^{(i)}(u_0(\tau),x), \quad i = 0,\pm1,...,$$
$$U^{(k)}(u_0(\tau),x_0) \equiv u_0(\tau)$$

Figure 2. Reference layer and reference response.

In this section we consider the case of strong coupling by assuming that $K=O(1)$, whereas in the next section we will consider the weakly coupled medium. First we consider standing waves in AZs (stop bands) of the periodic medium (Vakakis and King, 1995). Recognizing that these are synchronous oscillations, we adopt a nonlinear normal mode (NNM) approach (Vakakis et al., 1996) and introduce the following variable change,

$$u^{(i)}(x,\tau) = U^{(i)}(u_0(\tau),x), \quad i = 0,\pm1,..., \quad U^{(k)}(u_0(\tau),x_0) \equiv u_0(\tau) \quad (3)$$

where $u_0(\tau) \equiv u^{(k)}(x_0,\tau)$ is a reference response, defined as the response of the k-th layer at reference position $x = x_0$. This approach seeks nonlinear standing waves (NNMs) inside AZs of the nonlinear periodic medium, corresponding to trivial phase differences in the oscillations of any two material points of the i-th layer (b).

The plan is to introduce the change of the temporal independent variable $\tau \rightarrow u_0(\tau)$so that the explicit time dependence is eliminated from the equations of motion. First, we transform the time derivatives using the chain rule,

$$
\begin{aligned}
u^{(i)}(x,\tau) &= U^{(i)}(u_0(\tau),x), \quad i = 0, \pm 1, ..., \quad \Rightarrow \\
u_\tau^{(i)}(x,\tau) &= U_{u_0}^{(i)}(u_0(\tau),x)\, u_0'(\tau) \\
u_{\tau\tau}^{(i)}(x,\tau) &= U_{u_0 u_0}^{(i)}(u_0(\tau),x)\, u_0'^2(\tau) + U_{u_0}^{(i)}(u_0(\tau),x)\, u_0''(\tau)
\end{aligned}
\tag{4}
$$

where prime denotes differentiation with respect to τ. Given that the system under consideration is conservative we may employ (4) to express the velocity of the reference point using the expression of the total energy,

$$
\begin{aligned}
E = \sum_p \Big\{ &\int_0^1 \Big[(1/2)u_\tau^{(p)2}(x,\tau) + (1/2)u_x^{(p)2}(x,\tau) \Big]\, dx + \\
&(K/2)\left[u^{(p)}(0,\tau) - u^{(p-1)}(1,\tau) \right]^2 + O(\varepsilon) \Big\}
\end{aligned}
$$

so that,

$$
\begin{aligned}
&u_0'^2(\tau) = \\
&\frac{E - \sum_p \left\{ \int_0^1 \frac{U_x^{(p)2}}{2}dx + \frac{K}{2}\left[U^{(p)}(u_0(\tau),0) - U^{(p-1)}(u_0(\tau),1) \right]^2 + O(\varepsilon) \right\}}{\sum_p \int_0^1 \frac{U_{u_0}^{(p)2}(u_0(\tau),x)}{2}\, dx} \\
&\Rightarrow u_0'^2(\tau) \quad \equiv \quad F\left[u_0(\tau); E\right]
\end{aligned}
\tag{5}
$$

In turn, the second partial time derivative of the axial deformation can be expressed in terms of the reference response and the total (conserved) energy E as follows:

$$
u_{\tau\tau}^{(i)}(x,\tau) = U_{u_0 u_0}^{(i)}(u_0(\tau),x)\, F\left[u_0(\tau); E\right] + U_{u_0}^{(i)}(u_0(\tau),x)\, u_0''(\tau)
\tag{6}
$$

Employing the previous derivations the normalized equations of motion (1) are expressed as,

$$
U_{u_0 u_0}^{(i)} u_0'^2(\tau) + U_{u_0}^{(i)} u_0''(\tau) = U_{xx}^{(i)} + \varepsilon e U_x^{(i)2} U_{xx}^{(i)}, \quad i = 0, \pm 1, ...
\tag{7}
$$

with corresponding boundary conditions:

$$U_x^{(i)}(0, u_0(\tau)) + (\varepsilon e/3)\, U_x^{(i)3}(0, u_0(\tau)) = K\left[U^{(i)}(0, u_0(\tau))\right.$$
$$\left. -U^{(i-1)}(1, u_0(\tau))\right]$$
$$U_x^{(i)}(1, u_0(\tau)) + (\varepsilon e/3)\, U_x^{(i)3}(1, u_0(\tau)) = K\left[U^{(i+1)}(0, u_0(\tau))\right. \qquad (8)$$
$$\left. -U^{(i)}(1, u_0(\tau))\right]$$

Finally, expressing the reference acceleration and velocity in terms of the total conserved energy E and $U^{(i)}$, we transform the set of normalized governing equations (7) only in terms of $U^{(i)}$ and $u_0(\tau)$:

$$U_{u_0 u_0}^{(i)} F\left[u_0(\tau); E\right] +$$
$$U_{u_0}^{(i)}\left[U_{xx}^{(k)}(u_0(\tau), x_0) + \varepsilon e U_x^{(k)2}(u_0(\tau), x_0)\, U_{xx}^{(k)}(u_0(\tau), x_0)\right] = \qquad (9)$$
$$U_{xx}^{(i)}(u_0(\tau), x) + \varepsilon e U_x^{(i)2}(u_0(\tau), x)\, U_{xx}^{(i)}(u_0(\tau), x), \quad i = 0, \pm 1, \ldots$$

We note that at time instants of maximum potential energy satisfying $u_0(\tau) = u_0^*$ (where u_0^* is the maximum displacement of the reference response) the velocity of the reference response is zero, so it holds that $F[u_0^*; E] = 0$ and the coefficient of the highest derivative in (9) vanishes; it follows that these points represent regular singular points of the governing equations. Hence, we need to develop asymptotic approximations for $U^{(i)}(u_0(\tau), x)$ in open intervals $|u_0(\tau)| < u_0^*$, and then continue them up to the singularities by imposing the following 'orthogonality' condition valid at the intersections of the trajectory of the response (in the infinite dimensional configuration space) with the maximum equipotential energy surface (cf. Figure 3):

$$\left\{ U_{u_0}^{(i)}\left[U_{xx}^{(k)}(u_0(\tau), x_0) + \varepsilon e U_x^{(k)2}(u_0(\tau), x_0)\, U_{xx}^{(k)}(u_0(\tau), x_0)\right] = \right.$$
$$\left. U_{xx}^{(i)}(u_0(\tau), x) + \varepsilon e U_x^{(i)2}(u_0(\tau), x)\, U_{xx}^{(i)}(u_0(\tau), x) \right\}_{u_0 = u_0^*}, \qquad (10)$$
$$i = 0, \pm 1, \ldots$$

This procedure of analytical continuation of an asymptotic solution up to singular points by imposing boundary orthogonality conditions was first demonstrated in (Manevitch and Mikhlin, 1972) for discrete coupled nonlinear oscillators. Finally, we impose the obvious compatibility condition for the k-th layer (i.e., the reference layer):

$$U^{(k)}(u_0(\tau), x_0) = u_0(\tau) \qquad (11)$$

Summarizing, to compute standing waves in nonlinear AZs of the nonlinear periodic medium of Figure 2, we need to compute the modal functions

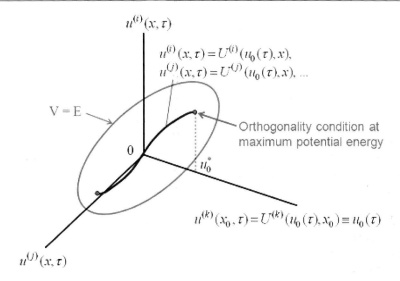

Figure 3. Trajectory of the nonlinear standing wave in infinite-dimensional configuration space and maximum equipotential surface $V = E$.

$U^{(i)}(u_0(\tau), x)$ governed by the functional equations (9) subject to the boundary conditions (8), the orthogonality conditions (10) and the compatibility condition (11). Once the asymptotic solutions for the modal functions are derived, the reference response $u_0(\tau)$ is computed by solving a nonlinear ordinary differential equation governing the oscillation of the reference layer at $x = x_0$,

$$u_{\tau\tau}^{(k)}(x_0, \tau) = u_{xx}^{(k)}(x_0, \tau) + \varepsilon\, e\, u_x^{(k)2}(x_0, \tau)\, u_{xx}^{(k)}(x_0, \tau) \Rightarrow$$
$$u_0''(\tau) = U_{xx}^{(k)}(u_0(\tau), x_0) + \varepsilon\, e\, U_x^{(k)2}(u_0(\tau), x_0)\, U_{xx}^{(k)}(u_0(\tau), x_0) \qquad (12)$$

subject to the initial conditions $u_0(0) = u_0^*$, $u_0'(0) = 0$. This completes the computation of the nonlinear standing waves.

The plan for the asymptotic solution of the problem (because of symmetric nonlinearity we only consider the solution in the half-interval $0 \leq u_0(\tau) \leq u_0^*$) is to first, develop asymptotic approximations $U^{(i)}(u_0(\tau), x)$, $i = 0, \pm 1, \dots$ that satisfy (8) and (9) in open intervals $|u_0(\tau)| < u_0^*$, and then to analytically continue these asymptotic approximations up the maximum potential energy level by satisfying the orthogonality conditions (10). The asymptotic solutions are sought in the form of successive approximations,

$$U^{(i)}(u_0(\tau), x) = \sum_{m=0}^{\infty} \varepsilon^m U^{(i)m}(u_0(\tau), x),$$
$$i = 0, \pm 1, ..., \quad 0 \le u_0(\tau) < u_0^* \tag{13}$$

with,

$$U^{(i)0}(u_0(\tau), x) = a_1^{(i)0}(x) u_0(\tau),$$
$$U^{(i)m}(u_0(\tau), x) = \sum_{q=1}^{\infty} a_q^{(i)m}(x) u_0^q(\tau) , \quad m \ge 1$$

This scheme recovers the linear dynamics as $\varepsilon \to 0$ (this holds since the system is weakly nonlinear). The various orders of approximation are computed by substituting the above expansions into the equations of motion (9), the boundary conditions (8), and the compatibility condition (10), and matching the coefficients of the various orders of ε.

A linear form is assumed for the leading order successive approximation $U^{(i)0}(u_0(\tau), x)$, since for $\varepsilon = 0$ the system (8-11) is linear and, hence, is separable in space and time; this does not hold, however, for higher order approximations which are nonlinear and not separable in space and time. The resulting near-field nonlinear standing waves have frequencies inside the linearized AZs (stop bands) of the system with $\varepsilon = 0$. The resulting equations are given by:

$$\left.\begin{array}{l} a_1^{(i)0''}(x) + \lambda^2 a_1^{(i)0}(x) = 0 \Rightarrow \\ a_1^{(i)0}(x) = A_i \cos \lambda x + B_i \sin \lambda x, \quad i = 0, \pm 1, ... \\ a_1^{(k)0}(x_0) = 1 \quad \Rightarrow \quad \lambda^2 = -a_1^{(k)0''}(x_0) \\ a_1^{(i)0'}(0+) = K \left[a_1^{(i)0}(0) - a_1^{(i-1)0}(1) \right] \\ a_1^{(i)0'}(1-) = K \left[a_1^{(i+1)0}(0) - a_1^{(i)0}(1) \right], \quad i = 0, \pm 1, ... \end{array}\right\} \Rightarrow$$

$$\lambda B_i - KA_i + KA_{i-1} \cos \lambda + KB_{i-1} \sin \lambda = 0$$
$$A_i(K \cos \lambda - \lambda \sin \lambda) + B_i(K \sin \lambda - \lambda \cos \lambda) - KA_{i+1} = 0, \tag{14}$$
$$i = 0, \pm 1, ...$$

This represents a linear set of homogeneous difference equations, so following (Mickens, 1987) we seek solutions:

$$A_i = \delta_1 \nu^i, \quad B_i = \delta_2 \nu^i \Rightarrow$$
$$\begin{bmatrix} -K + \dfrac{K \cos \lambda}{\nu} & \lambda + \dfrac{K \sin \lambda}{\nu} \\ K \cos \lambda - \lambda \sin \lambda - \nu K & K \sin \lambda + \lambda \cos \lambda \end{bmatrix} \begin{Bmatrix} \delta_1 \\ \delta_2 \end{Bmatrix} = 0 \Rightarrow \tag{15}$$
$$\cosh \mu = \cos \lambda - (\lambda/2K) \sin \lambda, \quad \nu = e^\mu, \quad \mu \in R$$

This expression relates the linearized wave number of the standing wave, λ, with the linearized propagation constant, μ, which provides a measure of spatial attenuation of the standing wave in the given AZ under consideration. Using these relations we obtain the O(1) linearized approximations of the AZs of the periodic medium which are depicted in Figure 4. The linearized standing waves in the AZs are then computed as,

$$a_1^{(i)0}(x) = \left[\frac{\lambda(\mu) + Ke^{-\mu} \sin \lambda(\mu)}{K(1 - e^{-\mu} \cos \lambda(\mu))} \cos \lambda(\mu)x_0 + \sin \lambda(\mu)x_0 \right]^{-1} e^{\mu(i-k)}$$
$$\times \left\{ \left[\frac{\lambda(\mu) + Ke^{-\mu} \sin \lambda(\mu)}{K(1 - e^{-\mu} \cos \lambda(\mu))} \right] \cos \lambda(\mu)x + \sin \lambda(\mu)x \right\}, \tag{16}$$

$$U^{(i)0}(u_0(\tau), x) = a_1^{(i)0}(x) \, u_0(\tau), \quad i = 0, \pm 1, \dots$$

where μ is a free parameter such that $|\cosh \mu| > 1$.

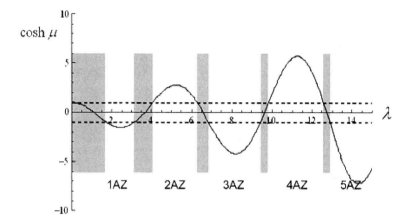

Figure 4. Leading order (linear) approximations to the AZs of the weakly nonlinear periodic system.

Proceeding to the next successive approximation,

$$U^{(i)}(u_0(\tau), x) = a_1^{(i)0}(x) \, u_0(\tau) + \varepsilon U^{(i)1}(u_0(\tau), x) + O(\varepsilon^2)$$
$$= a_1^{(i)0}(x) \, u_0(\tau) + \varepsilon \left\{ a_1^{(i)1}(x) \, u_0(\tau) + a_3^{(i)1}(x) \, u_0^3(\tau) + O(u_0^5(\tau)) \right\} \tag{17}$$
$$+ O(\varepsilon^2)$$

substituting into the equations of motion (9) we obtain the expressions,

$$a_3^{(i)1}(x) = \hat{A} \left[a_1^{(i)1''}(x) - a_1^{(k)1''}(x_0) \, a_1^{(i)0}(x) + \lambda^2 a_1^{(i)1}(x) \right] \tag{18}$$

where,

$$\hat{A} = \frac{\int_0^1 (\gamma \cos \lambda x + \sin \lambda x)^2 \, dx}{6u_0^{*2} \left\{ \int_0^1 (\lambda \cos \lambda x - \gamma \lambda \sin \lambda x)^2 \, dx + K \left[\gamma - e^{-\mu} (\gamma \cos \lambda x + \sin \lambda x) \right]^2 \right\}}$$

$$\gamma = K^{-1} \left(\lambda + K e^{-\mu} \sin \lambda \right) \left(1 - e^{-\mu} \cos \lambda \right)^{-1}, \quad i = 0, \pm 1, \dots$$

and the propagation constant μ is the free parameter. Substituting into the orthogonality conditions (10), the boundary conditions (8) and the compatibility condition (11), we derive the following 4th-order ordinary differential equation governing $a_1^{(i)1}(x)$,

$$a_1^{(i)1''''}(x) + F_1 a_1^{(i)1''}(x) + F_2 a_1^{(i)1}(x) = F_{3i}(x), \quad i = 0, \pm 1, \dots$$

with the boundary conditions,

$$\begin{aligned}
a_1^{(i)1'}(0) &= K \left[a_1^{(i)1}(0) - a_1^{(i-1)1}(1) \right] \\
a_1^{(i)1'}(1) &= K \left[a_1^{(i+1)1}(0) - a_1^{(i)1}(1) \right] \\
a_3^{(i)1'}(0) &= -(e/3) a_1^{(i)1'3}(0) + K \left[a_3^{(i)1}(0) - a_3^{(i-1)1}(1) \right] \\
a_3^{(i)1'}(1) &= -(e/3) a_1^{(i)1'3}(1) + K \left[a_3^{(i+1)1}(0) - a_3^{(i)1}(1) \right]
\end{aligned} \tag{19}$$

and the compatibility condition:

$$a_1^{(k)1}(x_0) = 0$$

The solution of the nonhomogeneous equation (19a) can be expressed as (King and Vakakis, 1995),

$$\begin{aligned}
a_1^{(i)1}(x) &= C_{i1} \cos \lambda x + C_{i2} \sin \lambda x + C_{i3} \cos \xi x + C_{i4} \sin \xi x \\
&\quad + C_{i5} \cos 3\lambda x + C_{i6} \sin 3\lambda x + C_{i7} x \cos \lambda x + C_{i8} x \sin \lambda x,
\end{aligned} \tag{20}$$

$$\xi = \left[3\lambda^2 + \left(\hat{A} u_0^* \right)^{-1} \right]^{1/2}, \quad i = 0, \pm 1, \dots$$

where C_{i5}, C_{i6} are explicitly computed in terms of the O(1) solution $a_1^{(i)0}(x)$, whereas C_{i7}, C_{i8} are yet undetermined since they depend on the solution on the k-th reference layer, $a_1^{(k)1}(x)$. Moreover, this solution holds only when $\xi \neq 3\lambda$, since otherwise a 1:3 internal resonance exists in each layer. Applying the four boundary conditions (19b) we obtain a set of four coupled linear (since these conditions are also linear) inhomogeneous difference equations in terms of C_{i1}, C_{i2}, C_{i3}, C_{i4}, which can be solved explicitly by taking into account the form of the inhomogeneous terms:

$$C_{ip} = c_{p1}ie^{i\mu} + c_{p2}e^{3i\mu}, \quad p = 1, 2, 3, 4 \tag{21}$$

Finally, imposing the compatibility condition at the k-th reference layer we evaluate all unknown terms and complete the solution.

Combining these results we express the leading-order nonlinear correction as follows:

$$
\begin{aligned}
a_1^{(i)1}(x) &= \left[c_{11}ie^{\mu(i-k)} + c_{12}e^{3\mu(i-k)}\right]\cos\lambda x \\
&+ \left[c_{21}ie^{\mu(i-k)} + c_{22}e^{3\mu(i-k)}\right]\sin\lambda x \\
&+ \left[c_{31}ie^{\mu(i-k)} + c_{32}e^{3\mu(i-k)}\right]\cos\xi x \\
&+ \left[c_{41}ie^{\mu(i-k)} + c_{42}e^{3\mu(i-k)}\right]\sin\xi x \\
&+ \left[c_{5}e^{3\mu(i-k)}\right]\cos 3\lambda x + \left[c_{6}e^{3\mu(i-k)}\right]\sin 3\lambda x \\
&+ \left[c_{71}e^{\mu(i-k)} + c_{72}e^{3\mu(i-k)}\right]x\cos\lambda x \\
&+ \left[c_{81}e^{\mu(i-k)} + c_{82}e^{3\mu(i-k)}\right]x\sin\lambda x \\
a_3^{(i)1}(x) &= \hat{A}\left[a_1^{(i)1''}(x) - a_1^{(0)}(x)a_1^{(k)1''}(x_0) + \lambda^2 a_1^{(i)1}(x)\right]
\end{aligned}
\tag{22}
$$

Hence, the $O(\varepsilon)$ linear spatial coefficient contains three wave numbers: λ identical to the $O(1)$ solution, and ξ and 3λ that are due to the material nonlinearities. Also, the $O(\varepsilon)$ solution possesses two propagation constants, μ and 3μ.

Combining the previous results, the responses of the layers of the nonlinear periodic medium for a nonlinear standing wave in an AZ are approximated as,

$$
\begin{aligned}
u^{(i)}(x,\tau) &= U^{(i)}(u_0(\tau), x) = \\
\left[a_1^{(i)0}(x) + \varepsilon a_1^{(i)1}(x)\right]&u_0(\tau) + \varepsilon a_3^{(i)1}(x)\, u_0^3(\tau) + O\left(\varepsilon u_0^5(\tau), \varepsilon^2\right), \\
i &= 0, \pm 1, \ldots \\
u^{(k)}(x_0, \tau) &= U^{(k)}(u_0(\tau), x_0) \equiv u_0(\tau)
\end{aligned}
\tag{23}
$$

As mentioned previously, this asymptotic analysis is valid only if $\xi \neq 3\lambda$; otherwise 1:3 internal resonance (IR) occurs in the periodic system, and nonlinear coupling between two linearized modes in each of the identical layers occurs. The system then performs multimode non-synchronouos oscillations that cannot be modeled by the previous analysis. The occurrence of IR is dictated by the strength of the coupling K between layers and the value of the propagation constant μ. Indeed, when $K = O(\varepsilon)$, the periodic system is weakly coupled and the leading order solution is the response of a free-free layer. The analysis of the nonlinear dynamics in that case requires a different approach which will be discussed in what follows. In Figure 5 we depict the previous approximations for $K = 1.0$, $e = 0.1$, $\varepsilon = 0.1$, $x_0 = 0.15$, $u_0^* = 1.0$ and $\lambda = 2.5$. We note that conditions for IR are satisfied at the boundaries between AZs and PZs, i.e., at natural frequencies of the free-free and fixed-fixed layer (b); also IR conditions are also satisfied at isolated points inside AZs.

The time dependence and period of a synchronous periodic motion inside an AZ are computed by analyzing the response of the reference point:

$$
\begin{aligned}
u_0''(\tau) + \left[\lambda - \varepsilon a_1^{(k)1''}(x_0)\right] u_0(\tau) + \\
\varepsilon \left[\varepsilon a_1^{(k)0'2}(x_0) - a_3^{(k)1''}(x_0)\right] u_0^3(\tau) + O(\varepsilon u_0^3(\tau), \varepsilon^2) = 0, \\
u_0(0) = u_0^*, \quad u_0'(0) = 0
\end{aligned}
\tag{24}
$$

The response of this reference oscillator can be either estimated asymptotically (Nayfeh and Mook, 1984), or explicitly by quadratures in terms of Jacobian elliptic functions (Byrd and Friedman, 1954).

Now we consider traveling waves in PZs of the strongly coupled nonlinear periodic medium of Figure 2. These are non-synchronous oscillations, so they cannot be studied by the previous NNM-based methodology. Nonlinear traveling waves can be studied by the method of multiple scales (Nayfeh and Mook, 1984), through the introduction of multiple temporal and spatial scales, $T_0 = \tau$, $T_1 = \varepsilon\tau, ...$ and $X_0 = x$, $X_1 = \varepsilon x, ...$, regarded as independent variables. Then, the normalized equations of motion (1) are expressed as,

$$
u_{T_0 T_0}^{(i)} + 2\varepsilon u_{T_0 T_1}^{(i)} - u_{X_0 X_0}^{(i)} - 2\varepsilon u_{X_0 X_1}^{(i)} - \varepsilon e\, u_{X_0}^{(i)2} u_{X_0 X_0}^{(i)} + O(\varepsilon^2) = 0,
\tag{25}
$$
$$
i = 0, \pm 1, \pm 2, ...
$$

where the responses are expressed in the following asymptotic form:

$$
u^{(i)}(T_0, T_1, ..., X_0, X_1, ...) = \sum_{q=0}^{\infty} \varepsilon^q u_q^{(i)}(T_0, T_1, ..., X_0, X_1, ...)
\tag{26}
$$

Upon substitution into the governing equations we obtain an hierarchy of linear problems (again, a signature of weak nonlinearity).

Considering the $O(1)$ approximation we obtain the following problem,

$$
u_{0 T_0 T_0}^{(i)} - u_{0 X_0 X_0}^{(i)} = 0 \Rightarrow
$$
$$
u_0^{(i)}(T_0, T_1, X_0, X_1) = A^{(i)}(T_1, X_1)\, e^{j(kX_0 - \lambda T_0)} + cc,
\tag{27}
$$
$$
j = (-1)^{1/2}, \quad i = 0, \pm 1, ...
$$

where k denotes the linearized wavenumber, λ the linearized frequency and $A^{(i)}(T_1, X_1)$ the complex (slowly modulated) amplitude of the traveling

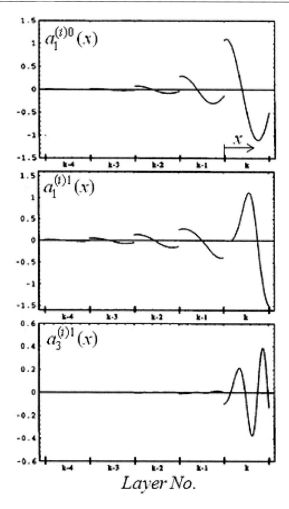

Figure 5. Leading order approximations for a nonlinear standing wave in an AZ of the strongly coupled periodic system of Figure 2.

wave; (cc) denotes complex conjugate. In addition, the linearized dispersion relation is given by $k^2 - \lambda^2 = 0$. The complexity of the amplitude
$A^{(i)}(T_1, X_1)$ incorporates the nontrivial phase differences between the oscillations of material points of the system in the traveling wave motion.

Considering the $O(\varepsilon)$ approximation, we evaluate the complex amplitudes $A^{(i)}(T_1, X_1)$ of the $O(1)$ approximation by eliminating secular terms
from the equations of this order:

$$u_{1T_0T_0}^{(i)} - u_{1X_0X_0}^{(i)} = -2u_{0T_0T_1}^{(i)} + 2u_{0X_0X_1}^{(i)} + e\left(u_{0X_0}^{(i)}\right)^2 u_{0X_0X_0}^{(i)} =$$
$$\left[2j\lambda A_{T_1}^{(i)} + 2jkA_{X_1}^{(i)} - ek^4 A^{(i)2}\bar{A}^{(i)}\right]e^{j(kX_0 - \lambda T_0)} + NST, \tag{28}$$
$$i = 0, \pm 1, \dots$$

Hence, obtain the modulation partial differential equations,

$$2j\lambda A_{T_1}^{(i)} + 2jkA_{X_1}^{(i)} - ek^4 A^{(i)2}\bar{A}^{(i)} = 0 \tag{29}$$

where overbar denotes complex conjugate. Seeking monochromatic waves
that possess a fixed wavenumber k, we set the partial derivative with respect to X_1 equal to zero and introduce the following polar transformation,
$A^{(i)}(T_1) = (1/2)a_i(T_1)e^{j\beta_i(T_1)}$. Separating real and imaginary parts in the
complex modulation equation we obtain two real modulation equations:

$$a_i'(T_1) = 0 \Rightarrow a_i(T_1) = a_{i0}$$
$$\beta_i'(T_1) = -\left(ek^4/8\lambda\right)a_{i0}^2 T_1 + \beta_{i0} \tag{30}$$

Waves with fixed frequency and slowly varying spatial waveform can be similarly analyzed. Combining the previous analytical results, we approximate
monochromatic waves in PZs of the nonlinear medium as:

$$u^{(i)}(x, \tau) = a_{i0}e^{j(kx+\beta_{i0})}e^{-j\left[\lambda+\left(\varepsilon ek^4/8\lambda\right)a_{i0}^2 + O(\varepsilon^2)\right]\tau} + O(\varepsilon), \tag{31}$$
$$i = 0, \pm 1, \dots$$

The nonlinear correction to the dispersion relation is given by, $\lambda^2 - k^2 = 0$
and $\omega = \lambda + \left(\varepsilon ek^4/8\lambda\right)a_{i0}^2 + O\left(\varepsilon^2\right)$, where ω is the frequency of the traveling
wave. We may express the response of the i-th layer as:

$$u^{(i)}(x, \tau) = a_{i0}e^{j(kx+\beta_{i0})}e^{-j\omega\tau} + O(\varepsilon) + cc \equiv$$
$$\Phi^{(i)}(x)e^{-j\omega\tau} + O(\varepsilon) + cc, \quad i = 0, \pm 1, \dots \tag{32}$$

We note that the spatial dependence of the amplitude is the same for all
layers and just differs by a phase shift, $\Phi^{(i)}(x) = C_i \cos kx + D_i \sin kx$, $C_i = a_{i0}e^{j\beta_{i0}}$, $D_i = ja_{i0}e^{-j\beta_{i0}}$. In fact, boundary displacements of neighboring

layers differ only by phase differences, and no spatial attenuation of the traveling wave occurs; this is the basic distinction between nonlinear waves in PZs and AZs. We now impose boundary conditions, and also introduce an additional relation relevant to traveling waves,

$$\Phi^{(i+1)}(0) = e^{j\mu}\Phi^{(i)}(0), \quad \Phi^{(i)}(1) = e^{j\mu}\Phi^{(i-1)}(1), \quad i = 0, \pm 1, \dots \quad (33)$$

where μ is the propagation constant computed by the relations:

$$\cos\mu = \cos\lambda - (\lambda/2K)\sin\lambda$$
$$e^{2j\beta_{i0}} = j\left(\lambda + Ke^{j\mu}\sin\lambda\right)\left(K - Ke^{-j\mu}\cos\lambda\right)^{-1} \quad (34)$$

The nonlinear PZs of the periodic medium are then approximated from the first of the above relations as,

$$\cos\mu = \pm 1 \Rightarrow \cos\lambda_b - (\lambda_b/2K)\sin\lambda_b = \pm 1$$
$$\omega_b = \lambda_b + \left(\varepsilon e \kappa^4/8\lambda_b\right) a_{i0}^2 + O(\varepsilon^2) \quad (35)$$

where the amplitudes a_{i0} are evaluated by the initial conditions of the problem. These relations show that the boundaries of the PZs depend on the amplitudes (energy), in contrast to linear theory.

This completes the analysis of the dynamics of the strongly coupled periodic system depicted in Figure 2. We found that the weakly nonlinear dynamics are perturbations of the linear dynamics (corresponding to $\varepsilon = 0$), and no essentially nonlinear dynamical phenomena are realized. This is not so for the weakly coupled periodic medium, since as shown in the next section, interesting nonlinear dynamics are realized in that case, having no counterparts in linear theory.

3 Nonlinear Modal Interactions Leading to Standing Breathers in a Weakly Coupled Periodic Medium

We consider the weakly coupled nonlinear periodic system (1,2) by scaling the linear coupling stiffness according to $K = \varepsilon k$, $0 < \varepsilon \ll 1$, and expressing the equations of motion in a slightly different form, that is, by incorporating the linear coupling forces between the nonlinear layers in (2) directly into the equations of motion,

$$u_{\tau\tau}^{(i)}(x,\tau) = u_{xx}^{(i)}(x,\tau) + \varepsilon e\, u_x^{(i)2}(x,\tau)\, u_{xx}^{(i)}(x,\tau) -$$
$$\varepsilon k\left[u^{(i)}(0,\tau) - u^{(i-1)}(1,\tau)\right]\delta(x) - \quad (36)$$
$$\varepsilon k\left[u^{(i)}(1,\tau) - u^{(i+1)}(0,\tau)\right]\delta(x-1)$$

and complemented by the simplifed boundary conditions

$$u_x^{(i)}(0, \tau) = u_x^{(i)}(1, \tau) = 0$$

with $i = 0, \pm 1, \dots$ Given the weak nonlinearity and weak coupling, in the
limit $\varepsilon \to 0$ this system degenerates into an infinite set of uncoupled linear
rods (cf. Figure 6). In addition, the coupling and nonlinear terms are now
of the same order.

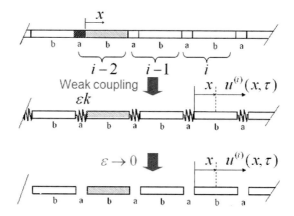

Figure 6. The weakly coupled nonlinear periodic system, and the limiting
linear system of uncoupled elastic layers in the limit $\varepsilon \to 0$.

First we consider the limiting linear system as $\varepsilon \to 0$:

$$\begin{aligned}
u_{\tau\tau}^{(i)}(x, \tau) &= u_{xx}^{(i)}(x, \tau), \quad i = 0, \pm 1, \dots \\
u_x^{(i)}(0, \tau) &= u_x^{(i)}(1, \tau) = 0
\end{aligned} \tag{37}$$

We then compute the complete and orthogonal basis of eigenfunctions and
eigenfrequencies of the generating problem:

$$\phi_m(x) = \begin{cases} 1, & m = 0 \\ \sqrt{2} \sin m\pi x, & m \neq 0 \end{cases} \tag{38}$$
$$\omega_m = m\pi$$

Using as basis the linear eigenfunctions $\phi_m(x)$ of the uncoupled rods, we
express the responses of the weakly coupled system as,

$$u^{(i)}(x, \tau) = \sum_{m=1}^{\infty} \phi_m(x) q_m^{(i)}(\tau) \tag{39}$$

and derive a bi-infinite set of coupled modal oscillators:

$$
\ddot{q}_l^{(i)} + (l\pi)^2 q_l^{(i)} = \varepsilon\, e_3 \sum_{m,n,p=0,1,\ldots} \Gamma_{lmnp} q_m^{(i)} q_n^{(i)} q_p^{(i)} -
$$
$$
\varepsilon k \left[\sum_{m=0,1,\ldots} \phi_l(0)\phi_m(0)q_m^{(i)} - \sum_{m=0,1,\ldots} \phi_l(0)\phi_m(1)q_m^{(i-1)} \right] -
$$
$$
\varepsilon k \left[\sum_{m=0,1,\ldots} \phi_l(1)\phi_m(1)q_m^{(i)} - \sum_{m=0,1,\ldots} \phi_l(1)\phi_m(0)q_m^{(i+1)} \right],
$$
$$
\Gamma_{lmnp} = \int_0^1 \phi_l \phi_m' \phi_n' \phi_p''\, dx, \quad \phi_l(0) = \left\{ \begin{array}{ll} 1, & l=0 \\ \sqrt{2}, & l\neq 0 \end{array} \right. , \tag{40}
$$
$$
\phi_l(1) = \left\{ \begin{array}{ll} 1, & l=0 \\ \sqrt{2}(-1)^l, & l\neq 0 \end{array} \right.
$$
$$
l = 0,1,\ldots, \quad i = 0,\pm 1,\ldots
$$

We employ again a NNM-based approach, whereby resonant oscillations of the discretized system (40) are sought. Note that countable infinities of internal resonances are possible in this system since it holds that $\omega_m/\omega_n = m/n, \quad m,n \in \in N$. To study nonlinear standing waves we introduce again a reference displacement, as the response of the s-th mode of the r-th (reference) layer, $q_0(\tau) \equiv q_s^{(r)}(\tau)$, assuming that is not identically equal to zero. Then, we express the response of the l-th mode of the i-th layer in terms of the following modal function, $q_l^{(i)}(\tau) = Q_l^{(i)}[q_0(\tau)]$, with the obvious compatibility condition $Q_s^{(r)}[q_0(\tau)] = q_0(\tau)$. An asymptotic methodology for computing the modal functions $Q_l^{(i)}[q_0(\tau)]$ is discussed in (Vakakis and King, 1997), and here we only provide a summary of the results.

In particular we will focus on solutions under conditions of 1:3 internal resonance between the s-th and $3s$-th modes in each layer. Substituting into the equations of motion we get,

$$
\frac{d^2 Q_l^{(i)}}{dq_0^2}\dot{q}_0^2 + \frac{dQ_l^{(i)}}{dq_0}\ddot{q}_0 + (l\pi)^2 Q_l^{(i)} =
$$
$$
\varepsilon\, e_3 \sum_{m,n,p=0,1,\ldots} \Gamma_{lmnp} Q_m^{(i)} Q_n^{(i)} Q_p^{(i)} -
$$
$$
\varepsilon k \left[\sum_{m=0,1,\ldots} \phi_l(0)\phi_m(0)Q_m^{(i)} - \right.
$$
$$
\left. \sum_{m=0,1,\ldots} \phi_l(0)\phi_m(1)Q_m^{(i-1)} \right] -
$$
$$
\varepsilon k \left[\sum_{m=0,1,\ldots} \phi_l(1)\phi_m(1)Q_m^{(i)} - \right. \tag{41}
$$
$$
\left. \sum_{m=0,1,\ldots} \phi_l(1)\phi_m(0)Q_m^{(i+1)} \right],
$$
$$
l = 0,1,\ldots, \quad i = 0,\pm 1,\ldots
$$

where the two time derivatives of the reference displacement are computed as follows. The acceleration of the reference point is computed by considering the equation of motion of the s-th mode of the r-th layer:

$$\ddot{q}_0 \equiv \ddot{q}_s^{(r)} = -(l\pi)^2 q_0 +$$
$$\varepsilon\, e_3 \sum_{m,n,p=0,1,...} \Gamma_{lmnp} Q_m^{(r)} Q_n^{(r)} Q_p^{(r)} -$$
$$\varepsilon k \left[\sum_{m=0,1,...} \left\{ \phi_s(0)\phi_m(0) + \phi_s(1)\phi_m(1) \right\} Q_m^{(r)} - \right.$$
$$\left. \sum_{m=0,1,...} \left\{ \phi_s(0)\phi_m(1)Q_m^{(r-1)} - \phi_s(1)\phi_m(0)Q_m^{(r+1)} \right\} \right] \tag{42}$$
$$\equiv F_1\left[Q_0^{(r)}, Q_1^{(r)}, ..., q_0 \right]$$

The velocity of the reference point is similarly computed by considering energy conservation,

$$\dot{q}_0^2 = 2 \int_{q_0}^{q_0^*} \left\{ (s\pi)^2 \xi - \right.$$
$$\varepsilon e_3 \sum_{m,n,p=0,1,...} \Gamma_{smnp} Q_m^{(r)}(\xi) Q_n^{(r)}(\xi) Q_p^{(r)}(\xi) +$$
$$\varepsilon k \sum_{m=0,1,...} [\phi_s(0)\phi_m(0) + \phi_s(1)\phi_m(1)] Q_m^{(r)} - \tag{43}$$
$$\left. \sum_{m=0,1,...} \left[\phi_s(0)\phi_m(1)Q_m^{(r-1)} + \phi_s(1)\phi_m(0)Q_m^{(r+1)} \right] \right\} d\xi$$
$$\equiv F_2[Q_0^{(r)}, Q_1^{(r)}, ..., q_0]$$

where q_0^* is the maximum value of the reference displacement (attained at the time instant of maximum potential energy). We note that $F_2[Q_0^{(r)}, Q_1^{(r)}, ..., q_0^*] = 0$ since $\dot{q}_0 = 0$ at the time instant of maximum potential energy.

Substituting (42) and (43) into (41) Hence, we obtain the following functional equations for the modal functions:

$$\frac{d^2 Q_l^{(i)}}{dq_0^2} F_2[Q_0^{(r)}, Q_1^{(r)}, ..., q_0] + \frac{dQ_l^{(i)}}{dq_0} F_1[Q_0^{(r)}, Q_1^{(r)}, ..., q_0] =$$
$$-(l\pi)^2 Q_l^{(i)} + \varepsilon\, e_3 \sum_{m,n,p=0,1,...} \Gamma_{lmnp} Q_m^{(i)} Q_n^{(i)} Q_p^{(i)} -$$
$$\varepsilon k \left[\sum_{m=0,1,...} \phi_l(0)\phi_m(0)Q_m^{(i)} - \sum_{m=0,1,...} \phi_l(0)\phi_m(1)Q_m^{(i-1)} \right] - \tag{44}$$
$$\varepsilon k \left[\sum_{m=0,1,...} \phi_l(1)\phi_m(1)Q_m^{(i)} - \sum_{m=0,1,...} \phi_l(1)\phi_m(0)Q_m^{(i+1)} \right],$$
$$l = 0, 1, ..., \quad i = 0, \pm 1, ...$$

complemented by the following orthogonality conditions at the time instants of maximum potential energy:

$$
\frac{dQ_l^{(i)}}{dq_0} F_1[Q_0^{(r)}, Q_1^{(r)}, ..., q_0]\Big|_{q_0=q_0^*} =
$$
$$
-(l\pi)^2 Q_l^{(i)} + \varepsilon\, e_3 \sum_{m,n,p=0,1,...} \Gamma_{lmnp} Q_m^{(i)} Q_n^{(i)} Q_p^{(i)} -
$$
$$
\varepsilon k \left[\sum_{m=0,1,...} \phi_l(0)\phi_m(0) Q_m^{(i)} - \right.
$$
$$
\sum_{m=0,1,...} \phi_l(0)\phi_m(1) Q_m^{(i-1)} \Big] -
$$
$$
\varepsilon k \left[\sum_{m=0,1,...} \phi_l(1)\phi_m(1) Q_m^{(i)} - \right.
$$
$$
\left. \sum_{m=0,1,...} \phi_l(1)\phi_m(0) Q_m^{(i+1)} \right]\Big|_{q_0=q_0^*},
$$
$$
l = 0, 1, ..., \quad i = 0, \pm 1, ...
$$
(45)

The solutions of this system of equations are sought in the following asymptotic form:

$$
Q_l^{(i)}(q_0) = \sum_{p=0,1,...} \varepsilon^p Q_{lp}^{(i)}(q_0) ,
$$
$$
Q_{lp}^{(i)}(q_0) = \sum_{j=1,3,5,...} c_{ljp}^{(i)} q_0^j(\tau), \quad l = 1, 2, ..., \quad i = 0, \pm 1, ...
$$
(46)

Summarizing, we seek asymptotic approximations,

$$
u^{(i)}(x, \tau) = \phi_s(x) q_s^{(i)}(\tau) + \phi_{3s}(x) q_{3s}^{(i)}(\tau) + ... =
$$
$$
\phi_s(x)\, Q_s^{(i)}[q_0(\tau)] + \phi_{3s}(x)\, Q_{3s}^{(i)}[q_0(\tau)] + ... =
$$
$$
\phi_s(x) [a_i q_0(\tau)] + \phi_{3s}(x) \left\{ b_i \left[q_0(\tau) - (4/3q_0^{*2}) q_0^3(\tau) \right] \right\} + ... ,
$$
$$
i = 0, \pm 1, ...
$$
(47)

involving 1:3 internal resonance between the s–th and the $3s$-th modes of each layer. The resulting motion is non-synchronous due to nonlinear modal interactions. As shown in (Vakakis and King, 1997) the coefficients a_i and b_i in (47) satisfy an infinite coupled set of nonlinear difference equations depending on a parameter Θ denoting the relative strength of the weak material nonlinearity of the layers with respect to the weak coupling between layers, $\Theta = e_3 s^4 \pi^4 q_0^{*2}/16k$. To derive analytical solutions for these coefficients we rescale $\Theta \to \Theta/\mu$, where μ is a small parameter, and write the sets of difference equations as:

$$
\mu \left\{ (-1)^s a_{r-1} - 2 + (-1)^s a_{r+1} \right\} a_i - \mu \left\{ (-1)^s a_{i-1} - \right.
$$
$$
2a_i + (-1)^s a_{i+1} \} +
$$
$$
3\Theta \left[a_i^2 - 1 \right] a_i + \Theta \left[a_i b_i + 4b_i^2 - b_r - 4b_r^2 \right] a_i = 0
$$
$$
\mu \left\{ (-1)^s b_{i-1} - 2b_i + (-1)^s b_{i+1} \right\} +
$$
$$
3\mu \left\{ (-1)^s a_{r-1} - 2 + (-1)^s a_{r+1} \right\} b_i +
$$
$$
2\mu \left\{ (-1)^s b_{r-1} - 2b_r + (-1)^s b_{r+1} \right\} b_i -
$$
$$
3\Theta \left[a_i^3 + 12a_i^2 b_i + 9b_i^3 + b_i - 5b_r b_i \right] = 0
$$
$$
i = 0, \pm 1, ...
$$
(48)

Analytical solutions are obtained either by direct analysis of the difference equations in the limit of small μ, or by a continuum approximation of the difference equations:

$$z_i \to z(i), \; z \equiv a, b,$$

$$(-1)^s z_{i-1} - 2z_i + (-1)^s z_{i+1} \to 2\left[(-1)^s - 1\right] z(i) + (-1)^s \frac{d^2 z(i)}{di^2} + \ldots$$

The approximations will hold as long as the modal amplitudes are slowly varying; higher order discreteness effects can be included by retaining higher order derivatives in the above continuum approximations. The resulting coupled system of odes are solved asymptotically (Vakakis and King, 1997).

First we consider in-phase (IP) continuum approximations governed by the following approximate ordinary differential equations:

- IP Continuous Approximation

$$
\begin{aligned}
& \mu \frac{d^2 a(i)}{di^2} + \big[3\Theta(-1)^s + \Theta(-1)^s b(r) + \\
& \quad 4\Theta(-1)^s b^2(r) - \mu \frac{d^2 a(r)}{di^2}\big] a(i) - \\
& 3\Theta(-1)^s a^3(i) - \Theta(-1)^s a^2(i) b(i) - 4\Theta(-1)^s b^2(i) a(i) \approx 0 \\
& \mu \frac{d^2 b(i)}{di^2} + \Big\{8\mu\left[1 - (-1)^s\right] + 3\mu \frac{d^2 a(r)}{di^2} + 4\mu\left[1 - (-1)^s\right] b(r) \\
& \quad +2\mu \frac{d^2 b(r)}{di^2} - 3\Theta(-1)^s + 15\Theta(-1)^s b(r)\Big\} b(i) - \\
& 3\Theta(-1)^s a^3(i) - 36\Theta(-1)^s a^2(i) b(r) - 27\Theta(-1)^s b^3(i) \approx 0 \\
& a(r) = 1
\end{aligned}
\tag{49}
$$

Asymptotic solutions of the IP equations can be derived in the limit of strong nonlinearity over coupling between layers; the first type of solutions is given by,

$$
\begin{aligned}
a^I(i;r) &= \sec h \left[\left(\frac{3\Theta}{2\mu}\right)^{1/2}(i-r)\right] + O(\mu) , \\
b^I(i;r) &= -\frac{\mu}{1.2} \sec h \left[\left(\frac{3\Theta}{2\mu}\right)^{1/2}(i-r)\right] + O(\mu^2)
\end{aligned}
\tag{50}
$$

where $\mu = 0.1$, $s = 1, 3, 5, \ldots \mu = 0.1$, and the selection of the reference layer r is a free parameter; it follows that there exists an infinity of solutions for $-\infty < r < \infty$. These are localized, time-periodic, standing breathers corresponding to single hump or multi-hump strongly localized in-phase localized standing oscillations of the weakly coupled nonlinear periodic medium. Indeed, Type I IP solutions belong to a

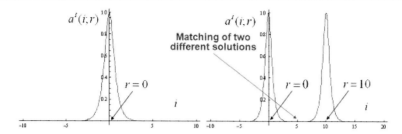

Figure 7. Type I IP solutions (in-phase localized standing breathers) of the weakly coupled system.

family of in-phase standing waves possessing countably infinite localized humps of every possible permutation of layers (b) (cf. Figure 7; these solutions can be rigorously categorized using symbolic algebra).

The second type of IP solutions of system (49) is given by,

$$
\begin{aligned}
a^{II}(i;d) &= \tan h \left[- \left(\frac{3\Theta}{2\mu} \right)^{1/2} (i - d) \right] + O(\mu) \ , \\
b^{II}(i;d) &= - \frac{\mu}{1.2} \tan h \left[- \left(\frac{3\Theta}{2\mu} \right)^{1/2} (i - d) \right] + O(\mu^2)
\end{aligned}
\tag{51}
$$

where $\mu = 0.1$, $s = 2, 4, 6, \ldots$ and the free parameter in this case is the position e of the localization of the slope. Type II solutions correspond to in-phase, time-periodic, standing waves with single or multiple slope localizations, as shown in Figure 8. Again, these solutions are members of a family of countable infinite breathers with arbitrary number of points of localized slope.

Similarly we study out-of-phase (OP) continuum approximations by introducing the out-of-phase coordinate transformations

$$
a_i \to (-1)^i a_i \ , \quad b_i \to (-1)^i b_i
$$

in (48), and performing a continuum approximation on the resulting equations. These transformations enable the study of oscillations where neighboring layers oscillate in an out-of-phase fashion with respect to each other:

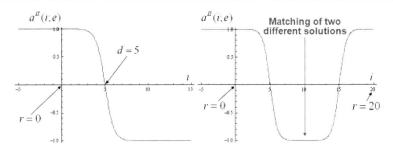

Figure 8. Type II IP solutions (in-phase breathers with localized slopes) of the weakly coupled system.

- OP Continuous Approximation

$$\mu\frac{d^2a(i)}{di^2} + \left\{ 4\mu\left[1-(-1)^{s+1}\right] + \mu\frac{d^2a(r)}{di^2} + \right.$$
$$\left. \Theta(-1)^{s+1}b^2(r)\right\}a(i) - 3\Theta(-1)^{s+1}\left[a^2(i)-1\right]a(i) - $$
$$\Theta(-1)^{s+1}a^2(i)b(i) - 4\Theta(-1)^{s+1}b^2(i)a(i) \approx 0$$

$$\mu\frac{d^2b(i)}{di^2} + \left\{ 8\mu\left[1-(-1)^{s+1}\right] + 3\mu\frac{d^2a(r)}{di^2} + \right. \tag{52}$$
$$4\mu\left[1-(-1)^{s+1}\right]b(r) + 2\mu\frac{d^2b(r)}{di^2} - 3\Theta(-1)^{s+1} + $$
$$\left. 15\Theta(-1)^{s+1}b(r)\right\}b(i) - 3\Theta(-1)^{s+1}a^3(i) - $$
$$36\Theta(-1)^{s+1}a^2(i)b(r) - 27\Theta(-1)^{s+1}b^3(i) \approx 0$$

$$a(r) = 1$$

As in the case of IP solutions, there are two additional types of time-periodic standing wave breathers, which we label them as Type III and Type IV PO solutions:

$$a^{III}(i;r) = \sec h\left[(3\Theta/2\mu)^{1/2}(i-r)\right] + O(\mu) \ ,$$
$$b^{III}(i;r) = -(\mu/1.2)\sec h\left[(3\Theta/2\mu)^{1/2}(i-r)\right] + O(\mu^2) \tag{53}$$
$$\mu = 0.1, \ s = 2,4,6,...$$

$$a^{IV}(i;e) = \tanh\left[-(3\Theta/2\mu)^{1/2}(i-e)\right] + O(\mu) \ ,$$
$$b^{IV}(i;e) = -(\mu/1.2)\tan h\left[-(3\Theta/2\mu)^{1/2}(i-e)\right] + O(\mu^2) \tag{54}$$
$$\mu = 0.1, \ s = 1,3,5,...$$

Although the OP solutions are identical in form to the IP expressions, they correspond to different classes of modes s.

This observation leads to to the conclusion that there exists an additional family of 'composite' standing breathers that are constructed by matching IP and OP solutions corresponding the same class of modes s (i.e., odd or even modes). That is, for $s = 1, 3, 5, ...$ Type I IP localized breathers can be matched with Type IV OP breathers with localized slopes, whereas for $s = 2, 4, 6, ...$ Type III OP localized breathers can be matched with Type II IP breathers with localized slopes. The construction of a composite breather is presented in Figure 9, shown a transition from an in-phase standing wave solution to an out-of-phase one with increasing i.

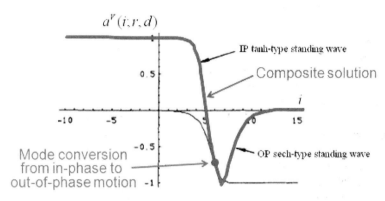

Figure 9. Composite standing wave solution, with transition from in-phase to out-of-phase motion with increasing i.

We note that additional families of nonlocalized syncronous standing waves exist in the weakly coupled periodic medium, together with other families of traveling localized waves. These solutions however, will not be discussed in this work. The analysis presented in this section provides a demonstration of the complexity of the nonlinear dynamics in the weakly coupled periodic medium, with the weakly and strongly localized families of breathers having no counterparts in the dynamics of the strongly coupled medium. This is in contrast to the strongly coupled system examined in the previous section, where the nonlinear dynamics were only perturbations of the linearized dynamics. These results hint on the effect that coupling may have on the nonlinear dynamics of spatially periodic systems. An open question regards the study of the bifurcations or smooth transitions of the different branches of standing or traveling wave solutions when the coupling over nonlinearity ratio increases from small to large values and the strongly nonlinear dynamics degenerate to weakly nonlinear dynamics.

In the next section we discuss some special techniques for studying the nonlinear dynamics of periodic systems. In a later section we will discuss the dynamics of a special class of strongly nonlinear discontinuous systems, namely ordered granular media and provide methodologies for studying their dynamics.

4 Special Techniques for Nonlinear Periodic Media

In this section we present two special methodologies for studying some classes of problems of nonlinear periodic media with weak or strong nonlinearities. The first approach is an analytical method that accounts exactly for discreetness effects in the dynamics of nonlinear elastic continua on a periodic array of discrete nonlinear supports. The method is based on non-smooth transformations of one of the independent variables of the problem, and leads to sets of smooth nonlinear boundary value problems (NLBVPs) in appropriately defined spaces. The NLBVPs can then be solved with standard asymptotic or perturbation methods of nonlinear dynamics. The basic elements of the method were originally conceived by Pilipchuk (1985, 1988) who was influenced by previous work of Zhuravlev (1976, 1977) on non-smooth coordinate transformations. The second approach (Pilipchuk et al., 1996) concerns the study of primary pulse transmission in an impulsively forced semi-infinite repetitive system of linear elastic layers which are coupled by means of strongly nonlinear coupling elements. This approach enables the study of maximum pulse penetration in periodic elastic media with even non-smooth coupling and can be applied to a broad class of problems of primary pulse transmission.

4.1 Analysis of Nonlinear Discreteness Effects Using Non-Smooth Coordinate Transformations

We demonstrate the analytical study of nonlinear discreteness effects in elastic continua by analyzing the dynamics of a forced linear string of infinite spatial extent, supported by a periodic array of nonlinear stiffnesses (Pilipchuk and Vakakis, 1996). The model is presented in Figure 10, from which we note that the dynamics possesses at least two characteristic spatial scales, namely, a long scale y and a short scale y/ε (modeling the spacing between adjacent supports). As in the method of multiple scales, these two spatial scales are treated as independent spatial variables. The governing partial differential equation of motion is given by,

$$\rho\frac{\partial^2 u}{\partial t^2} - T\frac{\partial^2 u}{\partial y^2} + 2f(u)\sum_{k=-\infty}^{\infty}\delta\left(\frac{y}{\varepsilon}-1-2k\right) = q\left(\frac{y}{\varepsilon},y,t\right) \quad (55)$$

$$0 < \varepsilon << 1, -\infty < y < \infty$$

where $u = u\left(y/\varepsilon,y,t\right)$ denotes the transverse displacement of the string, and $f(u)$ is the nonlinear characteristic of the discrete array of supporting springs; in addition, the forcing function q is assumed to be periodic in y/ε with normalized period equal to 4. No dissipation is assumed in the model, although dissipative effects can be conveniently incorporated in the analysis. The non-smooth effects due to the discrete supporting springs enter as a periodic set of Dirac functions in the governing equation (55).

Since the non-smooth (singular) terms in (55) are in terms of the short spatial scale, we introduce non-smooth coordinate transformations in order to remove the singularities and smoothen (exactly) the governing equation of motion. Indeed, we will show that by introducing a new non-smooth transformation of the short spatial scale, it is possible to remove the non-smooth effects and convert the system (55) to a nonlinear boundary value problem (NLBVP), which, however, is completely smooth and hence can be analyzed by known techniques of the theory of nonlinear oscillations. Hence, the proposed approach amounts to a smoothening of the problem by its conversion to a NLBVP in terms of the short spatial scale.

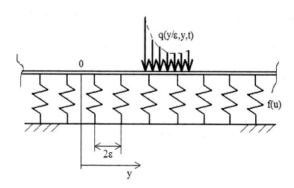

Figure 10. Forced linear string on discrete nonlinear supports.

Before we proceed to the analysis, we provide some basic elements of the method of non-smooth coordinate transformations as conceived by Pilipchuck

(1985, 1988) [but also see (Vakakis et al., 1996)]. Considering a generic independent variable ϕ, we introduce the following pair of non-smooth transformations,

$$\tau(\phi) = \frac{2}{\pi} \arcsin \left[\sin \left(\frac{\pi\phi}{2} \right) \right], \quad e(\phi) = \tau'(\phi) \tag{56}$$

where the derivative should be considered in the generalized sense. These
functions are depicted in Figure 11. Then we introduce the transformation
$\phi \to (\tau(\phi), e(\phi))$, in terms of which a general periodic function $g(\phi)$ with
normalized period $T=4$ in ϕ is transformed as follows,

$$g(\phi) = X(\phi) + e(\phi) Y(\phi) \tag{57}$$

where $X(\tau) = (1/2) [g(\tau) + g(2 - \tau)]$ and $Y(\tau) = (1/2) [g(\tau) - g(2 - \tau)]$.
Moreover, the derivative of this periodic function can also be transformed
in terms of the new variables as follows:

$$g'(\phi) = Y'(\tau) + eX(\tau) + e'Y(\tau) \tag{58}$$

We note that the last term in (58) is singular since,

$$e'(\phi) = \tau''(\phi) = 2 \sum_{k=-\infty}^{\infty} [\delta(z + 1 - 4k) - \delta(z - 1 - 4k)]$$

Requiring that the derivative $g'(\phi)$ be a smooth function of ϕ it is necessary
to eliminate the singularities that are localized at $\phi \in N = \{\phi/\tau(\phi) = \pm 1\}$
by imposing the following 'smoothening conditions':

$$Y|_{\phi \in N} = Y|_{\tau = \pm 1} = 0 \quad \Rightarrow \quad g'(\phi) = Y'(\tau) + eX'(\tau) \tag{59}$$

Higher derivatives are treated similarly; for example, $g''(\phi) = X''(\tau) +
eY''(\tau)$, provided that $X'|_{\phi \in N} = X'|_{\tau = \pm 1} = 0$. Similarly, a general smooth
function of the periodic function can be transformed as follows:

$$\begin{aligned} w[g(\phi)] &= R_w + eI_w, \\ R_w &= (1/2) [w(X + Y) + w(X - Y)], \\ I_w &= (1/2) [w(X + Y) - w(X - Y)] \end{aligned} \tag{60}$$

The terms R_w and I_w are termed the R- and I-components, respectively, of
the transform of $w[g(\phi)]$; note the resemblance to the real and imaginary
parts of a complex function (also note that $e^2 = 1$ in similarity to the relation
satisfied by the imaginary constant $j^2 = -1$). Hence, we have developed
the mathematical formulation in order to transform the governing partial

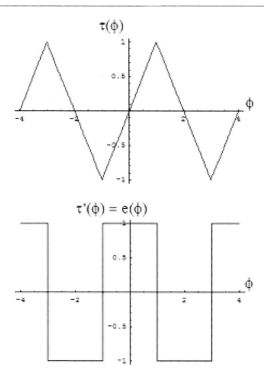

Figure 11. Non-smooth transformations $\tau(\phi)$ and $e(\phi) = \tau'(\phi)$.

differential equation of the problem and eliminate exactly the non-smooth nonlinear terms.

Considering again the non-smooth problem (55), we introduce the non-smooth transformations of the short spatial scale, $y/\varepsilon \to \{\tau(y/\varepsilon),\ e = e(y/\varepsilon)\}$, and express the solution and the forcing as:

$$u = U(\tau, y, t) + eV(\tau, y, t), \quad q = Q(\tau, y, t) + eP(\tau, y, t) \qquad (61)$$

Hence, we seek a solution that is periodic in terms of the short spatial scale, and replace the short scale with the pair of the previously introduced non-smooth variables. Note that the two spatial scales y and y/ε are treated as independent variables. It is interesting to note the the non-smooth term in (55) can be expressed in terms of the second derivative of the non-smooth variable τ:

$$2f(u) \sum_{k=-\infty}^{\infty} \delta\left(\frac{y}{\varepsilon} - 1 - 2k\right) = -f(u)\, sgn(\tau)\, \tau'' \tag{62}$$

Substituting (61) and (62) into (55), transforming the nonlinear function $f(u) = f(U + eV) = R_f + eI_f$ and the partial derivatives using the previous formulation, and taking into account that $\tau'\tau'' = 0$ [since in a generalized sense it is the derivative of the identity $e^2 = 1$ (Maslov and Omenianov, 1981)], we set separately equal to zero the R- and I-components of the resulting transformed expressions, we obtain the following two-point NLBVPs in terms of τ,

$$
\begin{aligned}
\rho\frac{\partial^2 U}{\partial t^2} - T\left[\frac{\partial^2 U}{\partial y^2} + \frac{2}{\varepsilon}\frac{\partial^2 V}{\partial y \partial \tau} + \frac{1}{\varepsilon^2}\frac{\partial^2 U}{\partial \tau^2}\right] &= Q(\tau, y, t) \\
\rho\frac{\partial^2 V}{\partial t^2} - T\left[\frac{\partial^2 V}{\partial y^2} + \frac{2}{\varepsilon}\frac{\partial^2 U}{\partial y \partial \tau} + \frac{1}{\varepsilon^2}\frac{\partial^2 V}{\partial \tau^2}\right] &= P(\tau, y, t) \\
-\left(\frac{1}{\varepsilon}\frac{\partial V}{\partial y} + \frac{1}{\varepsilon^2}\frac{\partial U}{\partial \tau}\right)\Big|_{\tau=\pm 1} &= \mp\frac{1}{T}R_f \\
\frac{\partial V}{\partial y}\Big|_{\tau=\pm 1} &= V|_{\tau=\pm 1} = 0
\end{aligned}
\tag{63}
$$

where $-1 \leq \tau \leq 1$, $-\infty < y < \infty$ and $0 \leq t < \infty$. The boundary conditions in (63) are, in essence, smoothening conditions ensuring smoothness of the derivatives of the transformed variables, and elimination of non-smooth terms from the governing equation of motion.

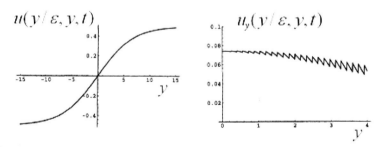

Figure 12. Snapshot of the analytical solution for the transverse displacement of the string showing non-smooth effects in the slope due to the discrete nature of the periodic array of supports; snapshot corresponds at the time instant of maximum potential energy (Pilipchuck and Vakakis, 1998).

The transformed system has double the dimension of the original problem, but is completely smooth. As mentioned previously it is in the form of a NLBVP in terms of variable τ, and its solution is carried only in the interval $\tau \in [-1,1]$. This provides an additional simplification from an analytical point of view, since, the analytical solution can be expressed in terms of a regular perturbation expansion in terms of τ without introducing secular terms. The plan is to analytically approximate the solution of the transformed problem using asymptotic or perturbation techniques. Then, the discreteness effects can be analytically accounted for through inverse transformations.

The solution of the transformed problem (63) is sought in the following regular series form,

$$U(\tau, y, t) = \sum_k \varepsilon^k U_k(\tau, y, t), \quad V(\tau, y, t) = \sum_k \varepsilon^k V_k(\tau, y, t) \qquad (64)$$

and the analysis is carried out in a series of subproblems in ascending powers of ε. At $O(1)$ we analyze the following problem:

$$\frac{\partial^2 U_0}{\partial \tau^2} = 0 \Rightarrow U_0(\tau, y, t) = A_0(y, t)\tau + B_0(y, t)$$
$$\frac{\partial^2 V_0}{\partial \tau^2} = 0 \Rightarrow V_0(\tau, y, t) = C_0(y, t)\tau + D_0(y, t) \qquad (65)$$

Taking into account the $O(1)$ boundary (smoothening) conditions, we evaluate the solution as,

$$\left. \frac{\partial U_0}{\partial \tau} \right|_{\tau = \pm 1} = 0 \,, \quad V_0 \big|_{\tau = \pm 1} = 0 \;\Rightarrow$$
$$U_0(\tau, y, t) = B_0(y, t) \,, \qquad (66)$$
$$V_0(\tau, y, t) = 0$$

where the function $B_0(y,t)$ is determined at a next order.

At $O(\varepsilon)$ we obtain a trivial NLBVP, so preceeding to the $O(\varepsilon^2)$ problem we obtain,

$$\frac{\partial^2 U_2}{\partial \tau^2} = -2\frac{\partial^2 V_1}{\partial y \partial \tau} + \frac{\rho}{T}\frac{\partial^2 U_0}{\partial t^2} - \frac{\partial^2 U_0}{\partial y^2} - \frac{Q(\tau, y, t)}{T}$$
$$\frac{\partial^2 V_2}{\partial \tau^2} = -2\frac{\partial^2 U_1}{\partial y \partial \tau} + \frac{\rho}{T}\frac{\partial^2 V_0}{\partial t^2} - \frac{\partial^2 V_0}{\partial y^2} - \frac{P(\tau, y, t)}{T} \qquad (67)$$
$$\left. \frac{\partial U_2}{\partial \tau} \right|_{\tau = \pm 1} = \mp \frac{1}{2T}\left[f(U_0 + V_0) + f(U_0 - V_0) \right] \,, \quad V_2 \big|_{\tau = \pm 1} = 0$$

The solution of this set of NLBVPs is given by,

$$U_2(\tau, y, t) = -\frac{\tau^2}{2T}f(B_0(y, t)) + \frac{\tau^2 + 2\tau}{4T}\int_{-1}^{1} Q(\xi, y, t)\, d\xi$$
$$+ \frac{1}{T}\int_{-1}^{\tau}(\xi - \tau)Q(\xi, y, t)\, d\xi + B_2(y, t) \qquad (68)$$
$$V_2(\tau, y, t) = \frac{1}{T}\int_{-1}^{\tau}(\xi - \tau)P(\xi, y, t)\, d\xi - \frac{\tau + 1}{2\tau}\int_{-1}^{1}(\xi - 1)P(\xi, y, t)\, d\xi$$

where the function $B_0(y,t)$ is governed by the following partial differential equation (this is derived by satisfying one of the boundary conditions):

$$\frac{\partial^2 B_0(y, t)}{\partial t^2} - \frac{T}{\rho}\frac{\partial^2 B_0(y, t)}{\partial y^2} + \frac{f(B_0(y, t))}{\rho} = \frac{1}{2\rho}\int_{-1}^{1} Q(\xi, y, t)\, d\xi,$$
$$-\infty < y < \infty \qquad (69)$$

To obtain the partial differential equation for $B_2(y,t)$ we must consider selected parts of the subproblems at $O(\varepsilon^3)$ and $O(\varepsilon^4)$. Then we obtain the following partial differential equation governing the function $B_2(y,t)$:

$$\rho \frac{\partial^2 B_2(y,t)}{\partial t^2} - T \frac{\partial^2 B_2(y,t)}{\partial y^2} + f'(B_0(y,t))B_2(y,t) =$$
$$\frac{\rho}{6T} \left[\frac{\partial^2 f(B_0(y,t))}{\partial t^2} - \frac{1}{2} \int_{-1}^{1} \frac{\partial^2 Q}{\partial t^2}(\xi,y,t)\, d\xi - \right.$$
$$\left. 3 \int_{-1}^{1} \int_{-1}^{\gamma} (\xi - \gamma) \frac{\partial^2 Q}{\partial t^2}(\xi,y,t)\, d\xi d\gamma \right] -$$
$$\frac{1}{6} \left[\frac{\partial^2 f(B_0(y,t))}{\partial t^2} - \frac{1}{2} \int_{-1}^{1} \frac{\partial^2 Q}{\partial y^2}(\xi,y,t)\, d\xi - \right. \tag{70}$$
$$\left. 3 \int_{-1}^{1} \int_{-1}^{\gamma} (\xi - \gamma) \frac{\partial^2 Q}{\partial y^2}(\xi,y,t)\, d\xi d\gamma \right] +$$
$$\frac{f'(B_0(y,t))}{2T} \left[f(B_0(y,t)) - \int_{-1}^{1} \left(\xi - \frac{1}{2} \right) Q(\xi,y,t)\, d\xi \right],$$
$$-\infty < y < \infty$$

Combining the previous results we derive the following analytical approximation for the transverse displacement of the forced string:

$$u(y/\varepsilon, y, t) = \left\{ B_0(y,t) + \varepsilon^2 \left[-\frac{\tau^2}{2T} f(B_0(y,t)) + \right. \right.$$
$$\frac{\tau^2 + 2\tau}{4T} \int_{-1}^{1} Q(\xi,y,t)\, d\xi +$$
$$\frac{1}{T} \int_{-1}^{\tau} (\xi - \tau) Q(\xi,y,t)\, d\xi + B_2(y,t) \bigg] \bigg\} + \tag{71}$$
$$e\,\varepsilon^2 \left[\frac{1}{T} \int_{-1}^{\tau} (\xi - \tau) P(\xi,y,t)\, d\xi - \right.$$
$$\frac{\tau+1}{2\tau} \int_{-1}^{1} (\xi - 1) P(\xi,y,t)\, d\xi \bigg] + O(\varepsilon^3)$$

We note that only the R-component of the excitation affects the leading-order approximation of the response of the string. The computed approximations are spatially periodic with respect to the short spatial scale y/ε, but not with respect to the long spatial scale y. The discreteness effects enter through the non-smooth variables τ and e , and are accounted only up to $O(\varepsilon^2)$ in the asymptotic analysis; these produce discontinuities in the first derivative of the envelope of the transverse oscillation of the string. The C^0 (but not C^1) differentiability of the asymptotic results is ensured by the imposed smoothness conditions. A variety of localized or nonlocalized, standing or traveling wave solutions and of, time periodic or transient solutions can be computed through the functions $B_0(y,t)$ and $B_2(y,t)$ computed by the equations (69) and (70), respectively.

As an application we consider the forced vibrations of a string with hardening stiffness supports possessing cubic nonlinearity, $f(u) = a_1 u + a_3 u^3$, and the time-harmonic excitation $q(y/\varepsilon, y, t) = Z \sec hy \cos(\pi y/2\varepsilon) \cos \omega t$. We considered standing, time-periodic solutions with localized slope distri-

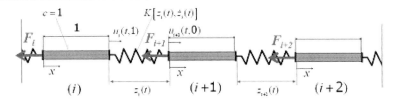

Figure 13. Primary pulse transmission in a nonlinear periodic medium; note the definition of a local spatial variable $x \in [0,1]$ for each of the layers.

bution. To the order of approximation considered in the analysis, discreteness effects appear at the envelope of the vibration, as depicted in Figure 12 (Pilipchuck and Vakakis, 1998). Extension of the methodology for nonlinear discrete supports with clearance stiffness nonlinearities (Salenger and Vakakis, 1998) or dry friction is possible.

In conclusion, the discussed approach can be used to study linear or nonlinear elastic media with nonlinear supports; it smoothens 'exactly' the dynamics my removing singularities from the equations of motion and transforming the problems to smooth NBVPs, albeit of double dimension. The algebra governing the non-smooth coordinate transformations resembles complexification, with the non-smooth variable e corresponding to the imaginary constant j. The methodology has been applied to strongly nonlinear (also non-smooth, e.g., vibro-impact) dynamical problems, and is one of the few available analytical methods enabling the asymptotic study of strongly nonlinear dynamics (since it uses non-smooth generating functions instead of harmonic ones).

4.2   Primary Pulse Transmission in a Nonlinear Periodic Medium

The second methodology was developed to study transmission of stress waves in an impulsively forced semi-infinite repetitive system of linear layers coupled by means of strongly nonlinear stiffness elements. Only primary pulse transmission and reflection at each nonlinear element is considered, and this permits the reduction of the problem to an infinite set of first-order strongly nonlinear ordinary differential equations. These equations can then be solved either analytically and numerically (Pilipchuck et al., 1996). This approach can be used to study primary pulse transmission in nonlinear periodic media.

In particular, we consider a periodic system of linearly elastic one di-

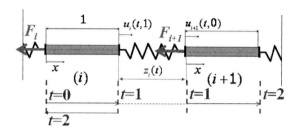

Figure 14. Schematic showing the time instants of arrival and reflection of propagating pulses in two layers of the periodic medium.

mensional layers in axial vibration, coupled by means of strongly nonlinear stiffnesses (cf. Figure 13). Dissipative effects in the coupling stiffnesses are taken into account. We aim to study wave transmission in this system when a shock $F_1(t)$ is applied to the left boundary of the first layer ($i = 1$). Moreover, we assume that the external pulse $F_1(t)$ is of an impulsive nature, that is, it has finite time duration of application in the time interval $0 \leq t \leq \varepsilon \ll 1$. Considering the axial vibration of the i-th layer,

$$\frac{\partial^2 u_i}{\partial t^2} - \frac{\partial^2 u_i}{\partial x^2} = 0, \quad 0 \leq x \leq 1, \quad t \geq 0, \quad i = 0, 1, 2, \ldots \quad (72)$$

we assume zero initial conditions $u_i(0, x) = \partial u_i(0, x)/\partial t = 0$, where $t = 0$ is the instant when the pulse arrives at the left boundary $x = 0$ of this layer (in fact, we consider a local spatial variable x for each of the layers of the periodic medium). For $i > 1$ the force acting on the left boundary of the i-th layer is computed by $F_i(t) = K[z_{i-1}(t), \dot{z}_{i-1}(t)]$, so the boundary conditions are given by:

$$\begin{aligned} \frac{\partial u_i(t,0)}{\partial x} &= F_i(t), \\ \frac{\partial u_i(t,1)}{\partial x} &= K[z_i(t), \dot{z}_i(t)] \equiv K(t), \\ z_i(t) &= u_{i+1}(t,0) - u_i(t,1) \end{aligned} \quad (73)$$

Similarly, considering the oscillation of the $(i + 1)$-th layer, the pulse arrives at its left boundary at $t = 1$ and its dynamics is governed by (cf. Figure 13),

$$\frac{\partial^2 u_{i+1}}{\partial t^2} - \frac{\partial^2 u_{i+1}}{\partial x^2} = 0, \quad 0 \leq x \leq 1, \quad t \geq 1 \\ \frac{\partial u_{i+1}(0,t)}{\partial x} = K(t) \equiv F_{i+1}(t), \quad u_{i+1}(1,x) = \frac{\partial u_{i+1}(1,x)}{\partial t} = 0 \quad (74)$$

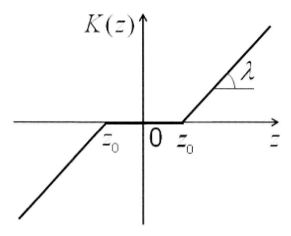

Figure 15. Clearance stiffness nonlinearity.

We now consider waves propagating in the i-th layer,

$$
\begin{aligned}
\frac{\partial u_i}{\partial x} &= F_i(t-x), \quad 0 < t < 1 \\
\frac{\partial u_i}{\partial x} &= F_i(t-x) + \Phi_i(t+x), \quad 1 < t < 2 \\
\frac{\partial u_{i+1}}{\partial x} &= F_{i+1}(t-x) = K(t-x), \quad 1 < t < 2
\end{aligned}
\tag{75}
$$

where the time instants of arrival and reflection of propagating pulses in layers i and $(i+1)$ of the periodic medium are depicted in the schematic of Figure 14. Considering the primary propagating pulse, the reflected wave in layer i is computed by the boundary condition at $x = 1$,

$$
F_i(t-1) + \Phi_i(t+1) = K(t) \;\Rightarrow\; \Phi_i(\xi) = -F_i(\xi-2) + K(\xi-1) \tag{76}
$$

so the dynamics of layers layers$i$ and $(i+1)$ for primary pulse propagation are computed by:

$$
\begin{aligned}
\frac{\partial u_i}{\partial x} &= F_i(t-x) + F_i(t+x-2) + K(t+x-1) \;\Rightarrow \\
u_i(t,x) &= u_i(t,0) - \\
\int_t^{t-x} F_i(\xi)d\xi &- \int_t^{t+x} F_i(\xi-2)d\xi + \int_t^{t+x} K(\xi-1)d\xi, \quad 1 < t < 2 \\
\frac{\partial u_{i+1}}{\partial x} &= K(t-x) \;\Rightarrow \\
u_{i+1}(t,x) &= u_{i+1}(t,0) - \int_t^{t-x} K(\xi)d\xi \\
& 1 < t < 2
\end{aligned}
\tag{77}
$$

Evaluating the responses of the layers at $x = 1$ and for $1 < t < 2$, we obtain,

$$
\left.
\begin{array}{c}
u_i(t,1) = u_i(t,0) - \int_t^{t-1} F_i(\xi)d\xi - \int_t^{t+1} F_i(\xi-2)d\xi + \int_t^{t+1} K(\xi-1)d\xi \\
u_{i+1}(t,1) = u_{i+1}(t,0) - \int_t^{t-1} K(\xi)d\xi
\end{array}
\right\} \Rightarrow
$$

$$
z_i(t) = u_{i+1}(t,0) - u_i(t,1) =
$$
$$
\int_t^{t-1} K(\xi)d\xi - \left[C - \int_t^{t-1} F_i(\xi)d\xi - \int_t^{t+1} F_i(\xi-2)d\xi + \int_t^{t+1} K(\xi-1)d\xi \right],
$$
$$
1 < t < 2
$$

(78)

Assuming that $u_i(t,0)$ is a constant, and differentiating the last of the expressions (78) with respect to t we obtain:

$$
\dot{z}_i(t) = 2K(t-1) - 2K(t) + 2F_i(t-1) - F_i(t) - F_i(t-2),
$$
$$
1 < t < 2
$$

(79)

Taking into account the impulsive nature of the applied excitation to the first layer of the medium, it holds that $K(t-1) = F_i(t) = F_i(t-2) = 0$ in the time interval $1 < t < 2$. Hence, we derive the relation,

$$
(1/2)\dot{z}_i(t) + K\left[z_i(t), \dot{z}_i(t)\right] = F_i(t),
$$
$$
z_i(0) = 0, \quad i > 1, \quad 0 < t < 1
$$

(80)

where the time shift $t - 1 \to t$ is imposed, so that the time of arrival of the primary pulse at an arbitrary layer is always set to $t = 0$. Moreover, the profile of the transmitted force pulse in the next layer $(i + 1)$ is computed as,

$$
F_{i+1}(t) = K\left[z_i(t), \dot{z}_i(t)\right] \Rightarrow
$$
$$
F_{i+1}(t) = F_i(t) - (1/2)\dot{z}_i(t),
$$
$$
i > 1, \quad 0 < t < 1
$$

(81)

where the previous time shift was imposed (in essence, a local normalized temporal variable $t \in [0,1]$ and a local normalized spatial variable $x \in [0,1]$ are defined for each elastic layer of the periodic medium).

Based on these results, we develop an iterative scheme to compute the primary pulse transmission through the nonlinear periodic medium. For $i = 1$ we solve (80) for prescribed applied impulsive excitation $F_1(t)$ and compute the relative response $z_1(t)$. This enables the computation of the force $F_2(t)$ by means of (81), representing the force applied to the left boundary of the second layer of the medium due to arrival of the primary pulse. The iterative computation of primary pulse propagation through the system is continued for the arbitrary layer i using the relations (80) and (81). This algorithm solves iteratively the problem of primary pulse transmission through the periodic medium by computing, (i) the profile $F_i(t)$ of the transmitted primary pulse at the left boundary of each layer, and (ii) the relative response $z_i(t)$ between neighboring layers. Note that due to the previous shift in time, the time of arrival of the primary pulse at an arbitrary layer is always set equal to zero, so the time domain of interest is always normalized to $0 \leq t \leq 1$. In addition to direct numerical integrations of these relations, continuum approximations can be developed that enable the study of maximum pulse penetration in the periodic medium.

As an application we consider a clearance nonlinearity for the coupling element with no dissipation, with $K = K(z)$ depicted in Figure 15. In this case the basic equations for the iteration scheme are expressed as,

$$\dot{z}_i + \lambda z_i + \phi(z_i) = F_i(t)$$
$$F_{i+1}(t) = F_i(t) - \dot{z}_i/2, \quad z_i(0) = 0, \quad 0 < t < 1 \tag{82}$$

where,

$$\phi(z) = \begin{cases} -\lambda z, & |z| \leq z_0 \\ -\lambda z_0 sgn(z), & |z| > z_0 \end{cases} \tag{83}$$

We consider the applied impulsive excitation,

$$F_1(t) = \begin{cases} Pt^2(t - \varepsilon)^2 \sin(t/\varepsilon), & 0 \leq t \leq \varepsilon \\ 0, & t > \varepsilon \end{cases} \quad , \quad P = 5.2 \times 10^7, \quad \varepsilon = 0.1$$

and study primary pulse arrest (entrapment) in the nonlinear periodic system with by depicting the primary pulse penetration in the first 15 layers (i.e., the transmitted forces $\tilde{F}_i(t)$, $i = 1, ..., 15$) for $\lambda = 10.0$ and three values of the clearance, $z_0 = 0$, 0.1 and 0.2. The results are depicted in Figure 16, from which we conclude that for sufficiently large clearances pulse arrest is achieved, that is, the primary pulse does not penetrate beyond a specific layer of the nonlinear periodic medium. For $z_0 = 0.1$ pulse arrest is realized at the 7th layer, whereas for $z_0 = 0.2$ at the 4th layer. From a practical point of view it is of interest to be able to predict when pulse arrest is realized in order to design the nonlinear periodic medium as shock mitigator.

To study maximum penetration of the transmitted primary pulse in the periodic medium of Figure 13, we reconsider the infinite set of equations of the previous iterative scheme,

$$(1/2)\dot{z}_{i+1}(t) + K\left[z_{i+1}(t), \dot{z}_{i+1}(t)\right] = F_{i+1}(t), \quad z_{i+1}(0) = 0,$$
$$F_{i+1}(t) = K\left[z_i(t), \dot{z}_i(t)\right], \quad i > 0, \quad 0 < t < 1 \tag{84}$$

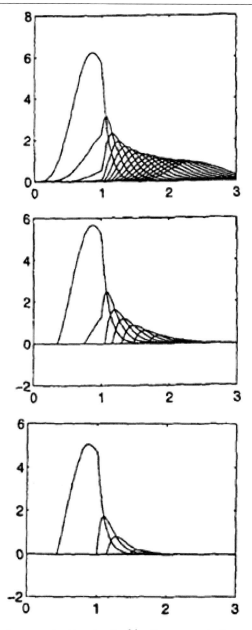

Figure 16. Pulse penetration in the nonlinear periodic medium with clearance nonlinearities: (a) $z_0 = 0$ (linear system), (b) $z_0 = 0.1$, and (c) $z_0 = 0.2$; note pulse arrest in (b) and (c).

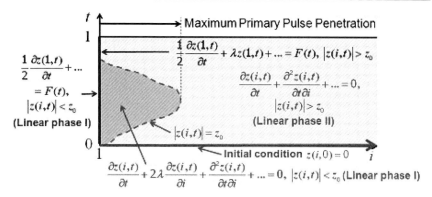

Figure 17. Schematic showing the two linear subproblems and the matching line for solving the problem of maximum pulse penetration in the impulsively forced nonlinear periodic medium with clearance nonlinearities.

and assume that $z_i(t)$ does not vary much between subsequent coupling elements, so that we can impose the following continuum approximations:

$$
\begin{aligned}
z_i(t) \to z(i,t), \quad z_{i+1}(t) &\to z(i+1,t) = z(i,t)+ \\
\frac{\partial z(i,t)}{\partial i} &+ \frac{1}{2!}\frac{\partial^2 z(i,t)}{\partial i^2} + \dots \\
K\left[z_{i+1}(t), \dot{z}_{i+1}(t)\right] &\to K\left[z(i,t), \frac{\partial z(i,t)}{\partial t}\right] + \\
\frac{\partial K[z,\partial z/\partial t]}{\partial z}\frac{\partial z(i,t)}{\partial i} &+ \frac{\partial K[z,\partial z/\partial t]}{\partial(\partial z/\partial t)}\frac{\partial^2 z(i,t)}{\partial t\partial i} + \dots
\end{aligned}
\tag{85}
$$

Hence, we may replace the infinite set of equations (84) by the following approximate partial differential equation,

$$
\frac{\partial z(i,t)}{\partial t} + 2\frac{\partial K[z,\partial z/\partial t]}{\partial z}\frac{\partial z(i,t)}{\partial i} +
\left\{1 + \frac{2\partial K[z,\partial z/\partial t]}{\partial(\partial z/\partial t)}\right\}\frac{\partial^2 z(i,t)}{\partial t\partial i} + \dots = 0
\tag{86}
$$

which must be solved in the semi-infinite strip:

$$
S = \left\{(t,i) \in R^2 / \ 0 \le t \le 1, \ 1 \le i < +\infty\right\}
$$

with boundary conditions,

$$
\frac{1}{2}\frac{\partial z(1,t)}{\partial t} + K\left[z(1,t), \partial z(1,t)/\partial t\right] + \dots = F_1(t) \equiv F(t), \quad 0 \le t \le 1
$$
$$
z(i,0) = 0, \quad 1 \le i \le +\infty
$$

One way to proceed would be to apply Laplace transform with respect to the spatial index i, and then solve the resulting initial value problem with respect to the temporal variable t. For the case of clearance nonlinearity we can formulate an interesting problem in order to study maximum pulse penetration.For the case of coupling elements with clearance nonlinearity it holds that,

$$\frac{\partial K\left[z, \partial z / \partial t\right]}{\partial z} = \begin{cases} \lambda, & |z| \leq z_0 \\ 0, & |z| > z_0 \end{cases}$$

$$\frac{\partial K\left[z, \partial z / \partial t\right]}{\partial \left(\partial z / \partial t\right)} = 0$$

and the problem of primary pulse transmission is formulated in terms of two linear phases, which are matched by imposing the condition $|z(i,t)| = z_0$. Using the continuum approximation we can study maximum penetration of the primary pulse in the periodic system with clearances. As shown in the schematic of Figure 17, the problem is formulated in terms of two linear phases, that are matched at the previous condition.

In summary, we developed an analytic approach for studying primary pulse transmission in a periodic medium with nonlinear coupling elements. The iterative approach was valid for impulsive external excitations, since a basic assumption of the analysis was that the force is applied over a short duration [of $O(\varepsilon)$]. Moreover, secondary stress wave scattering at the nonlinear interfaces was neglected in the primary pulse propagation analysis.The approach can be applied to strongly nonlinear periodic media (e.g., media with periodic arrays of clearances) and to interfaces with dissipation, and examples of application of the asymptotic analysis can be found in (Pilipchuck et al., 1996).

In the next section we consider the dynamics of ordered homogeneous and heterogeneous granular media, which represent a strongly nonlinear and discontinuous class of periodic media. We provide analytical techniques for studying standing and traveling waves in these media, and demonstrate the capacity of heterogeneous granular chains for passive pulse attenuation.

5 Dynamics of Homogeneous and Heterogeneous Ordered Granular Media

Ordered granular media represent a strongly nonlinear (non-linearizable) periodic systems in the form of spherical (or differently shaped) beads in contact (Nesterenko, 2001; Daraio et al., 2005, 2006). Assuming Hertzian law interaction and no applied pre-compression between beads these periodic media are strongly nonlinear, whereas applied pre-compression leads to gradual linearization of the dynamics. In this work we will be concerned only with one-dimensional granular media (homogeneous, with isolated intruders, and heterogeneous), and study their intrinsic dynamics.We will be especially interested in issues of adaptivity of stress wave transmission due to the essential nonlinearity of the granular medium.We will discuss special analytical techniques for studying the strongly nonlinear intrinsic dynamics (standing waves, traveling waves and solitary waves), and focus on the interplay between smooth / non-smooth, and strongly nonlinear / weakly nonlinear dynamical regimes of these systems. We note that uncompressed granular media completely lack linearized structural acoustics, and possess zero speed of sound [hence their characterization as 'sonic vacua' (Nesterenko, 2001)]; yet, as shown in this and the next sections, granular media possess very complex nonlinear dynamics, including solitary and traveling waves, as well as nonlinear resonances and anti-resonances.

5.1 Solitary and Traveling Waves in Homogeneous Granular Media

Strongly nonlinear solitary waves in uncompressed, one-dimensional homogeneous chains of beads in Hertzian contact were analytically, numerically and experimentally studied (Nesterenko, 2001; Daraio et al., 2005; Sen et al., 2008). In (Theocharis et al., 2009) the interplay between nonlinearity and disorder in strongly compressed one-dimensional chains with Hertzian contact was addressed, and the excitation of localized nonlinear modes, their stability and bifurcations were studied.Granular media are highly complex dynamical systems, since their dynamics is highly tunable. Indeed, their dynamics can be either weakly nonlinear in oscillatory regimes with high pre-compression or strongly nonlinear for weak or no pre-compression (Nesterenko, 2001). Hence, linear features such as PZs and AZs due to the periodicity of the medium may be induced in granular chains, simply by applying pre-compression; we will show later, however, that pass and stop bands exist even in the strongly nonlinear granular medium, i.e., with no pre-compression (Jayaprakash et al., 2011a). An added complication is that the dynamics of granular systems can be either smooth or non-smooth, depending on the occurrence or not of separation between beads which lead to non-smooth collisions and exchange of energy and momentum at fast time scales. It follows that the dynamics of granular media can be highly complex as they may exhibit alternating strongly / weakly nonlinear and smooth / non-smooth oscillatory regimes.

The propagation of solitary waves in granular media in Hertzian contact is a subject of intense current study. The existence of a family of single-hump solitary waves in these media was first proved by Nesterenko (2001). It is well-known that these solitary waves propagate without distortion despite rather violent dispersion effects. It is the balance of dispersion and nonlinearity that leads to spatially localized, coherent and shape preserving, propagating solitary waves with speed of propagation being directly proportional to the amplitude (energy). The importance the single-hump solitary wave in the homogeneous ordered granular chain derives from the fact that this wave is typically formed in granular chains excited by impulse excitation in one of their beads. Numerical, experimental and analytical studies have shown that, typically, as the pulse propagates through the granular chain this solitary wave develops in space and time. It follows that this solitary wave is a fundamental dynamical mechanism for energy transfer through homogeneous granular media.

Hence, we start our study of the dynamics of one-dimensional homogeneous granular chains by considering an infinite periodic chain composed of identical, spherical, and linearly elastic beads in contact with no precompression. The governing equations of motion can be expressed in normalized form (Starosvetsky and Vakakis, 2010) as follows,

$$\frac{d^2 z_n}{d\tilde{t}^2} = A \left[(z_{n-1} - z_n)_+^{3/2} - (z_n - z_{n+1})_+^{3/2} \right],$$
$$-N \le n \le N, \quad N \to \infty \tag{87}$$

where $z_n(\tilde{t})$ is the displacement of the n-th bead,

$$A = E(2R)^{1/2} / \left[3m(1 - \mu^2) \right]$$

where E is modulus of elasticity; R radius of each bead; μ Poisson's ratio; and m mass of each bead, the $(3/2)$ stiffness law models Hertzian contact between beads in compression, and the subscript $(+)$ on the right-hand-sides denotes that only positive arguments are allowed in the fractional powers, with the arguments being set equal to zero otherwise. This accounts explicitly for the possibility of separation between adjacent beads of the chain in the absence of compressive forces. Introducing the normalizations,

$$u = \frac{z}{R}, \quad t = \frac{\tilde{t}}{2R/C} \left(\frac{\pi(1 - \mu^2)}{\sqrt{2}} \right)^{1/2}, \quad C = (E/\rho)$$

we express the equations of motion in the following normalized form:

$$\frac{d^2 u_n}{dt^2} = (u_{n-1} - u_n)_+^{3/2} - (u_n - u_{n+1})_+^{3/2}, \quad -N \le n \le N, \quad N \to \infty$$

Hence, in the normalized equations there is complete absence of parameters, so our analysis will be general and applied to general granular chains, as long as the assumptions of purely elastic Hertzian interactions between adjacent spherical beads hold. These assumptions are that, (i) the contact surface area is small in comparison to the granular dimension, and (ii) sufficiently low strains are developed and beads with high modulus materials are used.

We seek a traveling wave solution of (87b) whose waveform remains unchanged as it propagates through the granular chain. By introducing the notion of time delay T between the displacements of adjacent beads it is possible to reformulate this problem in terms of a nonlinear delay differential equation (NDDE). It will be helpful at this point if we use as coordinates the relative displacements $\delta_n = (u_n - u_{n+1})$ between adjacent beads, and transforming (87b) as,

$$\ddot{\delta}_n = (\delta_{n-1})_+^{3/2} - 2(\delta_n)_+^{3/2} + (\delta_{n+1})_+^{3/2}, \quad -N \leq n \leq N, \quad N \to \infty \quad (88)$$

where dot denotes differentiation with respect to the normalized time t. We seek time-periodic solutions of (88) with nontrivial time delay(or equivalently, phase difference) between the relative displacements of adjacent pairs of beads, corresponding to a traveling wavein the infinite granular chain. These solutions are computed by reformulating (88) as a nonlinear delay differential equation (NDDE),

$$\ddot{\delta} = (\delta(t - T))_+^{3/2} - 2(\delta)_+^{3/2} + (\delta(t + T))_+^{3/2}$$
$$\delta(t) = \delta(t + NT), \quad N \to \infty \quad (89)$$

where T is the constant (phase) time shift of the traveling wave between any adjacent beads. Following the solitary solution developed by Nesterenko (2001), we will develop a Padé approximation of the solution of (89) in the form of a localized motion that decays at infinity. Indeed, assuming a localized solitary structure for the solution (single hump) we introduce the restrictions $\delta(0) \geq \delta(t)$, $\delta(t \to \pm\infty) = 0$. Performing another change of variables, $\delta_n = \delta \equiv \varphi^2$ and introducing the normalizations $\tau = t/T$, $\varphi = \tilde{\varphi}/T^2$, the NDDE (89) transforms to:

$$2(\tilde{\varphi}\tilde{\varphi}'' + \tilde{\varphi}'^2) = \tilde{\varphi}(\tau - 1)^3 - 2\tilde{\varphi}(\tau)^3 + \tilde{\varphi}(\tau + 1)^3 \quad (90)$$

In (90) primes denote differentiations with respect to normalized time τ. Because of its non-causal nature (90) represents an infinite-dimensional dynamical system that cannot be solved numerically by specifying initial conditions (guesses) at a single point in τ, or even at a single interval of τ.

Rather, an initial estimate over its entire waveform needs to be made before any kind of iteration or asymptotic analysis can be performed. Additional features that complicate the analysis even more are the strongly nonlinear (nonlinearizable) nature of the NDDE and the lack of a small parameter that could be used for perturbation analysis.

Hence, we resort to an analysis based on Pade' approximations, and seek the solution for the solitary wave by adopting the following simple *ansatz*:

$$\tilde{\varphi}(\tau) = \frac{1}{q_0 + q_2\tau^2 + q_4\tau^4 + q_6\tau^6 + q_8\tau^8 + \ldots} \tag{91}$$

We note that the *ansatz* (91) is non-unique as other forms of Padé approximants could be used in order to approximate the solitary wave [e.g., (Sen et al., 2008)]. Substituting (91) into (90) and collecting like powers of τ we derive a system of nonlinear algebraic equations in terms of the coefficients q_k, $k = 1, 2, \ldots$ which we truncate and solve numerically. A numerical convergence study is then performed in order to determine the limit of the truncation that guarantees convergence of the solution. For example, by retaining only four terms in the Padé approximation we derive the following system of equations for the coefficients,

$$
\frac{2}{q_0^3} - \frac{2}{(q_0+q_2)^2} - \frac{4q_2}{q_0^3} = 0
$$
$$
\frac{36q_2^2}{q_0^4} - \frac{16q_2^2}{(q_0+q_2)^2} + \frac{2}{(q_0+q_2)}\left[\frac{4q_2^2}{(q_0+q_2)^4} + \frac{2q_2(3q_2-q_0)}{(q_0+q_2)^4}\right] - \tag{92}
$$
$$
2q_2\left[\frac{(3q_2-q_0)}{(q_0+q_2)^5}\right] - \frac{60q_2}{q_0^4} = 0
$$

which has to be solved numerically. At each level of approximation the resulting algebraic system for the coefficients should be solved simultaneously for all unknowns and cannot be successively parameterized by q_0 (this is due to the presence of time delay in (28) which brings all the unknowns at the very first level of approximation.

Table 1. Convergence of the coefficients of the Pade' approximation (91).

Padé Approx.	q_0	q_2	q_4	q_6	q_8	Max. error
Two-term	1.14	0.17	-	-	-	0.2606
Three-term	0.9313	0.3057	0.0392	-	-	0.0834
Four-term	0.8604	0.3512	0.0696	0.0075	-	0.0311
Five-term	0.8357	0.3669	0.0831	0.0123	0.0011	0.0128

In Table 1 we present a convergence study for the coefficients of the Padé approximation (91) (Starosvetsky and Vakakis, 2010). Satisfactory convergence is achieved by using a five term Padé approximation which gives an error squared of $O(10^{-2})$. In this convergence study the maximum error is estimated by computing the supremum of the difference between the Padé approximation and the exact solution at certain selected time instants. In Figure 18 we present the convergence of the Padé approximations to the exact solitary wave for increasing number of terms. In these plots the exact solitary wave was computed by direct numerical simulations of the a homogeneous chain with N=200 beads subject to an impulse applied at the first bead at the free left boundary of the chain. In this system a propagating solitary wave is fully formed after the 5th bead so its waveform can be accurately estimated. Furthermore, of significant practical importance is the relationship between the amplitude of the solitary wave and the time shift, $T = \sqrt{\bar{\varphi}(0)}\delta(0)^{-1/4}$; the normalized initial condition $\delta(0)$ on the right-hand-side provides a measure of the energy (amplitude) of the solitary wave. The solitary wave is the fundamental mechanism for transferring energy in the one-dimensional homogeneous granular chain without pre-compression, in the sense that an arbitrary shock excitation applied at any bead of the chain leads eventually to the formation of a train of solitary waves of the type shown in Figure 18. In fact, the formation of the solitary wave is rather robust and represents a preferred way for energy transmission in the granular chain.

We now show that one-dimensional homogeneous granular chains support a countable infinity of families traveling waves that in a certain limit accumulate to the previously studied solitary wave. Our study of nonlinear traveling and standing waves in homogeneous granular chains is based on the analysis of the dynamics of reduced-order homogeneous granular systems with periodic boundary conditions (Starosvetsky and Vakakis, 2010). Equivalently, these reduced systems may be regarded as finite-degree-of-freedom (DOF) granular systems with cyclicity (cyclic symmetry). By studying time-periodic orbits [or nonlinear normal modes – NNMs (Vakakis et al., 1996)] of these reduced cyclic systems we can investigate the existence of traveling or standing waves with specific spatial periodicity (wavenumber) in homogeneous granular chains of infinite spatial extent. We start with the general definition of the reduced-order cyclic granular chain of N identical beads with periodic boundary conditions and no pre-compression:

$$\begin{aligned}
\ddot{u}_1 &= (u_N - u_1)_+^{3/2} - (u_1 - u_2)_+^{3/2} \\
\ddot{u}_2 &= (u_1 - u_2)_+^{3/2} - (u_2 - u_3)_+^{3/2} \\
&\quad \cdots \\
\ddot{u}_N &= (u_{N-1} - u_N)_+^{3/2} - (u_N - u_1)_+^{3/2} \\
&\quad u_0 \equiv u_N, \ \ u_{N+1} \equiv u_1
\end{aligned} \tag{93}$$

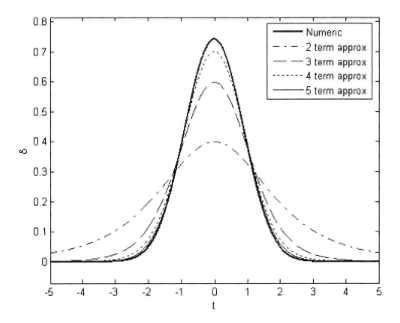

Figure 18. Convergence of the Padé approximations to the exact solution for the solitary wave (Starosvetsky and Vakakis, 2010).

System (93) will be termed the 'reduced granular system with N-bead cyclicity.' It should be clear that the dynamics of (93) can be directly extended to the corresponding granular chain of infinite spatial extent, since any solution of the N-bead reduced cyclic system represents also a spatially periodic solution of the infinite homogeneous granular chain which repeats itself every N-beads (i.e., with N-bead cyclicity). Hence, we can extend the solutions of the N-bead cyclic system to the corresponding infinite chain through the relation $u_{p+mN}(t) = u_p(t)$ for $p = 1, ..., N$, and $m = 1, 2, ...$

Focusing now in the strongly nonlinear system (93), this admits two independent first integrals of motion corresponding to conservation of energy (E) and linear momentum (G) in the absence of external excitations:

$$E = \frac{\dot{u}_1^2}{2} + \frac{\dot{u}_2^2}{2} + ... + \frac{\dot{u}_N^2}{2} + \frac{2}{5}\left[(u_N - u_1)_+^{5/2} + (u_1 - u_2)_+^{5/2} + \right.$$
$$\left. ... + (u_{N-1} - u_N)_+^{5/2}\right] \tag{94}$$
$$G = \dot{u}_1 + \dot{u}_2 + ... + \dot{u}_N = const$$

In addition to the above true first integrals of motion we will also impose the restriction that the initial velocity of center of mass of a cyclic chain is equal to zero, i.e., $G = 0$. This restriction is necessary in order to study time periodic motions in terms of bead displacements. This, without loss of generality, leads to the condition $u_1 + u_2 + ... + u_N = 0$. Given that the reduced system possesses place N DOF and there exist only two first integrals of motion, it is non-integrable for N>2, and, hence, it is expected to possess complex dynamics (including chaotic orbits).

Time-periodic solutions of (93) with nontrivial time delay (or equivalently, phase difference) T between the displacements of adjacent beads correspond to traveling waves (with spatial periodicity equal to the cyclicity index N) in the corresponding infinite granular chain. These solutions are computed by reformulating (93) as a nonlinear delay differential equation (NDDE),

$$\ddot{u} = (u(t - (N-1)T) - u(t))_+^{3/2} - (u(t) - u(t-T))_+^{3/2}$$
$$u(t) = u(t + NT) \tag{95}$$

where T is the constant time shift between the motions of neighboring beads for a traveling wave solution. The NDDE (95) represents an infinite dimensional dynamical system, and by introducing the rescalings $\tau = t/T$, $\bar{u}(\tau) = T^4 u(\tau T)$ we normalize it in the form,

$$\bar{u}'' = (\bar{u}(\tau - 2(N-1)) - \bar{u}(\tau))_+^{3/2} - (\bar{u}(\tau) - \bar{u}(\tau - 1))_+^{3/2}$$
$$\bar{u}(\tau) = \bar{u}(\tau + N) \tag{96}$$

where primes denote differentiations with respect to the rescaled time τ. Clearly, time-periodic solutions of (96) correspond to spatially- and time-periodic solutions of (93) corresponding to traveling waves with N-bead periodicity. Assuming the existence of a time-periodic solution $\bar{\gamma}(\tau)$ of (96) with constant amplitude $\bar{A}$, we obtain a general relationship $T^4 A = \bar{A} = const$, where A denotes the constant amplitude of the traveling wave in terms of the original (unscaled) coordinates. This relation enables us to compute the amplitude – time shift relationship between all members of a particular family of traveling waves once a single member of that family has been computed.

In Figure 19a we depict a traveling wave with $N = 3$ periodicity, with period T_p equal to exactly three times the time shift, $T_p = 3T$. Indeed, the 3-bead reduced cyclic system is the lowest order such system that admits a family of traveling waves. Members of this family are parametrized by energy (i.e., the wave amplitude). Considering the waveform of the traveling wave, we note its clear *asymmetry*. This is a consequence of the fact that

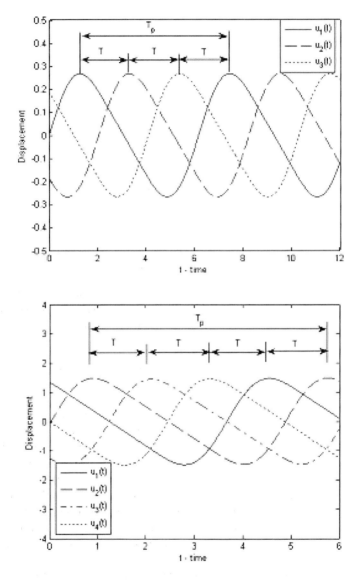

Figure 19. Traveling waves in the homogeneous granular chain at energy level $h = 0.1$ with, (a) $N = 3$, and (b) $N = 4$ periodicity.

during traveling wave propagation there exist two distinct regimes in the motion of each bead; a regime of 'smooth dynamics' during which each bead is in contact and interacts through Hertzian compressive forces with its neighbors, and a regime of 'non-smooth dynamics' where separation between beads takes place, and the bead in 'free flight' before colliding with its neighbor. These two regimes continuously interchange during the propagation of the traveling wave through the homogeneous granular chain, and give rise to multi-phase strongly nonlinear dynamical effects. We make the interesting observation that this family (as well as the other families with higher periodicity N) exists in a system that is characterized as 'sonic vacuum' (Nesterenko, 2001), since it possesses zero speed of sound and complete absence of linearized acoustics! The waveform of the traveling wave for the system with 4-bead periodicity is depicted in Figure 19b. The period of this wave family is exactly equal to four times the time shift, $T_p = 4T$, and the previous general relation, $T^4 A = \bar{A} = const$, can be used to find the characteristic amplitude – time shift (or amplitude-speed) relationship for this family of traveling waves. Comparing the waveforms of the traveling waves for 3- and 4-bead periodicities we conclude that the asymmetry increases with increasing cyclic periodicity index N. This is a direct consequence of the fact that with increasing periodicity index the regime of 'free flight' (or bead separation) is of longer duration compared to the regime of beads in compression. This raises the obvious question related to the asymptotic limit of the waveform of the traveling wave in the limit $N \to \infty$. This issue was addressed in (Starosvetsky and Vakakis, 2010) and it was shown that as the periodicity index N of the families of traveling waves increases, they approach the family of solitary waves studied by Nesterenko (2001) and discussed previously in this section. This results shows that the well studied family of solitary waves is the accumulation of a countable infinity of families of traveling waves with increasingly higher index of spatial periodicity.

In summary, we showed that the one-dimensional homogeneous uncompressed granular chain, although a 'sonic vacuum', supports a countable infinity of families of traveling waves. The non-smooth nature of the traveling waves prevent the use of a continuum approach (in contrast to the limiting family of solitary waves where no bead separation occurs and, hence, it can be analyzed by homogenization); instead one needs to consider directly the discrete equations of motion by imposing periodic boundary conditions, or equivalently, cyclic symmetry. As shown in (Starosvetsky and Vakakis, 2010) the speed of propagation of the traveling waves is smaller than the corresponding speed of the solitary wave studied by Nesterenko (2001), and strongly depends on energy as they are strongly nonlinear.

Finally, we demonstrate the significant effect that a light intruder can have on the dynamics of a forced homogeneous granular chain. To this end, we consider an uncompressed granular chain of identical beads in Hertzian contact with a single intruder depicted in Figure 20. An initial impulse applied to the free left boundary of the chain leads to the formation of a solitary wave that propagates undispersed until it impedes on the intruder. Then a strongly nonlinear interaction between the intruder and the chain takes place leading to oscillations of the intruder. An example of this interaction is depicted in the plot of Figure 21 where the displacements of the light intruder and its left and right neighbors are shown for a unit impulse applied to the left free end of a 201-bead granular chain. For this simulation the mass of the intruder was placed between the 100th and 102st beads of the chain and its mass was chosen as 5% of the mass of its neighboring beads (all masses and stiffness constants of the chain are normalized to unity by appropriate rescalings) (Starosvetsky et al., 2011). Moreover, the time window of integration was chosen so that no secondary reflected pulses originating from the boundaries of the granular chain enter into the picture. Two modes of intruder-chain interaction are inferred from this plot.

Figure 20. One-dimensional homogeneous granular chain with a light mass intruder.

The first mode of interaction corresponds to 'fast' oscillation of the light intruder under heavy compression from its left and right neighbor beads which themselves undergo 'slow' motions. There is no separation between these three beads so the dynamics is smooth and governed by the Hertzian law interaction between beads under compression. The second mode of interaction corresponds to separation between the intruders and its neighboring beads and the dynamics is non-smooth. The ensuing (elastic) collisions between the intruder and its left and right neighbors give rise to a localized *transient breather* (Job et al., 2009) at the site of the intruder, in the form of a 'fast' (non-smooth) oscillation of the intruder with varying amplitude and frequency and 'slow' motions of its neighboring beads. It will be shown that the excitation of this breather acts in essence as 'energy trap' since it acts as a high-frequency energy scatterer. Under certain conditions the

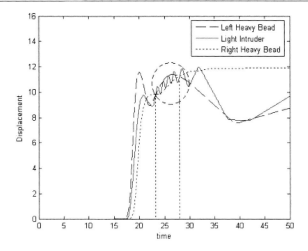

Figure 21. Excitation of the light intruder with 5% mass compared to the beads of the granular chain, showing the two distinct modes of intruder-chain interaction.

excitation of such transient breathers can drastically reduce the amplitude of a pulse propagating through a granular medium.

In Figure 22 we depict the formation of transient breathers in a granular chain composed of 40 beads with unit mass, with the intruder being the 11th bead. We apply an impulse excitation $F(t) = 5\delta(t)$ at the left free boundary of the chain and depict the response of the intruder and its neighboring beads for varying mass ε of the intruder. We note that for very light intruders (cf. Fig. 22a) a series of transient breathers is excited as the neighboring heavy beads separate and then compress the light intruder. For increased mass of the intruder (cf. Figs. 22b,c), however, a single transient breather of longer duration is excited, since the intruder prevents its compression from its neighboring beads after the first separation; in this case there is an interesting slow amplitude modulation of the fast oscillations of the intruder as it repeatedly collides with its neighbors. Finally, as $\varepsilon \to 1$ the fast oscillation of the intruder becomes progressively slower until the upper and lower amplitude modulations of the breather coalesce and the well known solitary wave (Nesterenko, 2001) is formed.

In Figure 23 we consider the velocity time series of the intruder for the same system with $\varepsilon = 0.01$ (corresponding to the plot of Figure 22b) and depict its wavelet spectrum. This provides us with the temporal evolution

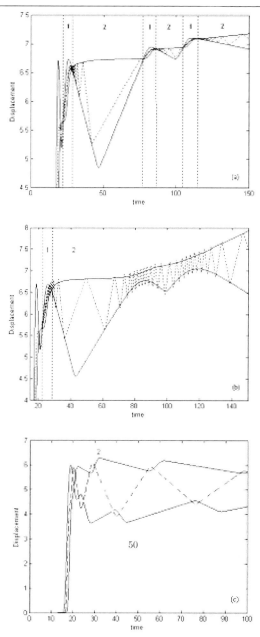

Figure 22. Transient breathers in granular chains with a single light intruder of mass, (a) $\varepsilon = 0.0001$, (b) $\varepsilon = 0.01$, and (c) $\varepsilon = 0.5$; ——— intruder, ------ neighboring beads (the two modes of intruder-chain interaction are indicated by labels 1 and 2).

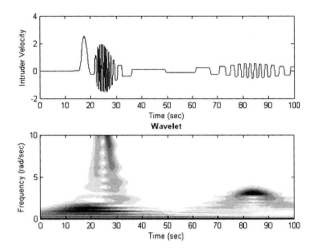

Figure 23. Velocity of the intruder for $\varepsilon = 0.01$ during the formation of the transient breather: (a) time series, (b) wavelet spectrum.

of the various harmonic components in the velocity response of the intruder. The excitation of the transient breather causes high-frequency scattering of pulse energy, especially in time intervals where frequent collisions occur between the intruder and its neighboring beads. In essence the intruder acts as an energy trap by scattering energy from low to high frequencies. It is expected that when dissipative effects are taken into account the effectiveness of the intruder as energy trap will be enhanced since it will cause the dissipation of a considerable part of the energy pulse during its repetitive collisions with its neighbors; in addition, energy gets more effectively dissipated at higher frequencies.

The results discussed in this section provide a hint on the complexity of the intrinsic dynamics of uncompressed ordered granular chains. This should be expected given the strongly nonlinear, and even non-smooth, nature of the dynamics of this class of dynamical systems. In the next section we show that homogeneous granular media possess propagation and attenuation zones (PZs) and (AZs) in similarity to linear and weakly nonlinear periodic systems.

5.2 Propagation and Attenuation Zones of Homogeneous Granular Media

We now examine standing waves (nonlinear normal modes – NNMs) in band zones in finite granular chains composed of spherical granular beads in Hertzian contact, with fixed boundary conditions. In the limit of infinite number of beads we find that attenuation and propagation zones (AZs and PZs), i.e., pass and stop bands in the frequency – energy plane of these dynamical systems, and classify the essentially nonlinear responses that occur in these bands. Moreover, we show how the topologies of these bands significantly affect the forced dynamics of these granular media subject to narrowband excitations.

We start our analysis (Jayaprakash et al., 2011a) by considering a three-bead chain with fixed-fixed boundary conditions and normalized equations of motion given by:

$$\ddot{u}_1 = (-u_1)_+^{3/2} - (u_1 - u_2)_+^{3/2}$$
$$\ddot{u}_2 = -(u_2 - u_3)_+^{3/2} + (u_1 - u_2)_+^{3/2} \qquad (97)$$
$$\ddot{u}_3 = -(u_3)_+^{3/2} + (u_2 - u_3)_+^{3/2}$$

First we compute the in-phase nonlinear normal mode (NNM) of this system by employing a shooting method by assuming zero initial velocities and initial displacements in the form,

$$u_3(0) = (5h/2)^{2/5}, \ u_2(0) = \alpha_2^{(3)}(5h/2)^{2/5}, \ u_1(0) = \alpha_1^{(3)}(5h/2)^{2/5}$$

where h is the (conserved) energy of the system, and, as previously, the coefficients $\alpha_{1,2}^{(3)}$ characterize the asymmetry in the initial deformation of the in-phase NNM. These coefficients are computed as, $0 < \alpha_1^{(3)} \approx 0.374 < \alpha_2^{(3)} \approx 0.744 < 1$ for all values of the energy h. In Figure 24 we depict the in-phase NNM; we infer that there are domains where beads become motionless and offset from their zero equilibrium positions. In addition, non-smooth effects in the dynamics are clearly noted, and a high non-synchronicity between bead oscillations is deduced. The most important (and unique) features of this mode are the patterns of separation and loss of contact between beads and between the end beads and the rigid walls. Indeed, the initial conditions required for realization of the in-phase NNM are such that except for one of the beads, all other beads are detached from each other and with the walls. It is this feature that prevents the study of this mode using continuum approximation approaches in higher dimensional systems with increased number of beads. Hence, the in-phase NNM is the mode most affected and influenced by the discrete nature of the granular system.

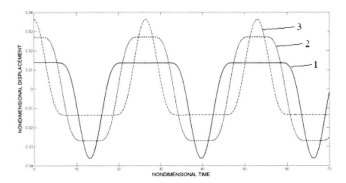

Figure 24. Bead displacements for the in-phase NNM (pseudo-traveling wave) of the three-bead granular system at normalized energy $h = 0.0001$.

An even more peculiar feature of the in-phase NNM (labeled as 'NNM 1') is that it resembles a traveling wave propagating back and forth in the granular chain. As discussed below, this becomes more apparent in higher-dimensional granular media, but a first hint is provided by studying the corresponding velocity profiles of the beads which are in the form of single-hump 'pulses'; half of the velocity profiles of the end beads match exactly that of the central bead, whereas the other half is strongly influenced by the interaction of each end bead with the wall. Moreover, there is a constant time delay between the transmission of velocity 'pulses' in neighboring beads, so the in-phase NNM resembles a traveling wave. This result can be generalized for in-phase NNMs of higher dimensional systems corresponding to velocity profiles of all beads (except for the two end beads) that are identical but for a constant time shift. Hence, although the in-phase NNM is in actuality a time-periodic standing wave, the motion of each bead is followed by an extended period where it settles to an offset stationary position until the bead executes a motion in the reverse direction after a time interval equal to the half period of the NNM. Since each bead (except for the end ones) executes an identical motion but for a constant time shift the in-phase NNM appears indeed as a traveling wave. Based on these observations we will refer from here on the in-phase NNM as a *pseudo-wave*.

The second NNM (labeled as 'NNM 2') of the three-bead system corresponds to out-of-phase motions between neighboring beads (Jayaprakash et al., 2011a). This mode is the highest frequency NNM of the three-bead system, and as such, it forms the upper bound of the domain of periodic motions in the frequency – energy plane. The third mode (labeled as 'NNM 3') of the three-bead system corresponds to stationarity of the central bead (bead 2) for all times. Then, the dynamics of the system can be partitioned into two phases. In the first phase the end beads share equally the energy of the system, which at $t = 0$ is purely elastic. At the end of the first phase the energy is completely transformed to kinetic and the two end beads have opposite velocities. Since the central bead is stationary, for this NNM it acts as a virtual wall. Hence, the dynamics of the system in the second phase is quite similar to the first one, and the equations of motion decouple throughout enabling us to solve for the bead responses in closed form. The last type of NNM (labeled as 'NNM 4') supported by the three-bead system is localized, with one of the end beads interacting with the wall and oscillating with an amplitude that is much larger (about twice) than the corresponding amplitudes of the other two beads (cf. Figure 25). In addition, neighboring beads oscillate in an out-of-phase fashion. It is clear that due to the symmetry of the system this mode is degenerate as it may be realized in an alternative symmetric configuration where the motion is localized to the other end bead. It is interesting to note that this type of nonlinear localization occurs in the homogeneous granular system, and in complete absence of pre-compression.

In addition to the four NNMs the system supports a countable infinity of subharmonic orbits (Jayaprakash et al., 2011a). These motions can be represented in a frequency – energy plot (FEP) of Figure 26. As discussed in (Vakakis et al., 2008), by depicting the dynamics of a system in a FEP it is possible to study the influence of these modes on the forced and damped dynamics; moreover, it would be possible to better relate the dynamics of the granular system to the dynamics of higher dimensional granular systems that will be considered below. From the discussion above the number of NNMs in this system exceeds its degrees of freedom, but this is possible in essentially nonlinear discrete oscillators (Vakakis et al., 1996). The frequency-energy curve of the out-of-phase NNM provides the upper limit of time-periodic orbits for this system, so no periodic oscillations can be realized in the upper region of the FEP, which is labeled as *'prohibited' band*. The lowest frequency mode is the in-phase NNM 1, which as discussed previously is a pseudo-wave. We note that the families of the four NNMs and subharmonic orbits are defined over the entire energy range and are represented by smooth curves that bifurcate from the origin of the FEP. It is of interest to study how the topological structure of the FEP changes as we increase the number of beads, and in particular, how the 'prohibited' band changes as the number of beads tends to infinity and the granular chain becomes of infinite extent. It is found (Jayaprakash et al., 2011a) that

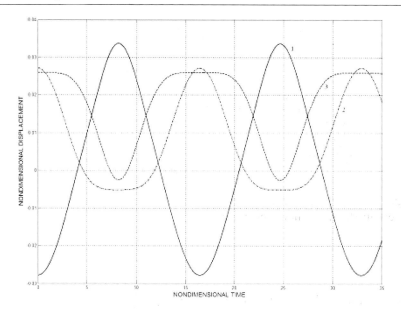

Figure 25. Bead displacements for the localized NNM of the three-bead granular system at normalized energy $h = 0.0001$.

with increasing number of beads the upper boundary (corresponding to the out-of-phase NNM) moves toward higher frequencies, whereas the curve corresponding to the in-phase (pseudo-wave) NNM moves towards lower frequencies.We conclude by noting that no time-periodic orbits (NNMs or subharmonic orbits) can occur in the 'prohibited' band.

Considering homogeneous granular chains of higher dimensionality, we identify regions in the FEP where spatial transmission of disturbances (i.e., energy) is facilitated or prohibited by the intrinsic dynamics of the granular medium. This information is of practical significance when such media are designed as passive mitigators of shocks or other types of unwanted disturbances. Since the out-of-phase and in-phase NNMs represent the highest and lowest frequency NNMs of the granular medium, respectively, all other NNMs (localized or non-localized) are realized in the frequency – energy band defined by these 'bounding' NNMs, irrespective of the dimensionality of the granular medium. Hence, we can only focus on these two NNMs and investigate how the topology of the band of realization of NNMs (and its complementary high-frequency 'prohibited' band) changes with increasing

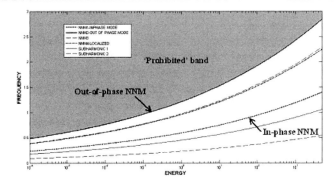

Figure 26. Representation of different NNMs of the three-bead granular system in the frequency – energy plot (FEP).

number of beads in the granular chain. In Figure 27 we depict the in-phase and out-of-phase NNMs in the FEP for systems composed of two to seven beads. We note that as the number of beads increases the out-of-phase NNM converges (accumulates) to a definite upper bounding curve, whereas the in-phase NNM makes a similar transition towards lower frequencies. No such quick convergence is noted for the in-phase NNM, but rather, as the number of beads tends to infinity it tends towards the zero frequency axis. This raises an interesting question concerning the physics of the dynamics of the infinite granular system as the in-phase NNM approaches the zero-frequency limit. Namely, as discussed previously due to bead separation the in-phase NNM resembles a traveling wave corresponding to a single-hump velocity disturbance propagating back and forth through the granular system. Moreover, the 'silent' period for each bead (corresponding to the time period where the bead remains motionless at an offset position) progressively increases with increasing dimensionality of the system. As discussed previously, in actuality the in-phase NNM is a stable standing wave, but for any specific bead the recurrence of the disturbance happens after a 'silent' period equal to half the period of the in-phase NNM; in turn, this period increases with increasing number of beads, as is the corresponding 'silent' period of the in-phase NNM. Following this argument one step further, one might deduce that in the limit of infinite number of beads the period of the NNM tends to infinity (and its frequency tends to zero as it approaches the energy axis in the FEP).

In the limit of infinite number of beads the FEP is partitioned into two

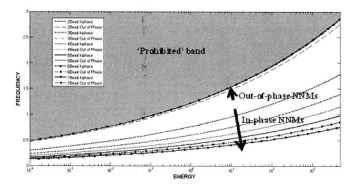

Figure 27. Representation of the in-phase and out-of-phase NNMs in the frequency – energy plot (FEP) for granular systems from two to seven beads; in the limit of infinite number of beads the FEP is partitioned into a propagation and an attenuation band.

regions, namely a propagation zone (PZ) and an attenuation zone (AZ). Inside the PZ the infinite granular medium possesses the continuous families of traveling waves of the type studied in the previous section; these are families of stable traveling waves in the form of propagating multi-hump velocity pulses with arbitrary wavelengths. These families of waves are parameterized by the 'silent' regions of zero velocity that separate successive maxima of the propagating velocity pulses. As these 'silent' regions increase, the corresponding wavelengths and periods of the traveling waves also increase and the frequencies of the waves decrease. As discussed in the previous section, in the limit of infinite 'silent' region the wave ceases to be periodic and its frequency becomes zero; this asymptotic limit of single-hump solitary waves is the solitary wave studied by Nesterenko (2001). In addition to these families of traveling periodic waves additional motions can occur inside the PZ, including subharmonic motions, standing waves with recurring localization features (i.e., periodically spaced humps) and chaotic orbits.

Considering the dynamics of the infinite granular system inside the attenuation zone (AZ) of the FEP, no spatially-periodic standing or traveling waves can occur for frequency – energy ranges in that zone, so no spatial transfer of energy in the granular medium is possible for motions inside the AZ. Rather, near-field motions occur inside the AZ, corresponding to spatially periodic oscillations of the beads about different positive offset positions with spatially decaying envelopes; overall, the motion of the granular medium is a standing wave with decaying envelope, with each bead performing a time periodic oscillation about its own (positive offset) equilibrium position. Indeed, the response of the granular medium inside the AZ resembles the responses of unforced linear periodic media (Mead, 1975; Vakakis and Cetinkaya, 1996; Norris, 1993), which possess similar decaying standing waves [and even decaying 'complex' waves when more than one coupling coordinates between individual substructures exist (Cetinkaya et al., 1995)] in well defined attenuation bands.

We conclude by demonstrating the influence of the intrinsic dynamics of the *unforced* granular system on the *forced* dynamics of the same system under narrowband excitation. We expect that the partitioning of the FEP in terms of propagation and attenuation zones will affect in a significant way the capacity of the granular medium to transmit or attenuate disturbances through it. This, in turn has obvious implications on the capacity of the granular chain to act as passive mitigator of unwanted disturbances. The influence of the intrinsic dynamics of the unforced chain on the forced dynamics is inferred from our previous observation that disturbances initiated inside the PZ of the FEP can spatially transfer energy within the granular medium (through excitation of traveling waves), whereas motions initiated inside the AZ are near-field solutions that cannot transfer energy through the medium. To numerically verify this theoretical prediction we excited a 50-bead granular chain by imposing a harmonic excitation on its left boundary in the form $y_0 = A \sin \omega t$, where ω is the cyclic frequency of the excitation in terms of the normalized time t. Although this is a finite-dimensional medium, we expect that, with the exception of certain end effects, its intrinsic dynamics will be close to the dynamics of the corresponding infinite chain. In order to eliminate high-frequency components in the response of the chain that result due to the excitation of low-amplitude chaotic motions we added weak viscous dissipative forces in the interactions between neighboring beads (Jayaprakash et al., 2011a). In Figure 28 we depict the forced response of the granular system for excitation parameters $A = 1.5$ and $\omega = 1$. This corresponds to dynamics that occur inside the PZ of the infinite chain, and this is confirmed by the spatially extended dynamical response of the system. We note that the initial disturbance generated at the left boundary of the 50-bead chain is transmitted throughout the chain; in addition, the total energy in the system gradually builds up until it forms an oscillation about a high level.

A qualitatively different picture for the dynamics is inferred from the results of Figure 29 depicting the forced response of the chain inside the

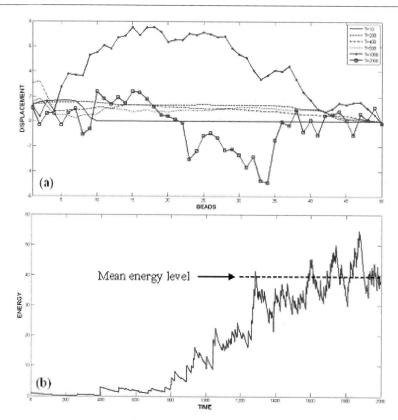

Figure 28. Forced response of the 50-bead granular chain for harmonic wall excitation $y_0 = 1.5 \sin \tau$: (a) Snapshots of chain deformation at selected time instants, (b) evolution of total energy in the chain.

AZ of the FEP. In this case the excitation parameters are selected as for $A = 0.3$ and $\omega = 1$; so we keep the same frequency with the previous simulation by decrease the amplitude of the excitation, or equivalently the energy input in the system. We note that in this case no energy transmission occurs through the chain, but rather the input energy is confined in a region close to the left boundary where it is originally generated by the oscillating left rigid wall. In this case the intrinsic dynamics of the granular chain does not enable the transmission of disturbances. In addition, from the snapshot plots of Figure 29a we conclude that with increasing time the motion inside the AZ settles into a near-linear decaying configuration where the beads nearest to the excitation have the greatest offsets from the trivial equilibrium, whereas the ones furthest have negligible offsets. This indicates that away from the excitation point there is effective pre-compression in the granular chain so the dynamics in the AZ can be studied by adopting a weakly nonlinear approach, whereby we may expand the response of each bead in Taylor series about its mean offset position and retain only the leading-order nonlinear terms. Comparing the temporal evolution of the total energy in the system (i.e., Fig. 28b and 29b) we note the for excitation frequency inside the AZ the energy decays with time and reaches a zero steady state value, whereas for excitation frequency inside the PZ the energy reaches a steady state where it fluctuates about a nonzero mean value. This is a clear demonstration of the capacity of the intrinsic dynamics of the granular medium to attenuate energy inside the AZ of the FEP.

We conclude that the topological structure of the PZ and AZ of the frequency-energy plot affects significantly the narrowband forced dynamics of the granular chain. Indeed, depending on the frequency-energy content of the excitation, the intrinsic dynamics of the medium either facilitates or hinders the spatial propagation of disturbances within the granular medium, allowing the propagation or dissipating certain frequency components. In the next section we review the dynamics of a nonlinear periodic system that is representative of heterogeneous periodic granular media, namely, a dimer granular chain with no pre-compression. We prove the existence of strongly nonlinear resonances in these system, which may lead to strong passive attenuation of propagating pulses, and effective shock mitigation designs.

5.3   Resonances in Dimer Granular Media

In this final section we consider a diatomic ("dimer") granular chain composed of alternating 'heavy' and 'light' spherical beads with no pre-compression, and show that this periodic medium can exhibit countable

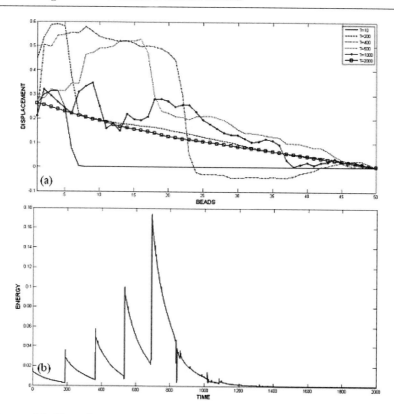

Figure 29. Forced response of the 50-bead granular chain for harmonic wall excitation $y_0 = 0.3 \sin \tau$: (a) Snapshots of chain deformation at selected time instants, (b) evolution of total energy in the chain.

infinities of nonlinear resonances. We study the nonlinear resonance mechanism and show that it can lead to drastic attenuation of pulses propagating in the dimer.

We consider one-dimensional granular dimers governed by the following equations of motion,

$$m_i \frac{d^2 z_i}{dt^2} = (4/3)E_* \sqrt{R} \left[(z_{i-1} - z_i)_+^{3/2} - (z_i - z_{i+1})_+^{3/2} \right]$$
$$m_{i+1} \frac{d^2 z_{i+1}}{dt^2} = (4/3)E_* \sqrt{R} \left[(z_i - z_{i+1})_+^{3/2} - (z_{i+1} - z_{i+2})_+^{3/2} \right] \quad (98)$$
$$i = \pm 1, \pm 3, \pm 5, \ldots$$

where $R_i = R_1$ and $R_{i+1} = R_2$; $m_i = m_1 = (4/3)\pi R_1^3 \rho_1$ and $m_{i+1} = m_2 = (4/3)\pi R_2^3 \rho_2$; $E_* = E_1 E_2 / \left[E_2(1 - \mu_1^2) + E_1(1 - \mu_2^2) \right]$; E_1 (E_2) is the elastic modulus, R_1 (R_2) the radius, μ_1 (μ_2) the material density, and μ_1 (μ_2) the Poisson's ratio of bead 1 (2). Introducing the normalizations,

$$x_i = \frac{z_i}{R_1} \ , \ t = \left[\frac{E_*}{\pi R_1^2 \rho_1} \left(\frac{R_2}{R_1 + R_2} \right)^{1/2} \right]^{1/2} \tilde{t}$$

we obtain the system of non-dimensional equations,

$$\ddot{x}_i = (x_{i-1} - x_i)_+^{3/2} - (x_i - x_{i+1})_+^{3/2}$$
$$\varepsilon \ddot{x}_{i+1} = (x_i - x_{i+1})_+^{3/2} - (x_{i+1} - x_{i+2})_+^{3/2} \quad (99)$$
$$i = \pm 1, \pm 3, \pm 5, \ldots$$

where overdots denote differentiations with respect to non-dimensional time t. In (3) the only non-dimensional parameter is $\varepsilon = \rho_2 R_2^3 / \rho_1 R_1^3$ denoting the mass ratio between the light and heavy beads. Hence, there is a single parameter governing the intrinsic dynamics of the general dimer system in the elastic range. In the following asymptotic analysis we will regard ε as the small parameter in the range $0 < \varepsilon << 1$.

In (Jayaprakash et al., 2011b) numerical experiments were performed to study pulse attenuation in a free-fixed dimer composed of 85 beads (excluding a fixed light bead at the right boundary), by applying a unit impulse excitation to the left end, and recording the maximum force transmitted to the right fixed boundary. Since the dimer system (99) is re-scalable with respect to the magnitude of the applied impulse (energy), there is no loss of generality by considering unit impulse excitation, so the results are applicable to other input levels. Clearly, the lower the force transmitted to the right end is, the higher is the capacity of the dimer to attenuate propagating pulses through scattering of the pulse by its inherent dynamics. The plot of transmitted force versus normalized mass ratio ε

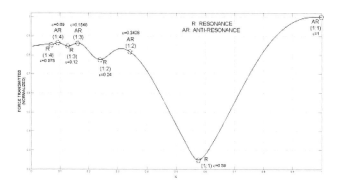

Figure 30. Force transmitted to the right end of the normalized dimer of 85 beads of for unit applied impulse at the left end (Jayaprakash et al., 2011b).

is reproduced in Figure 30 (Jayaprakash et al., 2011b). At the discrete values $\varepsilon_{AR} \approx 0.3428,\ 0.1548,\ 0.0901,\ ...$solitary waves (anti-resonances) are realized, evidenced as peaks (local maxima) of the plot. This result which might seem counterintuitive (since, it is not typical for nonlinear heterogeneous periodic media to support solitary waves), indicates that solitary waves is a basic mechanism for effective transmission of energy through the dimer, similarly to the Nesterenko solitary wave for homogeneous granular chains (see section 5.1). On the contrary, at the discrete values $\varepsilon_R \approx 0.59,\ 0.24,\ 0.12,\ 0.075,\ ...$valleys (local minima) of the plot of transmitted force of Figure 30 exist, corresponding to locally optimal scattering of the energy of the pulse by the dimer (with the exception of the first valley at $\varepsilon_R \approx 0.59$ which represents a global minimum). This hints on the existence of resonances at these values of the normalized mass ratio, which cause maximum scattering of a propagating pulse and attenuation of its magnitude.

To understand the mechanism for nonlinear resonance in the dimer (99), in Figure 31 we examine the waveforms of the light and heavy bead responses for one of the valleys of the plot corresponding to $\varepsilon_R \approx 0.12$; the waveforms were shifted so that $t = 0$ lies at the center of the combined plot. We note that the velocity waveforms are not symmetric, but instead exhibit maximum break of symmetry; that is, the fast component of the velocity waveform of the light bead has a phase difference approximately equal to

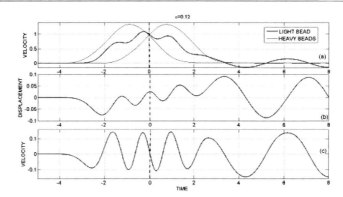

Figure 31. Resonance at $\varepsilon_R \approx 0.12$: (a) Velocity profiles of a light bead and its neighboring heavy beads, (b,c) fast components of the displacement and velocity, respectively, of the light bead.

$\pi/2$ compared to the corresponding waveforms of its adjacent heavy beads. The fast component of the dynamics corresponds to high-frequency, small-amplitude oscillations that are superimposed to the slowly varying (averaged) dynamics. Furthermore, the velocity of the light bead does not decay to zero, but rather attains a finite non-zero velocity towards the completion of the *'squeeze mode'*, i.e., of the phase of the motion when the light bead is continuously compressed by its heavy bead neighbors (during the arrival of the propagating primary pulse). Similarly, the velocities of the heavy beads decay to nonzero values, which is an indication that the primary pulse looses a portion of its energy as it propagates through the pairs of dissimilar beads in the dimer under consideration. In addition, contrary to anti-resonances where the light bead has zero displacement and nonzero velocity at the axis of symmetry ($t = 0$) (Jayaprakash et al., 2011b), in resonances the light bead has nonzero displacement and zero velocity at the same time instant.

At the end of squeeze mode the response of the light bead does not decay to zero as in the case for anti resonance, but rather separates from its heavy bead neighbors and collides repeatedly with them in a strongly nonlinear *'collision mode.'* This mode generates the oscillating 'tails' that appear in the trail of the primary pulse transmission in the dimer. These are strongly nonlinear waves that propagate in the opposite direction with respect to the primary pulse and continuously radiate energy from the pulse to the far field. During resonance there is locally maximum (with respect to ε)

amplification of these radiating waves, resulting in effective dispersion and attenuation of the propagating pulse.In fact, during resonance the primary pulse looses a part of its energy due to scattering at each interface between heavy and light beads, and significantly attenuates as it propagates through the dimer. Hence, the resonance mechanism for strong pulse dispersion is the opposite from the anti-resonance mechanism that leads to a countable infinity of families of solitary waves that exist at discrete values of ε and are parametrized by energy.

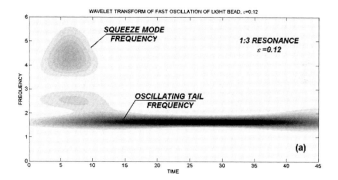

Figure 32. Wavelet transform spectra of the fast component of the oscillation of a light bead for$\varepsilon \approx 0.12$ (1:3 resonance).

It is expected that nonlinear resonances are realized when commensurable relationships between two or more characteristic frequencies of the dynamics of a system occur. Indeed, this is exactly what happens during resonance in the granular dimer, when one considers the characteristic frequencies of the light beads in each of the two phases of their motion; that is, in the initial squeeze mode during the arrival of the primary pulse (when the light beads are in a state of continuous compression from their neighboring heavy beads), and in the following collision mode (after the primary pulse has propagated) when the light beads collide with their neighboring heavy beads. These two characteristic frequencies, although attributed to two different dynamical behaviors determine the type of nonlinear resonance realized in the dimer. During the squeeze mode the dynamics is linearized and the corresponding characteristic frequency can be asymptotically approximated, whereas during the collision mode the dynamics is strongly nonlinear and the characteristic frequency of the bead collisions can only be estimated numerically. Nonlinear resonance occurs whenever these two frequencies become commensurate, and in the case of the dimer this frequency ratio is equal to $1 : m$ where m is the integer that determines the order of the particular resonance. Hence, a countable infinity of nonlinear resonances is anticipated in the elastic dimer chain.

In Figure 32 we depict the wavelet spectra of the fast components of the oscillations of the light beads for the resonance depicted in Figure 31 with $\varepsilon_R \approx 0.12$; in this particular resonance the ratio of characteristic frequencies is approximately equal to 3 signifying a 1:3 resonance, and a multi-scale partition of the dynamics can be performed, since the time scale of the squeeze mode is well separated from the time-scale of the collision mode. However, most important from a practical point of view is the case $\varepsilon_R \approx 0.59$, with a frequency ratio of near unity and 1:1 resonance. In this case the time scale separation between the squeeze mode and the oscillating tail that exists for smaller values of ε is lost, so this strong resonance corresponds to mixed-scale dynamics.In this case it is not possible to apply a multiple-scales approach, but still analytical treatement can be carried out based on binary model approximations (Jayaprakash et al., 2011c). From a physical point of view, for this large value of ε the dynamics of the light beads cannot be regarded as being driven by the dynamics of their neighboring heavy beads anymore, but rather there is strongly nonlinear interaction between them. Yet, as the plot of Figure 30 indicates, 1:1 resonance corresponds to the strongest possible pulse attenuation in the dimer (of about 75%) compared to the homogeneous chain composed only of heavy beads.

It is interesting that in the limit of small normalized mass ratio ε we may derive analytical approximations for the amplitudes of the strongly nonlinear traveling waves that are radiated from the propagating pulse. This was performed in (Jayaprakash et al., 2011c) to which the interested reader can refer. Moreover, based on these estimates we will be able to analytically estimate pulse attenuation in the dimer chain. Resonances are a feature of the intrinsic dynamics of ordered heterogeneous periodic granular media and can lead to strong dispersion and attenuation of propagating pulses. In that context, 1:1 resonance was found to provide the optimal results for the dimer system. The resulting effective pulse attenuation is attributed solely to redistribution of the energy of the propagating pulse to strongly nonlinear modes of the dimer and to traveling waves radiating energy to the far field.

The area of granular periodic media is a promising area of research with numerous applications, ranging from passive shock mitigation designs (Jayaprakash et al., 2011c) to designs of metamaterials, acoustic bullet formation and nonlinear acoustic lens (Spandoni and Daraio, 2010). They represent an interesting class of nonlinear periodic media that, depending on the state of its dynamics can be either weakly or strongly nonlinear and smooth or non-smooth. Of particular interest is the use of nonlinear granular interfaces in layered elastic or viscoelastic media for metamaterial designs (Starosvetsky and Vakakis, 2011).

References

Anderson P.W., Absence of Diffusion in Certain Random Lattices, *Phys. Rev.*, 109, 1492-1505, 1958.

Asfar O.R., and Nayfeh A.H., The Application of Multiple-scales to Wave Propagation in Periodic Structures, *SIAM Rev.*, 25, 455-480, 1981.

Aubrecht J., and Vakakis A.F., Localized and Non-Localized Nonlinear Normal Modes in a Multi-Span Beam With Geometric Nonlinearities, *J. Vib. Acoust.*, 118, 533-542, 1996.

Boechler N., and Daraio C., An Experimental Investigation of Acoustic Band Gaps and Localization in Granular Elastic Chains, *Proc. ASME 2009 International Design Engineering Technical Conferences – IDETC 2009*, San Diego, CA, Aug. 30 – Sept. 2, 2009.

Brillouin L., *Wave Propagation in Periodic Structures*, Dover Publications, New York, 1946.

Byrd P.F., and Friedman M.D., *Handbook of Elliptic Integrals for Engineers and Physicists*, Springer Verlag, Berlin and New York, 1954.

Cetinkaya C., *Axisymmetric Stress Wave Propagation in Weakly Coupled Layered Structures: Analytical and Computational Studies*, PhD Thesis, University of Illinois at Urbana – Champaign, 1995.

Cetinkaya C., Vakakis A.F., and El-Raheb M., Axisymmetric Elastic Waves in Weakly Coupled Layered Media of Infinite Radial Extent,*J. Sound Vib.*, 182(2), 283-302, 1995.

Daraio C., Nesterenko V.F., Herbold E.B., and Jin S., Strongly Nonlinear Waves in a Chain of Teflon Beads, *Phys. Rev. E* 72, 016603, 2005.

Daraio C., Nesterenko V.F., Herbold E.B., and Jin S., Energy Trapping and Shock Disintegration in a Composite Granular Medium, *Phys. Rev. Lett.*, 96, 058002, 2006.

Daraio C., Ngo D., Nesterenko V.F., and Fraternali F., Highly Nonlinear Pulse Splitting and Recombination in a Two-dimensional Granular Network, *Phys. Rev. E*, 82, 036603, 2010.

El-Raheb M., Transient Elastic Waves in Finite Layered Media: One Dimensional Analysis, *J. Acoust. Soc. Am.*, 94(1), 172-184, 1993.

Flytzanis N., Pnevmatikos S., and Remoissenet M., Kink, Breather and Asymmetric Envelope or Dark Solitons in Nonlinear Chains, *J. Phys. Solid State Physics*, 18, 4603-4620, 1985.

Goodman R.H., Holmes P.J., and Weinstein M.I., Strong NLS Soliton–Defect Interactions, *Physica D*, 192, 215–248, 2004.

Hodges C.H., Confinement of Vibration by Structural Irregularity, *J. Sound Vib.*, 82, 411-424, 1982.

Jayaprakash K.R., Starosvetsky Y., Vakakis A.F., Peeters M., and Kerschen G., Nonlinear Normal Modes and Band Zones in Granular Chains with no Pre-compression, *Nonl.Dyn.*, 63, 359-385, 2011a.

Jayaprakash K.R., Starosvetsky Y., and Vakakis A.F., New Family of Solitary Waves in Granular Dimer Chains with no Pre-compression, *Phys. Rev. E*, 83, 036606, 2011b.

Jayaprakash K.R., Starosvetsky Y., and Vakakis A.F., Nonlinear Resonances Leading to Strong Shock Attenuation in Granular Dimer Chains, *Physica D* (submitted), 2011c.

Job S., Santibanez F., Tapia F., and Melo F., Wave Localization in Strongly Nonlinear Hertzian Chains with Mass Defect,arXiv:0901.3532v1, 2009.

Knopoff L., Matrix Methods for Elastic Waves, *Bull. Seism. Soc. Am.*, 54, 431-438, 1964.

Lomdahl P.S., Solitons in Josephson Junctions: An Overview, *J. Stat. Phys.* 39, 551-561, 1985.

Mal A.K., Wave Propagation in Layered Composite Laminates Under Periodic Surface Loads, *Wave Motion*, 10, 257-266, 1988.

Manevitch L.I., and Mikhlin Yu.V., On Periodic Solutions Close to Rectilinear Normal Vibration Modes, *PMM*, 36(6), 1051-1058, 1972.

MaslovV.P., and Omelianov G.A., Asymptotic Soliton-like Solutions of Equations with Small Dispersion, *Advances in Mathematical Sciences – Usp.Mat.Nauk.* 36(3), 63-126, 1981.

Mead D.J., Wave Propagation and Natural Modes in Periodical Systems: I – Mono-coupled Systems, and II – Multi-coupled Systems With and Without Damping, *J. Sound Vib.*, 40(1), 1-39, 1975.

Mead D.J., A New Method of Analyzing Wave Propagation in Periodic Structures, *J. Sound Vib.*, 104(1), 9-27, 1986.

Mickens R.E., *Difference Equations*, Van Nostrand Reinhold, New York, 1987.

Nayfeh A.H., and Mook D.T., *Nonlinear Oscillations*, WileyInterscience, New York, 1984.

Nesterenko V.F., *Dynamics of Heterogeneous Materials*, Springer-Verlag, New York, 2001.

Nesterenko V.F., Daraio C., Herbold E.B., and Jin S., Anomalous Wave Reflection at the Interface of Two Strongly Nonlinear Granular Media, *Phys. Rev. Lett.*, 95, 158702, 2005.

Norris A.W., Waves in Periodically Layered Media: A Comparison of Two Theories, *SIAM J. Appl. Math.*, 53, 1195-1209, 1993.

Pierre C , and Dowell E.H., Localization of Vibrations by Structural Irregularity, *J. Sound Vib.*, 114(3), 549-564, 1987.

Pilipchuck V.N., The Calculation of Strongly Nonhnear Systems Close to Vibration-impact Systems, *PMM* 49(5), 572-578, 1985.

Pilipchuk V.N., A Transformation for Vibrating Systems Based on a Non-smooth Periodic Pair of Functions, *Doldady AN Ukr. SSR*, Ser. A, 4, 37-40, 1988 (in Russian).

Pilipchuck V.N., Azeez M.A.F., and Vakakis A.F., Primary Pulse Transmission in a Strongly Nonlinear Periodic System, *Nonl.Dyn.*, 11, 61-81, 1996.

Pilipchuck V.N., and Vakakis A.F., Study of the Oscillations of a Nonlinearly Supported String Using Nonsmooth Transformations, *J. Vib. Acoust.*, 120, 434-440, 1998.

Potapov A.I., Pavlov I.S., Gorshkov K.A., and Maugin G.A., Nonlinear Interactions of Solitary Waves in a 2D Lattice, *Wave Motion*, 34, 83-95, 2001.

Rizzi S.A., and Doyle J.F., A Spectral Element Approach to Wave Motion in Layered Solids, *J. Vib. Acoust.*, 114, 569-577, 1992.

SalengerG., and Vakakis A.F., Discreteness Effects in the Forced Dynamics of a String on a Periodic Array of Nonlinear Supports, *Int. J. Nonlinear Mech.*, 33(4), 659-673, 1988.

Schmidt H., and Jensen F., A Full Wave Solution for Propagation in Multilayered Viscoelastic Media with Application to Gaussian Beam Reflection at Fluid-Solid Interfaces, *J. Acoust. Soc. Am.*, 77(3), 813-825, 1985.

Sen S., Hong J., Bang J., Avalos E., and Doney R.,Solitary Waves in the Granular Chain, *Phys. Rep.* 462, 21, 2008.

Spadoni A., and Daraio C., Generation and Control of Sound Bullets with a Nonlinear Acoustic Lens, *Proc. Nat. Ac. Science*, doi/10.1073/pnas.1001514107, 2010.

Starosvetsky Y., and Vakakis A.F., Traveling Waves and Localized Modes in One-dimensional Homogeneous Granular Chains with no Pre-compression, *Phys. Rev. E*, 82, 026603, 2010.

Starosvetsky Y., and Vakakis A.F., Primary Wave Transmission in Systems of Elastic Rods with Granular Interfaces, *Wave Motion* (in press), 2011.

Starosvetsky Y., Jayaprakash K.R., and Vakakis A.F., Scattering of Solitary Waves and Excitation of Transient Breathers in Granular Media by Light Intruders and no Pre-compression,*J. Appl. Mech.* (in press), 2011.

Theocharis G., Kavousanakis M., Kevrekidis P.G., Daraio C., Porter

M.A., Kevrekidis I.G., Localized Breathing Modes in Granular Crystals with Defects, arXiv:0906.4094v1 [nlin.PS], 2009.

Toda M., *Theory of Nonlinear Lattices*, Springer Verlag, Berlin and New York, 1989.

Vakakis A.F., Relaxation Oscillations, Subharmonic Orbits and Chaos in the Dynamics of a Linear Lattice with a Local Essentially Nonlinear Attachment, *Nonl.Dyn.*, 61, 443-463, 2010.

Vakakis A.F. and Cetinkaya C., Dispersion of Stress Waves in One-dimensional Semi-infinite, Weakly Coupled Layered Systems,' *Int. J. Solids Str.* 33 (28), 4195-4213, 1996.

Vakakis A.F., Nayfeh T., and King M.E., A Multiple-Scales Analysis of Nonlinear, Localized Modes in a Cyclic Periodic System, *J. Appl. Mech.*, 60, 388-397, 1993.

Vakakis A.F., and King M.E., Nonlinear Wave Transmission in a Mono-coupled Elastic Periodic System, J. Acoust. Soc. Am., 98(3), 1534-1546, 1995.

Vakakis A.F., Manevitch L.I., Mikhlin Yu.,Pilipchuck V.N., Zevin A.A., *Normal Modes and Localization in Nonlinear Systems*, Wiley and Sons, New York, 1996.

Vakakis A.F., and King M.E., Resonant Oscillations of a Weakly Coupled, Nonlinear Layered System, *Acta Mech.*, 128, 59-80, 1998.

Vakakis A.F., Gendelman O., Bergman L.A., McFarland D.M., Kerschen G., Lee Y.S., *Nonlinear Targeted Energy Transfer in Mechanical and Structural Systems*, Springer Verlag, Berlin and New York, 2008.

Zhou J., Huang H.B., Qi G.X., Yang P., and Xie X., Communication with Spatial Periodic Chaos Synchronization, *Phys. Lett.A*, 335(2-3), 191-196, 2005.

Zhuravlev V.F., A Method of Analyzing Vibro-impact Systems Using Special Functions, *Izv.Akad.Nauk SSSR MTI* 2, 30–34, 1976.

Zhuravlev V.F., Investigation of Some Vibro-impact Systems by the Method of Non-Smooth Transformations, *Izv. Akad.Nauk SSSR MTT* 6, 24–28, 1977.